THz: Research Frontiers for New Sources, Imaging and Other Advanced Technologies

THz: Research Frontiers for New Sources, Imaging and Other Advanced Technologies

Editors

Stefano Lupi
Akinori Irizawa
Augusto Marcelli

MDPI • Basel • Beijing • Wuhan • Barcelona • Belgrade • Manchester • Tokyo • Cluj • Tianjin

Editors

Stefano Lupi
Sapienza University of Rome
Italy

Akinori Irizawa
Osaka University
Japan

Augusto Marcelli
Istituto Nazionale di Fisica
Nucleare—Laboratori Nazionali
di Frascati
Italy

Editorial Office

MDPI
St. Alban-Anlage 66
4052 Basel, Switzerland

This is a reprint of articles from the Special Issue published online in the open access journal *Condensed Matter* (ISSN 2410-3896) (available at: https://www.mdpi.com/journal/condensedmatter/special_issues/Terahertz_spectroscopy).

For citation purposes, cite each article independently as indicated on the article page online and as indicated below:

LastName, A.A.; LastName, B.B.; LastName, C.C. Article Title. *Journal Name* **Year**, *Volume Number*, Page Range.

ISBN 978-3-0365-2163-3 (Hbk)
ISBN 978-3-0365-2164-0 (PDF)

Cover image courtesy of Stefano Lupi.

Contents

About the Editors

Stefano Lupi, Since 1992, Stefano Lupi has been involved in condensed matter research by using infrared and terahertz synchrotron radiation. His main research areas include: High-Tc Superconductors, strongly correlated materials, graphene, and topological materials. He developed the SISSI infrared beamline at Elettra Sincrotrone Trieste and the TERASPARC beamline at the Free Electron Laser SPARC at the INFN-LNF laboratory. Since 2015, he has cooperated with Osaka University on THz radiation and with the University of Science and Technology at Hefei on 3D Graphene and other aerogel materials. He was the visiting professor at the Elettra Sincrotrone Trieste and the University of Hamburg in 2010, and at the University of Milan in 2015. He is responsible for the SapienzaTerahertz laboratory work at Sapienza University on high-intensity THz radiation and pump-probe spectroscopy.

Akinori Irizawa's main research areas are: infrared spectroscopy, photoemission spectroscopy, and solid-state physics. Since 2001, he has been involved in synchrotron radiation research, and since 2010, he has participated in free electron laser research at the Institute of Scientific and Industrial Research (ISIR) of Osaka University.

Augusto Marcelli, As a senior scientist at the Laboratori Nazionali di Frascati of the INFN since early 1980s, Augusto Marcelli has been involved in synchrotron radiation research. He contributed to the construction of beamlines at ADONE, DAΦNE, and ESRF. In particular, he proposed and built the first Italian infrared/THz beamline and until 2006, he was the scientist responsible for its operation. In 2019, he was appointed as a scientific expert on bilateral policies and activities for the internationalization of scientific and technological research of the Italian Ministry of Foreign Affairs. His main research areas include: correlation phenomena in X-ray absorption spectroscopy, circular magnetic X-ray dichroism investigations, X-ray absorption in elements of geophysical interest, dust and aerosol characterization, and ultra-trace detection for indoor and outdoor environmental research.

Preface to "THz: Research Frontiers for New Sources, Imaging and Other Advanced Technologies"

Terahertz (THz = 10^{12} Hz) radiation, alternatively referred to as T rays, is the electromagnetic radiation lying between the microwave and infrared portions of the spectrum. There is no standard definition for the THz band, but it has most often come to refer to the region of the electromagnetic spectrum from 0.1 to 20 THz (3–600 cm^{-1}, 3 mm–15 μm wavelength, 0.3–80 meV). The lack of readily available THz sources and detectors has led to the relatively unexplored region of the electromagnetic spectrum between microwaves and infrared being termed the "THz gap".

This is a frontier area for research in Physics, Chemistry, Biology, Materials Science and Bio-Medicine. Indeed, THz radiation can be used to investigate and characterize small molecules rotating at THz frequencies, while biologically-important collective modes of proteins DNA and RNA vibrate at THz frequencies, frustrated rotations and collective modes causing polar liquids such as water absorb at THz frequencies and gaseous and solid-state plasmas oscillate at THz frequencies. Also electrons in semiconductors and their nanostructures resonate at THz frequencies; electrons in highly-excited atomic Rydberg states orbit at THz frequencies and superconducting energy gaps fall at THz frequencies.

The publication of this volume has been made possible thanks to the support of the Italian Ministry of Foreign Affairs and International Cooperation in the framework of the Bilateral Cooperation Agreement between Italy and Japan (Major relevance Project N. PGR00728).

Stefano Lupi, Akinori Irizawa, Augusto Marcelli
Editors

condensed matter

MDPI

Editorial

Terahertz as a Frontier Area for Science and Technology

Akinori Irizawa [1], Stefano Lupi [2,3] and Augusto Marcelli [3,4,*]

1 The Institute of Scientific and Industrial Research (ISIR), Osaka University, 8-1 Mihogaoaka, Ibaraki, Osaka 5670047, Japan; irizawa@sanken.osaka-u.ac.jp
2 Department of Physics, Sapienza University, P.le Aldo Moro 5, 00185 Rome, Italy; stefano.lupi@roma1.infn.it
3 Istituto Nazionale di Fisica Nucleare-Laboratori Nazionali di Frascati (INFN-LNF), Via Enrico Fermi 54, 00044 Frascati, Italy
4 International Centre for Material Science Superstripes, RICMASS, Via dei Sabelli 119A, 00185 Rome, Italy
* Correspondence: augusto.marcelli@lnf.infn.it

Abstract: Recent theoretical and experimental research is triggering interest to technologies based on radiation in the region from ~0.1 to 20 Terahertz (THz). Today, this region of the electromagnetic (e.m.) spectrum is a frontier area for research in many disciplines. The technological roadmap of the THz radiation considers outdoor and indoor communications, security, drug detection, biometrics, food quality control, agriculture, medicine, semiconductors, and air pollution, and demands high-power and sub-ps compact sources, modern detectors, and new integrated systems. There are still many open questions regarding working at THz frequencies and with THz radiation. In particular, we need to invest in new methodologies and in materials exhibiting the unusual or exotic properties of THz. This book contains original papers dealing with some emerging THz applications, new devices, sources and detectors, and materials with advanced properties for applications in biomedicine, cultural heritage, technology, and space.

Keywords: THz; FEL; coherent synchrotron radiation; waveguides; THz detector; THz imaging

Citation: Irizawa, A.; Lupi, S.; Marcelli, A. Terahertz as a Frontier Area for Science and Technology. *Condens. Matter* **2021**, *6*, 23. https://doi.org/10.3390/condmat6030023

Received: 7 June 2021
Accepted: 22 June 2021
Published: 28 June 2021

1. Introduction

Photonics is the physical science describing the generation, the detection, the manipulation, and the application of many different devices of light. Photonics, combined with micro- and nano-electronics, biotechnology, and/or nanotechnology, is emerging as a strategic research and technological area as addressed by the United Nations which celebrated 2015 as the International Year of Light. Hence, the continuous development of photonic technologies needs a deeper understanding of light–matter interactions, the implementation of new concepts, and the synthesis of novel materials with tailored properties [1]. Following the development of laser-based Terahertz (THz) time-domain spectroscopy in the 1980s and 1990s, the field of THz science and technology expanded rapidly and now involves 'real world' applications [2–4]. The new approaches led to multidisciplinary research, combining photonics modeling with long-term strategic aims towards challenging applications such as spintronics, bio-photonics, and many unique devices that can be manufactured with the currently available technologies; others will soon appear, thanks to emerging technologies.

The spectacular advancements observed in the last decade has led to an increasing interest in THz technology in the effort to harness the power of thermal radiation in the region from ~100 GHz to 20 THz (3.3–670 cm^{-1} wavenumber, 3 mm–15 μm wavelength, 0.4–83 meV photon energy). The interest shares some uniquely attractive characteristics of the THz radiation with unique application possibilities [2,5].

The THz region of the electromagnetic spectrum is a frontier area for research in physics, chemistry, biology, materials science, and biomedicine. Indeed, small molecules rotate at THz frequencies; biologically important collective modes of proteins DNA and RNA vibrate at THz frequencies; frustrated rotations and collective modes cause polar liquids such as water to absorb at THz frequencies; electrons in semiconductors and their nanostructures resonate at THz frequencies; electrons in highly excited atomic Rydberg states orbit at THz frequencies; superconducting energy gaps fall at THz frequencies; and gaseous and solid-state plasmas also oscillate at THz frequencies, therefore this radiation can be used to study and control an extraordinary vast number of materials, fundamental systems, and phenomena [2,5–9].

The roadmap for the development of THz technologies [2,10,11] considers applications in the fields of outdoor and indoor communications, security, drug detection, biometrics, food quality control, agriculture, medicine, semiconductors, and air pollution. Their exploitation and realization demand high-power compact THz sources, more sensitive detectors, and more functional integrated THz systems.

2. Discussion

Sources of high-quality radiation in the THz region of the e.m. spectrum have been scarce [12], but this "THz gap", after continuous research efforts, has been filled by a wide range of new technologies ranging from accelerated relativistic electrons [13,14], to high-power femtosecond laser-based sources [15] and Quantum Cascade Lasers [16,17]. Thus, THz radiation is now available in both continuous wave (CW) and pulsed form, down to single-cycles and sub-s duration, with peak powers up to tens of MW; several THz facilities are distributed worldwide for fundamental experiments, user applications, and industrial R&D.

The rapid advance of THz technologies in terms of radiation generators, systems, and scientific or industrial applications has put a particular focus on compact sources with challenging performances. Free-electron-laser (FEL)-based sources are probably the best candidates to represent such versatility. There are a number of schemes that have been investigated over the years to generate coherent radiation from free electrons in the mm-wave and THz regions of the spectrum. The paper by Doria et al. [18] proposes new approaches for exploring the limits in the performance of radio frequency-driven free-electron devices in terms of ultra-short pulse duration, wide bandwidth operation, and energy recovery for near CW operation. This work demonstrates that a THz FEL is very powerful and flexible. This is possible by tailoring its spectral features with innovative design schemes that can be introduced for the realization of a compact FEL source, as discussed in this contribution.

Considering modern THz sources, TeraFERMI is the THz beamline at the FERMI free-electron-laser facility in Trieste (Italy). It uses super-radiant Coherent Transition Radiation emission to produce THz pulses from 10 to 100 μJ intensity over a spectral range, which can extend up to 12 THz. This facility can be used to perform non-linear, fluence-dependent THz spectroscopy and THz-pump/IR-probe measurements. Adhlakha et al. [19] describe in their contribution the optical set-up in use, and discuss the properties of the generated THz electric field profile. They showed that the THz emission from TeraFERMI is the combination of two components, and the essential role in the emission properties of this source played by CSR induced wakefields.

THz radiation is a very appealing band of the electromagnetic spectrum due to its practical applications. In this context, THz generation and manipulation are an essential part of the technological development. The demand of THz antennas is still high, as it is difficult to obtain directive, efficient, planar, low-cost, and easy-to-fabricate THz radiating systems. Tofani and Fuscaldo, in their contribution [20], focused on different classes of leaky-wave antennas, based on materials with tunable quasi-optical parameters to produce patterns with particularly high efficiency, taking into account costs and fabrication complexity. Design strategies have been also discussed and some examples have been presented and analyzed.

D'Arco et al., in their contribution [21] measured the complex refraction index of one HMQ-TMS (2-(4-hydroxy-3-methoxystyryl)-1-methylquinolinium 2,4,6-trimethyl benzene-sulfonate) single crystal from THz to UV, observing a remarkable anisotropy, strongly attenuated at long wavelengths. The HMQ-TMS is a recently discovered anisotropic organic crystal that can be exploited for the production of broadband high-intensity THz radiation through the optical rectification technique. It plays a central role in THz technology due to its broad transparency range, large electro-optic coefficient and coherence length, and excellent crystal properties. The authors extracted both the refraction index n and the extinction coefficient k between 50 and 35,000 cm^{-1}, and measured the THz radiation generated by optical rectification at different wavelengths and along the axes. Data highlighted well the remarkable anisotropic linear and non-linear optical behavior of these crystals, extending the knowledge of their properties from THz to UV wavelengths and suggesting applications in THz pump-probe experiments of exotic electronic systems.

The THz electromagnetic characterization of materials is also particularly important in many applications in accelerator physics. As, in storage rings and linear accelerators of new generation, very short bunches might be required, extending the exploration of the beam spectrum far in frequency is required to assess the coating material response up to hundreds of GHz. The paper of Passarelli et al. [22] describes a time domain method based on the waveguide spectroscopy to infer the coating properties at very high frequencies. The approach is extremely useful for the evaluation of the real part of the surface impedance, currently used for modeling the resistive wall component of the beam impedance in accelerators. In fact, the emergence of meta-materials opened new opportunities in many fields including THz. It has been shown that meta-material waveguides can guide a mode, even when the core diameter is more than 10 times smaller than the operating wavelength [9,23].

The incredible scientific and technological advances have recently allowed the production and the coherent detection of sub-ps pulses of THz radiation, making spectroscopy and imaging in biomedicine at very long wavelengths now possible [24–27]. THz pulsed imaging (TPI) was pioneered in the last decade and D'Arco et al., in their manuscript [28], presented a review of TPI. They discussed the basic principles and performances, and described state-of-the-art applications on biomedical systems. Concerning the THz radiation properties, they also discussed the major applications of TPI with ex vivo and in vivo studies and cases of histopathological imaging of cancers. Although THz technology is rapidly expanding as a medical imaging modality, its use requires more attention, and extended trials should be carried out to apply TPI systems in clinical fields and hospitals to transfer this already mature technique for research in a powerful tool for clinical tests [29].

InfraRed synchrotron radiation (IRSR) is a unique broadband light source covering a spectral region from THz to IR. Its brilliance represents the main advantage for microspectroscopy experiments, perfect to match the demands of extremely small samples that prevent the use of conventional thermal radiation sources [30]. One of the most promising applications of IRSR is cultural heritage, where materials to investigate are unique, delicate, and valuable. A non-destructive approach is mandatory, and extremely small pieces have to be investigated whenever is required, following the logic of preservation and conservation. Ikemoto et al. [31] analyzed several samples at the BL43IR beamline at SPring-8, showing an advantage of more than two orders of magnitude vs. thermal radiation sources and, in particular, in the low-wavenumber region. Iron oxides in bengala red pigment, biomineral distribution caused by lichen growing on a stone, adhesives of fibers of a traditional costume, and organic materials in overglaze layers on ceramics were characterized at high-spatial resolution.

As outlined above, accelerators show other fantastic properties that can be exploited, particularly in the THz domain. Coherent synchrotron radiation from an electron storage ring is observed in the THz spectral range when the bunch length is shortened to the sub-mm-range. Schade and co-workers, in their contribution [32], showed that increasing the stored current in the ring causes the bunch to become longitudinally unstable and modulates the THz emission in the time domain. Their results set the limits for the possible time resolution of pump-probe experiments with coherent THz synchrotron radiation from a storage ring. Ultra-short electron bunches, such as those delivered by a high-brightness photo-injector, are suitable to produce high peak power THz radiation, both broad and narrow band, with sub-picosecond down to femtosecond pulse shaping. The features of this kind of source are extremely appealing for frequency- and time-domain experiments in a wide variety of fields. The manuscript by Chiadroni et al. [33] will overview the method of generation and characterization of THz radiation produced by high-brightness electron beams. This radiation can be produced under different regimes, by a proper control of the bunch shaping and charge distribution. Different electron longitudinal distribution can be tailored and manipulated by properly setting the UV photocathode laser and the photo-injector parameters. Data collected at the SPARC_LAB accelerator at the Frascati National Laboratory of the INFN, with both ultra-short single bunch and multi-bunches electron comb beams, provided high energy per pulse and broad and narrow spectral bandwidth THz radiation, for non-linear and pump-probe experiments in solid-state physics and materials science.

Correlated with this research, the manuscript by Sakai and co-workers [34] is an original study that introduces a new technology to measure electron bunch length at high time resolutions. Using the coherent synchrotron radiation (CSR) emission, they evaluated the RMS bunch lengths in full-bunch mode and in burst mode. CSR was measured by using a narrow-band detector at the S-band linear accelerator facility of the Nihon University. The results can be applied to the precise evaluation of FEL gain vs. charge to develop high-power terahertz light sources.

The demand of new detectors is emerging, correlated with the emergence of new THz sources [35]. The realization of emitters and receivers is challenging as THz frequencies are too high for conventional electronics and the photon energies are too small for classical optics. At present, there exists a large variety of THz radiation sensors and detectors as traditional bolometers, and new devices based on different principles and materials. [36] In his contribution Rezvani et al. presented an innovative photon detector based on the proximity junction array device (PAD) working at long wavelengths. [37] The PAD has a relatively high photo-response, even at frequencies below the expected characteristic frequency, while its superconducting properties such, as the order parameter and the Josephson characteristic frequency, can be easily modulated via external fields to extend the detection band. Authors characterized the vortex dynamics of this proximity array device vs. external parameters, e.g., magnetic and electric fields, showing the occurrence of the transition from a Mott insulating to a metallic state. Advanced devices such as these represent promising detectors for new high-sensitivity long-wavelength systems and for THz imaging applications.

THz imaging is one of the many challenging and stimulating THz applications. Using the unique characteristics of the FEL, Irizawa et al. [38] successfully performed high-sensitivity spectral imaging of different materials in the THz and far-infrared (FIR) domain. Due to the intensity and directionality of the ISIR source at Osaka, they collected high-sensitivity transmission imaging of extremely low-transparency materials and three-dimensional objects in the 3–6 THz range. Using the FEL time structure, fast spectral imaging was demonstrated on different samples in various forms. Taking advantage of the available high power, they succeeded in the acquisition of transmission images of "opaque" solid materials and liquids for the first time in the THz/FIR domain. By accurately identifying the intrinsic absorption wavelength of organic and inorganic materials, they also succeeded in the mapping of spatial distribution of individual components. They also showed, again for the first time, that non-linear phenomenon can be observed by changing the FEL intensity, pointing out the huge opportunities of these coherent high-power, high brilliance sources for much interdisciplinary research.

The incredible opportunities offered by intense THz sources is well represented by the manuscript of Macis et al. [39] which showed the damage induced by an intense coherent THz beam on copper surfaces. This metallic surface, irradiated by multiple picosecond THz pulses generated by a FEL, exposed the sample surface to an electric field up to ~4 GV/m. No damage occurs at normal incidence, while images and spectroscopic analysis of the surface point out a clear dependence on the incidence angle, the electric field intensity, and polarization of the pulsed THz radiation. The experimental approach they introduced, with multiple fast irradiations, represents a new powerful technique useful to test the damage induced by an intense electric gradient on metallic surfaces in a reproducible way. This work shows for the first time the occurrence of an angular dependent reproducible damage induced by THz multiple high-intensity pulses on a metallic surface depending on incidence angle, number of shots, and fluence. The phenomenology behind the observed behavior can be understood in terms of a fast local temperature increase, which may induce ablation and melting on the surface. This original work clearly points out the need to consider electric-related effects, e.g., discharges and/or breakdowns occurring in the air due to the intense applied electric field [40]. The mechanism behind THz-radiation-induced morphological change is also unclear in soft materials. Changes in the intermolecular conformation of bulk materials without damaging the chemical structures have been already observed in bio-macromolecules. Consequently, the use of THz irradiation may constitute a new method also applicable to biological science, to design and characterize new functional materials [41].

3. Conclusions

There is still too much we do not know about working at THz frequencies and with THz radiation. We should continue to invest in basic science and technology, and to introduce new computational models for analyzing devices and operation in this energy range. We also need to invest in new methodologies [9] to better understand material properties [42–44] and meta-materials [45] at terahertz frequencies, as well as general terahertz phenomenology. This book deals with emerging applications in THz solid-state physics and devices such as sources, detectors, advanced materials, and light–matter interactions with theoretical, methodological, well-established, and validated examples: spectroscopic characterization of novel materials, imaging applications in biomedicine, cultural heritage, and earth science [46–49]. It intends to offer results and ideas for a very vast audience, from basic science to engineering, technology experts, newcomers, and students. Indeed, the published contributions describe, in our opinion, an up to date and reliable state-of-art of THz research and, we hope, their reading will help and promote research on and with THz radiation.

Author Contributions: A.I., S.L. and A.M. contributed equally to the manuscript. All authors have read and agreed to the published version of the manuscript.

Funding: We acknowledge the financial support of the Bilateral Cooperation Agreement between Italy and Japan of the Italian Ministry of Foreign Affairs and of the International Cooperation (MAECI) in the framework of the project of major relevance N. PGR00728.

Institutional Review Board Statement: Not applicable to the studies discussed in this text.

Informed Consent Statement: Not applicable to these studies discussed in this text.

Conflicts of Interest: The authors declare no conflict of interest.

References

1. Scalari, G.; Maissen, C.; Cibella, S.; Leoni, R.; Reichl, C.; Wegscheider, W.; Beck, M.; Faist, J. THz ultrastrong light-matter coupling. *Il Nuovo Saggiatore* **2015**, *31*, 4–14.
2. Dhillon, S.S.; Vitiello, M.S.; Linfield, E.H.; Davies, A.G.; Hoffmann, M.C.; Booske, J.; Paoloni, C.; Gensch, M.; Weightman, P.; Williams, G.P.; et al. The 2017 terahertz science and technology roadmap. *J. Phys. D Appl. Phys.* **2017**, *50*, 043001–043049. [CrossRef]
3. Kawano, Y. Terahertz waves: A tool for condensed matter, the life sciences and astronomy. *Contemp. Phys.* **2013**, *54*, 143–165. [CrossRef]
4. Tonouchi, M. Cutting-edge terahertz technology. *Nat. Photonics* **2007**, *1*, 97–105. [CrossRef]
5. Mittleman, D.M. Perspective: Terahertz science and technology. *J. Appl. Phys.* **2017**, *122*, 230901. [CrossRef]
6. Terahertz Optics. Available online: https://www.nature.com/collections/njscjcdjjr (accessed on 23 June 2021).

7. Wang, H.B.; Wu, P.H.; Yamashita, T. Terahertz Responses of Intrinsic Josephson Junctions in High T_C Superconductors. *Phys. Rev. Lett.* **2001**, *87*, 107002. [CrossRef] [PubMed]
8. Innocenzi, P.; Malfatti, L.; Marcelli, A.; Piccinini, M.; Sali, D.; Schade, U. Application of Terahertz spectroscopy to time-dependent chemical-physical phenomena. *J. Phys. Chem. A* **2009**, *113*, 9418–9423. [CrossRef] [PubMed]
9. Limaj, O.; Giorgianni, F.; Gaspare, A.D.; Giliberti, V.; Marzi, G.D.; Roy, P.; Ortolani, M.; Cunnane, D.; Xi, X.; Lupi, S. Superconductivity-Induced Transparency in Terahertz Metamaterials. *ACS Photonics* **2014**, *1*, 570–575. [CrossRef]
10. Zhang, X.C.; Shkurinov, A.; Zhang, Y. Extreme terahertz science. *Nat. Photonics* **2017**, *11*, 16–18. [CrossRef]
11. Ma, J.; Shrestha, R.; Adelberg, J.; Yeh, C.-Y.; Hossain, Z.; Knightly, E.; Jornet, J.M.; Mittleman, D.M. Security and eavesdropping in terahertz wireless links. *Nature* **2018**, *563*, 89–96. [CrossRef]
12. Gallerano, G.P. Overview of Terahertz Radiation Sources. In Proceedings of the 2004 FEL Conference, Trieste, Italy, 29 August–3 September 2004; pp. 216–221.
13. Chiadroni, E.; Bacci, A.; Bellaveglia, M.; Boscolo, M.; Castellano, M.; Cultrera, L.; Di Pirro, G.; Ferrario, M.; Ficcadenti, L.; Filippetto, D.; et al. The SPARC linear accelerator based terahertz source. *Appl. Phys. Lett.* **2013**, *102*, 094101. [CrossRef]
14. Perucchi, A.; Di Mitri, S.; Penco, G.; Allaria, E.; Lupi, S. The TeraFERMI terahertz source at the seeded FERMI free-electron-laser facility. *Rev. Sci. Instrum.* **2013**, *84*, 022702. [CrossRef]
15. Fülöp, J.A.; Tzortzakis, S.; Kampfrath, T. Laser-Driven Strong-Field Terahertz Sources. *Adv. Opt. Mater.* **2020**, *8*, 1900681. [CrossRef]
16. Williams, B.S. Terahertz quantum-cascade lasers. *Nat. Photonics* **2007**, *1*, 517–525. [CrossRef]
17. Vitiello, M.S.; Tredicucci, A. Physics and technology of Terahertz quantum cascade lasers. *Adv. Phys. X* **2021**, *6*, 1893809. [CrossRef]
18. Doria, A.; Gallerano, G.P.; Giovenale, E. Novel Schemes for Compact FELs in the THz Region. *Condens. Matter* **2019**, *4*, 90. [CrossRef]
19. Adhlakha, N.; Di Pietro, P.; Piccirilli, F.; Cinquegrana, P.; Di Mitri, S.; Sigalotti, P.; Spampinati, S.; Veronese, M.; Lupi, S.; Perucchi, A. The TeraFERMI Electro-Optic Sampling Set-Up for Fluence-Dependent Spectroscopic Measurements. *Condens. Matter* **2020**, *5*, 8. [CrossRef]
20. Tofani, S.; Fuscaldo, W. Fabry-Perot Cavity Leaky Wave Antennas with Tunable Features for Terahertz Applications. *Condens. Matter* **2020**, *5*, 11. [CrossRef]
21. D'Arco, A.; Tomarchio, L.; Dolci, V.; Di Pietro, P.; Perucchi, A.; Mou, S.; Petrarca, M.; Lupi, S. Broadband Anisotropic Optical Properties of the Terahertz Generator HMQ-TMS Organic Crystal. *Condens. Matter* **2020**, *5*, 47. [CrossRef]
22. Passarelli, A.; Koral, C.; Masullo, M.R.; Vollenberg, W.; Lain Amador, L.; Andreone, A. Sub-THz Waveguide Spectroscopy of Coating Materials for Particle Accelerators. *Condens. Matter* **2020**, *5*, 9. [CrossRef]
23. Atakaramians, S.; Afshar, S.; Monro, T.M.; Abbott, D. Terahertz dielectric waveguides. *Adv. Opt. Photonics* **2013**, *5*, 169–215. [CrossRef]
24. Choi, H.; Son, J.-H. Terahertz Imaging and Tomography Techniques, Chapt. 4. In *Terahertz Biomedical Science and Technology*; Son, J.-H., Ed.; CRC Press: Boca Raton, FL, USA, 2014; pp. 47–66.
25. Mittleman, D.M. Twenty years of terahertz imaging. *Opt. Express* **2018**, *26*, 9417. [CrossRef]
26. Baxter, J.B.; Guglietta, G.W. Terahertz Spectroscopy. *Anal. Chem.* **2011**, *83*, 4342–4368. [CrossRef] [PubMed]
27. McIntosh, A.I.; Yang, B.; Goldup, S.M.; Watkinson, M.; Donnan, R.S. Terahertz spectroscopy: A powerful new tool for the chemical sciences? *Chem. Soc. Rev.* **2012**, *21*, 8210. [CrossRef] [PubMed]
28. D'Arco, A.; Di Fabrizio, M.; Dolci, V.; Petrarca, M.; Lupi, S. THz Pulsed Imaging in Biomedical Applications. *Condens. Matter* **2020**, *5*, 25. [CrossRef]
29. Yu, C.; Fan, S.; Sun, Y.; Pickwell-MacPherson, E. The potential of terahertz imaging for cancer diagnosis: A review of investigations to date. *Quant. Imaging Med. Surg.* **2012**, *2*, 33–45. [CrossRef]
30. Marcelli, A.; Cinque, G. Infrared synchrotron radiation beamlines: High brilliance tools for IR spectromicroscopy. A practical guide to the characteristics of the broadband and brilliant non-thermal sources. In *Biomedical Applications of Synchrotron Infrared Microspectroscopy*; Moss, D., Ed.; Royal Society of Chemistry: London, UK, 2011; Chapter 3; pp. 67–104.
31. Ikemoto, Y.; Tanaka, M.; Higuchi, T.; Semba, T.; Moriwaki, T.; Kawasaki, E.; Okuyama, M. Infrared Synchrotron Radiation and Its Application to the Analysis of Cultural Heritage. *Condens. Matter* **2020**, *5*, 28. [CrossRef]
32. Schade, U.; Kuske, P.; Lee, J.; Marchetti, B.; Ortolani, M. Cross-Correlation of THz Pulses from the Electron Storage Ring BESSY II. *Condens. Matter* **2020**, *5*, 24. [CrossRef]
33. Chiadroni, E.; Cianchi, A.; Ferrario, M.; Mostacci, A.; Pompili, R.; Shpakov, V. A Versatile THz Source from High-Brightness Electron Beams: Generation and Characterization. *Condens. Matter* **2020**, *5*, 40. [CrossRef]
34. Sakai, T.; Hayakawa, K.; Tanaka, T.; Hayakawa, Y.; Nogami, K.; Sei, N. Evaluation of Bunch Length by Measuring Coherent Synchrotron Radiation with a Narrow-Band Detector at LEBRA. *Condens. Matter* **2020**, *5*, 34. [CrossRef]
35. Rogalski, A.; Sizov, F. Terahertz detectors and focal plane arrays. *Opto−Electron. Rev.* **2011**, *19*, 346–404. [CrossRef]
36. Sizov, F. THz radiation sensors. *Opto−Electron. Rev.* **2010**, *18*, 10–36. [CrossRef]
37. Rezvani, S.J.; Gioacchino, D.D.; Gatti, C.; Ligi, C.; Guidi, M.C.; Cibella, S.; Fretto, M.; Poccia, N.; Lupi, S.; Marcelli, A. Proximity Array Device: A Novel Photon Detector Working in Long Wavelengths. *Condens. Matter* **2020**, *5*, 33. [CrossRef]

38. Irizawa, A.; Fujimoto, M.; Kawase, K.; Kato, R.; Fujiwara, H.; Higashiya, A.; Macis, S.; Tomarchio, L.; Lupi, S.; Marcelli, A.; et al. Spatially Resolved Spectral Imaging by a THz-FEL. *Condens. Matter* **2020**, *5*, 38. [CrossRef]
39. Macis, S.; Tomarchio, L.; Tofani, S.; Rezvani, S.J.; Faillace, L.; Lupi, S.; Irizawa, A.; Marcelli, A. Angular Dependence of Copper Surface Damage Induced by an Intense Coherent THz Radiation Beam. *Condens. Matter* **2020**, *5*, 16. [CrossRef]
40. Agranat, M.B.; Chefonov, O.V.; Ovchinnikov, A.V.; Ashitkov, S.I.; Fortov, V.E.; Kondratenko, P.S. Damage in a Thin Metal Film by High-Power Terahertz Radiation. *Phys. Rev. Lett.* **2018**, *120*, 085704. [CrossRef] [PubMed]
41. Hoshina, H.; Suzuki, H.; Otani, C.; Nagai, M.; Kawase, K.; Irizawa, A.; Isoyama, G. Polymer Morphological Change Induced by Terahertz Irradiation. *Sci. Rep.* **2016**, *6*, 27180. [CrossRef]
42. Williams, C.R.; Andrews, S.R.; Maier, S.A.; Fernandez-Dominguez, A.I.; Martin-Moreno, L.; Garcia-Vidal, F.J. Highly confined guiding of terahertz surface plasmon polaritons on structured metal surfaces. *Nat. Photonics* **2008**, *2*, 175–179. [CrossRef]
43. D'Apuzzo, F.; Piacenti, A.R.; Giorgianni, F.; Autore, M.; Cestelli Guidi, M.; Marcelli, A.; Schade, U.; Ito, Y.; Chen, M.; Lupi, S. Terahertz and Mid-Infrared Plasmons in 3-Dimensional Nanoporous Graphene. *Nat. Commun.* **2017**, *8*, 14885. [CrossRef] [PubMed]
44. Sacchetti, A.; Cestelli Guidi, M.; Arcangeletti, E.; Nucara, A.; Calvani, P.; Piccinini, M.; Marcelli, A.; Postorino, P. Far-infrared absorption of $La_{1-x}Ca_xMnO_{3-y}$ at high pressure. *Phys. Rev. Lett.* **2006**, *96*, 035503. [CrossRef] [PubMed]
45. Pendry, J.B.; Holden, A.; Robbins, D.; Stewart, W. Magnetism from conductors and enhanced nonlinear phenomena. *IEEE Trans. Microw. Theory Tech.* **1999**, *47*, 2075–2084. [CrossRef]
46. Cinque, G.; Marcelli, A. Synchrotron Radiation InfraRed microspectroscopy and imaging in the characterization of archaeological materials and cultural heritage artefacts. In *EMU Notes in Mineralogy*; European Mineralogical Union and the Mineralogical Society of Great Britain & Ireland: London, UK, 2019; Volume 20, Chapter 12; pp. 411–444.
47. Della Ventura, G.; Bellatreccia, F.; Marcelli, A.; Cestelli Guidi, M.; Piccinini, M.; Cavallo, A.; Piochi, M. Application of micro-FTIR imaging in Earth Sciences. *Anal. Bioanal. Chem.* **2010**, *397*, 2039–2049. [CrossRef]
48. Della Ventura, G.; Marcelli, A.; Bellatreccia, F. SR-FTIR microscopy and FTIR imaging in the Earth Sciences. *Rev. Mineral. Geochem.* **2014**, *78*, 447–479. [CrossRef]
49. Valušis, G.; Lisauskas, A.; Yuan, H.; Knap, W.; Roskos, H.G. Roadmap of Terahertz Imaging 2021. *Sensors* **2021**, *21*, 4092. [CrossRef]

condensed matter

MDPI

Article

Broadband Anisotropic Optical Properties of the Terahertz Generator HMQ-TMS Organic Crystal

Annalisa D'Arco [1,*], Luca Tomarchio [1,2,*], Valerio Dolci [1,2], Paola Di Pietro [3], Andrea Perucchi [3], Sen Mou [1,2], Massimo Petrarca [1,4] and Stefano Lupi [2,5]

1 INFN-Section of Rome 'La Sapienza', P.Le Aldo Moro, 2, 00185 Rome, Italy; valerio.dolci@roma1.infn.it (V.D.); sen.mou@roma1.infn.it (S.M.); massimo.petrarca@uniroma1.it (M.P.)
2 Department of Physics, Sapienza University, P.le Aldo Moro, 5, 00185 Rome, Italy; stefano.lupi@roma1.infn.it
3 Elettra–Sincrotrone Trieste S.C.p.A., S.S.14 km 163.5, Basovizza, 34149 Trieste, Italy; paola.dipietro@elettra.eu (P.D.P.); andrea.perucchi@elettra.eu (A.P.)
4 SBAI Department, Sapienza University, Via A. Scarpa, 14-00161 Rome, Italy
5 INFN-LNF, Via Enrico Fermi 40, 00044 Frascati (RM), Italy
* Correspondence: annalisa.darco@roma1.infn.it (A.D.); luca.tomarchio@uniroma1.it (L.T.)

Received: 19 May 2020; Accepted: 11 July 2020; Published: 14 July 2020

Abstract: HMQ-TMS (2-(4-hydroxy-3-methoxystyryl)-1-methylquinolinium 2,4,6-trimethylbenzenesulfonate) is a recently discovered anisotropic organic crystal that can be exploited for the production of broadband high-intensity terahertz (THz) radiation through the optical rectification (OR) technique. HMQ-TMS plays a central role in THz technology due to its broad transparency range, large electro-optic coefficient and coherence length, and excellent crystal properties. However, its anisotropic optical properties have not been deeply researched yet. Here, from polarized reflectance and transmittance measurements along the x_1 and x_3 axes of a HMQ-TMS single-crystal, we extract both the refraction index n and the extinction coefficient k between 50 and 35,000 cm^{-1}. We further measure the THz radiation generated by optical rectification at different infrared (IR) wavelengths and along the two x_1 and x_3 axes. These data highlight the remarkable anisotropic linear and nonlinear optical behavior of HMQ-TMS crystals, expanding the knowledge of its properties and applications from the THz to the UV region.

Keywords: terahertz; THz spectroscopy; optical indices; nonlinear effects; optical rectification; organic crystals; HMQ-TMS

1. Introduction

THz radiation (1 THz$\sim$33 cm^{-1} or 4 meV photon energy) has gained over the years a considerable interest due to its broad variety of applications. Starting from fundamental scientific investigations, where THz can be used as a resonant probe for the plethora of excitations in condensed matter physics [1–3], its applications reach also to various industrial and biomedical activities [4–7], security applications [8–10], and particle-accelerator physics [11,12]. Following the growing interest, a rapid development of both THz generators and detectors has been made possible thanks to novel technologies that have become available in these last two decades, such as quantum cascade lasers, photoconductive antennas, Gunn lasers, and sources based on nonlinear optical (NLO) effects. The latter realm has been the starting point for the production of single cycle, high-intensity THz signals comparable to those obtained from free-electron facilities [13,14]. The process of difference frequency generation [15,16] or optical rectification (OR) [15,17–22] still holds the greatest interest due to its capabilities of reaching electric field magnitudes up to tens of MV/cm providing a broad THz spectral range going from nearly 0.1 THz up to 15 THz [1,23]. Due to these properties, novel NLO materials have been highly investigated in terms of THz transparency and linear and nonlinear optical responses. As already highlighted in literature, the production of THz radiation through OR process is highly dependent on the material properties, like the microscopic optical response functions of the crystal [15,24]. The real and imaginary parts of the refractive index, both in the optical and THz emission regions, give information about the phase matching condition and the absorption effects inside the crystal. Therefore, the knowledge of those optical parameters is of great importance in order to optimize the OR process and the emitted THz spectrum. Moreover, many efforts are also required in order to optimize future growth processes of new THz crystals [25,26].

Among the many materials already discovered, like inorganic NLO crystals such as ZnTe and GaP [27], organic NLO crystals offer the best platform, mainly due to their strong nonlinear optical response arising from the molecules hyperpolarizability and orientation inside the crystal [15]. Organic crystals like DAST, DSTMS, OH1 [28–30], 2-(4-hydroxy-3-methoxystyryl)-1-methylquinolinium 2,4,6-trimethylbenzenesulfonate (HMQ-TMS) [31] and BNA [32], are already widely used for THz photonics. Here, HMQ-TMS is an organic molecular crystal built upon HMQ (2-(4-hydroxy-3-methoxystyryl)-1-methylquinolinium) cations and TMS (2,4,6-trimethylbenzenesulfonate) counter anions. HMQ-TMS shows a polar axis oriented along the x_3 direction, as shown in Figure 1. Although the electromagnetic properties of HMQ-TMS have been partially studied [31,33,34], a complete investigation of its anisotropic optical properties is still missing. In recent works, Brunner et al. [33] estimated the crystal optical group index, with light polarized along the polar axis, through retardation of laser pulses [35], covering a range from 600 to 2000 nm. In addition, the absorption coefficient α_3 were also extracted from transmission measurements in the same spectral range and between 0.3 and 1.5 THz through THz time-domain spectroscopy. The same optical parameters have also been estimated for a broader THz spectral range (1.2–12 THz) in Reference [34]. In this paper, we extract from polarized reflectance and transmittance measurements, from THz to ultraviolet (UV), both the real (refraction index n) and the imaginary part (extinction coefficient k) of the complex refractive index $\tilde{n} = n - ik$, along the x_1 and x_3 (polar) axes of a HMQ-TMS single-crystal. We further measure the THz radiation generated by optical rectification at different infrared (IR) pumping wavelengths and along the two x_1 and x_3 axes. These data highlight the remarkable anisotropic linear and nonlinear optical behavior of HMQ-TMS crystal, as predicted from the crystallographic theory.

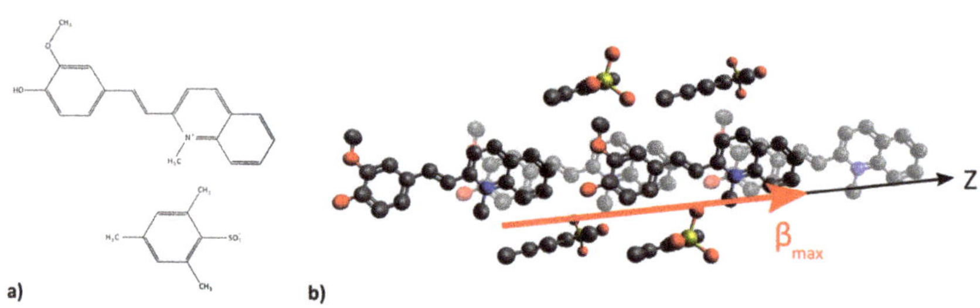

Figure 1. (**a**) Chemical structure of HMQ-TMS system. (**b**) Orientation inside the 2-(4-hydroxy-3-methoxystyryl)-1-methylquinolinium 2,4,6-trimethylbenzenesulfonate (HMQ-TMS) crystal of the HMQ and TMS molecular groups projected on the crystallographic b-axis. A massive hyperpolarizability is associated to the HMQ chromophores, which are aligned along the polar axis x_3 in such a way to define the maximum possible value of the order parameter $\langle \cos^3 \theta \rangle$ [15].

2. Experimental Methods

2.1. Linear Response Study

The Reflectance (R) and transmittance (T) at room temperature of a HMQ-TMS single crystal have been measured, from THz to UV (50–35,000 cm^{-1}), along the x_3 and x_1 axes. The crystal is characterised by a thickness of 190 μm (as measured by a micrometer) and lateral dimensions of 4 mm × 2.5 mm. The face of incidence coincides with the crystallographic *ac*-plane, parallel to the radiation polarization. THz and Mid-Infrared (MIR) regions have been investigated at the SISSI Infrared beamline in Elettra Synchrotron (Trieste) through a Bruker Vertex 70V Michelson interferometer [36–38]. The region going from NIR to UV has been studied at the Physics Department of the University of Rome "La Sapienza" through a JASCO V-770 spectrometer. A calibrated gold (aluminium) mirror in the THz/MIR (NIR/UV) has been used as a reference in reflectance experiments. The linear complex refractive index as a function of frequency, $n(\omega) - ik(\omega)$ (here ω is a wavenumber), has been obtained from $T(\omega)$ and $R(\omega)$ data by deriving the exact analytical solution to the inverse problem for a slab under the approximation $k^2 \ll n$ (no absorption at interfaces). The two indices can be expressed as [39]:

$$n(R,T,\omega) = \frac{1+R_F(R,T)}{1-R_F(R,T)} + \left\{ \frac{4R_F(R,T)}{[1-R_F(R,T)]^2} - \left(\frac{1}{2\omega d}\right)^2 \ln^2\left[\frac{R_F(R,T)T}{R-R_F(R,T)}\right] \right\}^{1/2} \tag{1}$$

$$k(R,T,\omega) = \frac{1}{2\omega d} \ln\left[\frac{R_F(R,T)T}{R-R_F(R,T)}\right], \tag{2}$$

where d is the slab thickness and the single interface reflectance R_F takes the form

$$R_F = \frac{2+T^2-(1-R)^2-\{[2+T^2-(1-R)^2]^2-4R(2-R)\}^{1/2}}{2(2-R)}$$

This method, based on both R and T, is independent from any major approximation. It is thus expected to be very precise in the determination of n and k values across the broad spectroscopic range.

2.2. Ir Pumping Scheme

In order to study the THz emission from the HMQ-TMS crystal, an optical apparatus has been developed based on a collinear optical parametric amplifier (OPA) from Light Conversion®, which permits the production of femtosecond pulses at tunable IR wavelengths, going from 1200 nm up to 1600 nm. The system is shown in Figure 2. A femtosecond high-intensity pulse at 780 nm pumps the OPA, while a minor intensity is used for detection of the THz signal through the electro-optic effect in a GaP 0.2 mm thick crystal. The signal emitted from the OPA is then used in order to pump the HMQ-TMS crystal at varying wavelengths. At constant fluence (4 mJ/cm^2), the wavelength range spans from 1300 nm to 1600 nm, and four different values have been compared for the THz generation: 1300, 1400, 1500 and 1600 nm, as suggested by previous literature [34].

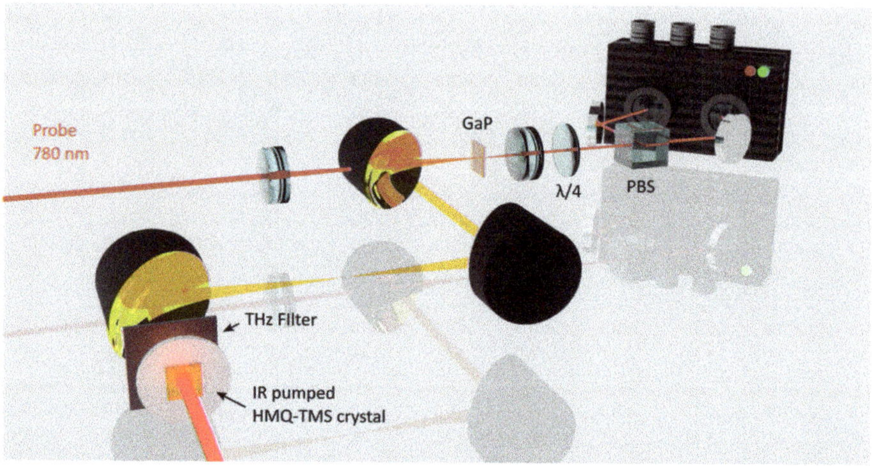

Figure 2. Infrared (IR) pumping scheme for THz generation in a HMQ-TMS crystal. An initial 40 fs pulse at 780 nm, generated from a Ti:Sapphire laser (COHERENT® Legend Elite), is injected into the optical parametric amplifier (OPA) for the generation of IR femtosecond and high-fluence pulses. The residual pump is sent to the electro-optical detection system after passing through a delay stage and a series of optical elements. A GaP crystal is used for the detection.

3. Results

3.1. Linear Optical Parameters

In Figure 3, $R(\omega)$ and $T(\omega)$ measurements for the HMQ-TMS crystal are reported between 50 and 35,000 cm^{-1}. The blue (red) solid-lines concern R data, while blue (red) dashed-lines refer to T data with light polarization along the x_1 and x_3 axes, respectively. From the T measurements, one can notice a broad transparent spectral region extending from the mid-IR to the VIS region (5000–20,000 cm^{-1}) for both axes, with a plateau at 83% along x_1 and up to 80% along x_3, respectively. This sligthly higher absorption is attributed to the major alignment of both HMQ cations and TMS anions along the x_3 axis (see Figure 1) [31]. The first electronic transition is approximately located around 20,000 cm^{-1} and corresponds to a strong reduction of transmittance along both axes, with a relatively low cut-off wavelength < 580 nm, in accordance with the estimation of Brunner et al. [33]. This electronic transition is related to the HMQ cations, which exhibit (in a methanol solution) an absorption maximum around 439 nm (nearly 22,000 cm^{-1}) [31], mapping a smaller modulation of the reflectance (Figure 3) along the x_1 axis. In the inset of Figure 3, a magnified plot of T (R) curves in the THz/MIR region is shown. Here, minima (maxima) correspond to both intra- and intermolecular (phonon) absorptions extending to the MIR region (see Figure 3).

In order to extract the real and imaginary parts of the refraction index from R and T data, the partial transparency of the HMQ-TMS single crystal in the MIR and VIS spectral region (see Figure 3) should be taken into account. Indeed, this transparency does not allow the use of Kramers-Kronig transformations. However, the complementary T and R data allow the derivation of an analytical method (see Equations (1) and (2) which considers Fresnel losses due to multiple reflections at the crystal surfaces [39]. The extracted optical parameters have been double-checked by using the RefFit constrained fitting program for a thin slab [40]. In Figure 4, the real (n) part of refraction index along x_1 (blue curves) and x_3 (red curves) axes are shown. For both axes, n is nearly constant from MIR to red, showing an average value of 1.6 (2.0) for the x_1 (x_3) axis. A strong modulation of n can be observed between 20,000 and 25,000 cm^{-1}, in correspondence to the electronic insulating gap. In the spectral range (5000–16,000 cm^{-1}), and along the x_3 axis, the refractive index behaves accordingly to already published data [33]. Moreover, the inset of the same figure shows n in the THz/MIR spectral region, which behavior is in accordance with previous works [34].

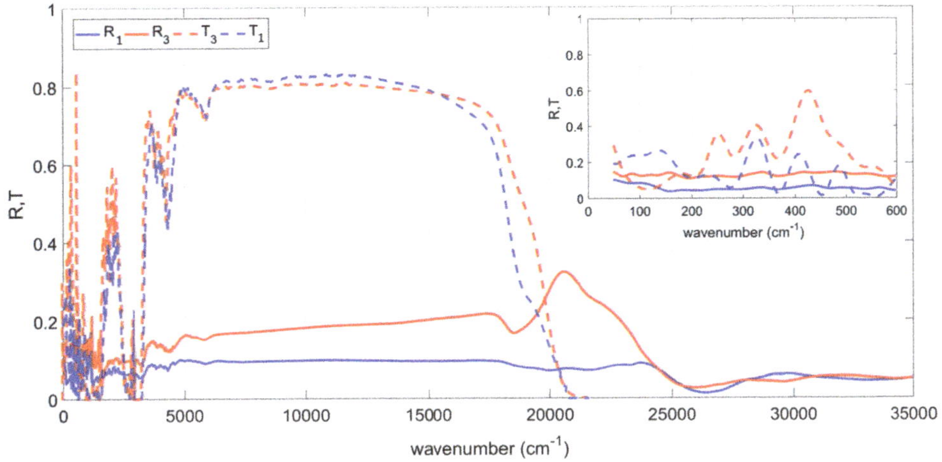

Figure 3. Polarized T and R data of HMQ-TMS single crystal at room temperature along the x_1 and x_3 axis, in the 50–35,000 cm^{-1} spectral region. T (R) data with light polarization along x_1 and x_3 are represented by dashed (solid) blue and red lines, respectively. In the inset, R and T data are plotted in the THz/MIR region.

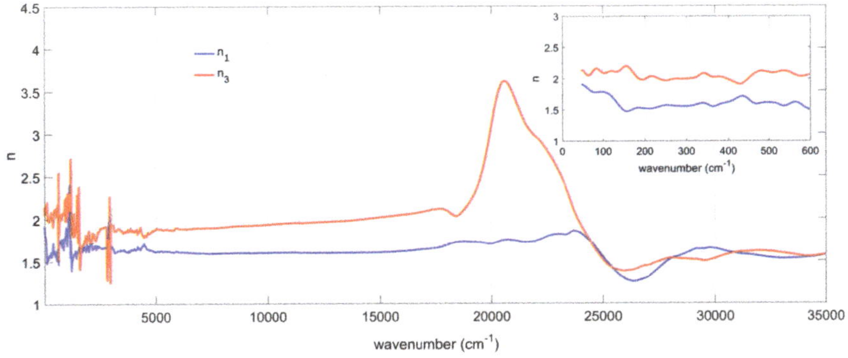

Figure 4. Real part of refraction index (n) of the HMQ-TMS crystal at room temperature. Solid-blue (red) line corresponds to the value along the x_1 (x_3) axis. In the inset, n is plotted in the THz/MIR spectral range. The strong variation of n around 20,000 cm^{-1} is generated by an electronic transition of the HMQ cations.

3.2. Spectral Analysis

The absorption coefficients, along both x_1 and x_3 axes, are calculated through the extinction coefficient k as $\alpha = 4\pi\omega k$ (ω is a wavenumber). They are reported in the spectral range 400–4000 cm^{-1}, where most of the vibrational excitations of HMQ and TMS chemical groups fall (see Figure 5a,b). Differently to the electronic transitions that show a remarkable anisotropy (see Figure 4), the two vibrational spectra have several peaks in common. A small anisotropy can be observed only below 1000 cm^{-1}, where ring-structure bending and lattice modes are located, and can be attributed to molecules orientation. In order to assign those peaks, one can observe that aromatic rings in the HMQ-TMS structure (see Figure 1) show several C-H and C=C-C vibrational modes. Specifically, the bending modes of quinolinium ring are present below 650 cm^{-1}. C-H out-of-plane and in-plane bending vibrations occur in the regions 670–900 cm^{-1} and 950–1225 cm^{-1}, respectively [41]. Along axis x_1, the band at 528 cm^{-1} is identified with the C-N-C and C-C-N in-plane bending modes. For axis x_3, in the region 450–600 cm^{-1}, two shoulders are distinguished at 470 and 609 cm^{-1}, and assigned to the symmetric and asymmetric bending vibrations of the -SO_3 group [42–44]. The peaks at about 1030 cm^{-1}, 1140 cm^{-1} and 1350 cm^{-1} can be assigned to the symmetric and asymmetric SO_3^- stretching, respectively [41]. Between 1260–1340 cm^{-1}, three weak shoulders can be associated to aromatic primary amine C-N stretching. The peaks at 1530 and 1590 cm^{-1} can be attributed to the vibrations of aromatic rings, while the absorptions around 1390, 1430 and 1480 cm^{-1} are due to trimethyl CH_3. The shoulder at 2652 cm^{-1} is related to the stretching vibration of C-CH_3, located in the trimethylbenzenesulfonate. The methylamino N-CH_3 vibrational band is located at 2760 cm^{-1}. Above 2800 cm^{-1}, the C-H bonds vibrate with the methyl C-H symmetric and asymmetric stretching at 2860 and 2960 cm^{-1}, respectively, and the methyl ether O-CH_3 and C-H stretching corresponding to the band at 2815 cm^{-1}.

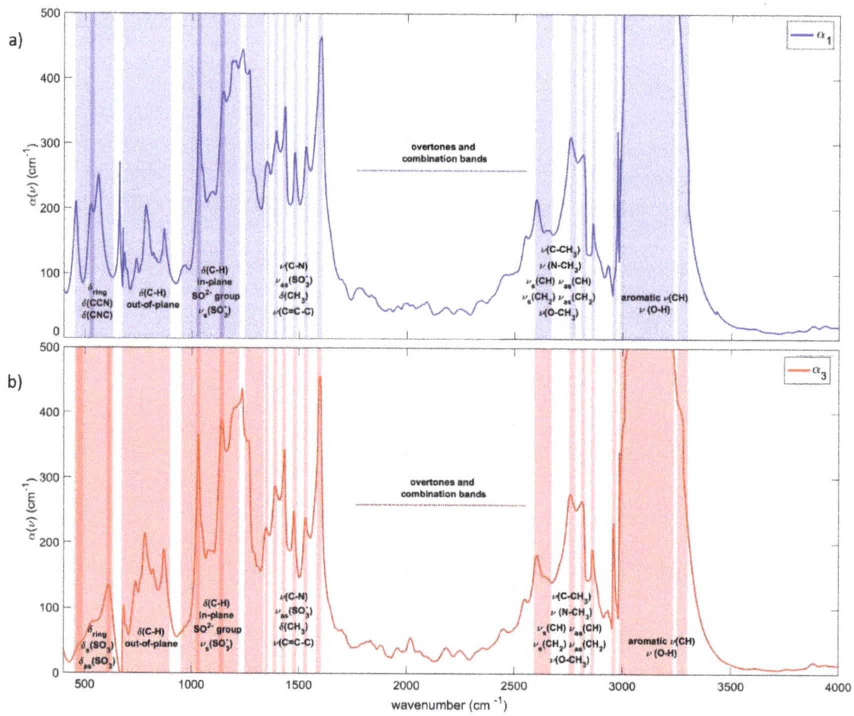

Figure 5. Absorption coefficients of HMQ-TMS crystal at room temperature in the vibrational spectral region 400–4000 cm^{-1} along the polar x_3 (**a**) and x_1 (**b**) axes. The shaded area pictorially assigns the absorption peaks (or vibrational regions) to specific vibrational modes of HMQ and TMS chemical groups. (The labels are: $\delta \rightarrow$ bending vibration, $\nu \rightarrow$ stretching vibration, $as \rightarrow$ asymmetric, $s \rightarrow$ symmetric, $ring \rightarrow$ quinolinium ring).

The narrow peak at 3000 cm^{-1} and the shoulder around 3010 cm^{-1} are attributed to C-H bonds around the aromatic rings [41]. The region between 3020–3230 cm^{-1}, related to aromatic C-H stretching and hydroxyl group vibrations, shows a very strong absorption that is nearly saturated. The small shoulder, located at 3250 cm^{-1}, can be attributed to O-H vibrational bonds. These general assignments are reported in Figure 5a,b.

3.3. Thz Generation

For completeness, the nonlinear properties in terms of THz generation vs. different pumping wavelengths, along the x_1 and x_3 axes, have also been measured. Referring to the scheme of Figure 2, and varying the timing overlap between the THz pulse and the 780 nm probe in a GaP detection crystal, it is possible to scan the THz electric field magnitude in order to compute the spectral amplitude. The amplitudes along the x_3 axis (coinciding with the maximum generation efficiency) at a fluence of 4 mJ/cm^2, and at different IR pumping wavelengths (1300, 1400, 1500 and 1600 nm), are shown in Figure 6a. The comparable magnitudes of the field vs. the pumping wavelength suggest a nearly flat THz efficiency of the HMQ-TMs crystal in the whole infrared range.

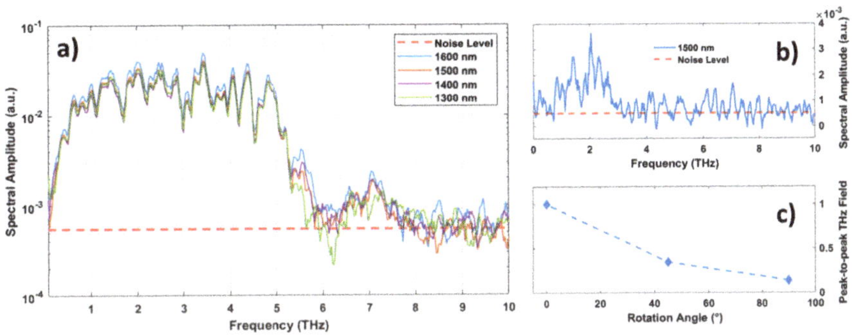

Figure 6. (**a**) Spectral amplitude of the THz field generated by a HMQ-TMS single crystal pumped by a fs optical pulse at different IR wavelengths (1300, 1400, 1500 and 1600 nm). A broad frequency THz generation is visible from 0.5 up to 6 THz, with a further contribution around 7 THz. Minima in the spectra are mainly due to the water vapour absorption. (**b**) Spectral amplitude of the THz field generated along the x_1 axis. The intensity and spectral bandwidth are strongly reduced in comparison to the x_3 axis. Red-dashed lines in (a) and (b) represent the noise level in our experiment. (c) Peak-to-peak THz field magnitude *vs* the angle between the x_3 axis and the pumping polarization. The THz field value strongly decreases for an increasing angle, indicating a strong reduction of the THz emission efficiency for a pump polarization along the x_1 axis.

The anisotropic THz emission properties of HMQ-TMS have been studied by varying the crystal orientation with respect to the linearly polarized pump. In particular, both the incidence OPA polarization and the GaP detection crystal orientation have been kept fixed while the crystal has been rotated. Although we cannot exclude that some THz intensity might come from orientation misalignment and polarization losses, a small THz emission can be observed along x_1 centered around 2 THz (Figure 6b) for a pumping wavelength at 1500 nm (similar data have been obtained at the other wavelengths). More specifically, the THz magnitude vs. the crystal orientation (Figure 6c) progressively decreases when the pumping polarization approaches x_1.

4. Conclusions

In this paper, we have measured the complex refraction index of a HMQ-TMS single crystal from terahertz to ultraviolet, both along the polar x_3 axis and the orthogonal x_1 axis on the crystallographic *ac*-plane. In the visible-ultraviolet region, we observe a remarkable anisotropy which is strongly attenuated in the infrared and terahertz range. The precise extraction of both the refractive indices and the absorption coefficients proposes an inverse problem approach for the THz generation study. Therefore, we have also measured the terahertz emission spectra along the same axes when pumping in the infrared through a fs-amplifier laser. As expected from theoretical grounds, the THz emission shows a huge intensity reduction when the pumping polarization is parallel to the x_1 axis. These data expand our knowledge of the HMQ-TMS optical properties across the broad spectral range from THz to UV, allowing a better understanding of its possible applications in THz pump-probe experiments of exotic electronic systems [45,46].

Author Contributions: Conceptualization, A.D., L.T., S.L. and M.P.; methodology, A.D. and L.T.; software, A.D. and L.T.; validation, S.L., A.D. and L.T.; formal analysis, A.D. and L.T.; investigation, V.D., L.T., S.M., P.D.P. and A.P.; resources, S.L. and M.P.; data curation, A.D. and L.T.; writing—original draft preparation, A.D., L.T. and S.L.; writing—review and editing, all authors; visualization, A.D. and L.T.; supervision, M.P. and S.L.; project administration, M.P. and S.L.; funding acquisition, S.L. All authors have read and agreed to the published version of the manuscript.

Funding: Ministero dell'Istruzione, dell'Università e della Ricerca (MIUR) (Rita Levi Montalcini); Italian Ministry of Foreign Affairs and International Cooperation (PGR00806).

Conflicts of Interest: The authors declare no conflicts of interest.

Abbreviations

The following abbreviations are used in this manuscript:

MDPI	Multidisciplinary Digital Publishing Institute
DOAJ	Directory of open access journals
TLA	Three letter acronym
LD	linear dichroism

References

1. Zhang, X.; Xu, J. *Introduction to THz Wave Phoronics*; Springer: New York, NY, USA, 2010.
2. Giorgianni, F.; Sakai, J.; Lupi, S. Overcoming the thermal regime for the electric-field driven Mott transition in vanadium sesquioxide. *Nat. Commun.* **2019**, *1*, 1159. [CrossRef] [PubMed]
3. Giorgianni, F.; Chiadroni, E.; Rovere, A.; Cestelli-Guidi, M.; Perucchi, A.; Bellaveglia, M.; Castellano, M.; Di Giovanale, D.; Di Pirro, G.; Ferrario, M.; et al. Strong nonlinear terahertz response induced by Dirac surface states in Bi2Se3 topological insulator. *Nat. Commun.* **2016**, *7*, 11421. [CrossRef] [PubMed]
4. Hochrein, T. Markets, Availability, Notice, and Technical Performance of Terahertz Systems: Historic Development, Present, and Trends. *J. Infrared Millim. Terahertz Waves* **2015**, *36*, 235–254. [CrossRef]
5. Tonouchi, M. Cutting-edge terahertz technology. *Nat. Photonics* **2007**, *1*, 97–105. [CrossRef]
6. Pickwell, E.; Wallace, V.P. Biomedical applications of terahertz technology. *J. Phys. D Appl. Phys.* **2006**, *39*, R301–R310. [CrossRef]
7. D'Arco, A.; Di Fabrizio, M.; Dolci, V.; Petrarca, M.; Lupi, S. THz Pulsed Imaging in Biomedical Applications. *Condens. Matter* **2020**, *5*, 25. [CrossRef]
8. Federici, J.F.; Schulkin, B.; Huang, F.; Gary, D.; Barat, R.; Oliveira, F.; Zimdars, D. THz imaging and sensing for security applications—Explosives. weapons and drugs. *Semicond. Sci. Technol.* **2005**, *20*, S266–S280. [CrossRef]
9. Ergün, S.; Sönmez, S. Terahertz Technology for Military Applications. *J. Manag. Inf. Sci.* **2015**, *3*, 13–16. [CrossRef]
10. Leahy-Hoppa, M.R.; Fitch, M.J.; Osiander, R. Terahertz spectroscopy techniques for explosives detection. *Anal. Bioanal. Chem.* **2009**, *395*, 247–257. [CrossRef] [PubMed]
11. Curcio, A.; Marocchino, A.; Dolci, V.; Lupi, S.; Petrarca, M. Resonant plasma excitation by single-cycle THz pulses. *Sci. Rep.* **2018**, *8*, 1052. [CrossRef] [PubMed]
12. Nanni, E.A.; Huang, K.-H.; Ravi, K.; Fallahi, A.; Moriena, G.; Miller, R.J.D.; Kartner, F.X. Terahertz-driven linear electron acceleration. *Nat. Commun.* **2015**, *6*, 8486. [CrossRef] [PubMed]
13. Chiadroni, E.; Bacci, A.; Bellaveglia, M.; Boscolo, M.; Castellano, M.; Cultrera, L.; Di Pirro, G.; Ferrario, M.; Ficcadenti, L.; Filippetto, D.; et al. The SPARC linear accelerator based terahertz source. *Appl. Phys. Lett.* **2013**, *102*, 094101. [CrossRef]

14. Perucchi, A.; Mitri, S.D.; Penco, G.; Allaria, E.; Lupi, S. The TeraFERMI terahertz source at the seeded FERMI free-electron-laser facility. *Rev. Sci. Instrum.* **2013**, *84*, 022702. [CrossRef] [PubMed]
15. Jazbinsek, M.; Puc, U.; Abina, A.; Zidansek, A. Organic Crystals for THz Photonics. *Appl. Sci.* **2019**, *9*, 882. [CrossRef]
16. Sutherland, R.L. *Handbook of Nonlinear Optics*; Dekker: New York, NY, USA, 2003.
17. Zhang, X.C.; Ma, X.F.; Jin, Y.; Lu, T.M.; Boden, E.P.; Phelp, P.D.; Stewart, K.R.; Yakymyshyn, C.P. Terahertz optical rectification from a nonlinear organic crystal. *Appl. Phys. Lett.* **1992**, *61*, 3080–3082. [CrossRef]
18. Winnewisser, C.; Jepsen, P.; Schall, M.; Schyja, V.; Helm, H. Electro-optic detection of THz radiation in $LiTaO_3$, $LiNbO_3$ and ZnTe. *Appl. Phys. Lett.* **1997**, *70*, 3069–3071. [CrossRef]
19. Vicario, C.; Jazbinsek, M.; Ovchinnikov, A.V.; Chefonov, O.V.; Ashitkov, S.I.; Agranat, M.B.; Hauri, C.P. High efficiency THz generation in DSTMS, DAST and OH1 pumped by Cr:forsterite laser. *Opt. Express* **2015**, *23*, 4573–4580. [CrossRef]
20. Dolci, V.; Cascioli, V.; Curcio, A.; Ficcadenti, L.; Lupi, S.; Petrarca, M. Intensity and phase retrieval of IR laser pulse by THz-based measurement and THz waveform modulation. *Nucl. Instrum. Methods Phys. Res. A* **2018**, *909*, 204–207. [CrossRef]
21. Curcio, A.; Dolci, V.; Lupi, S.; Petrarca, M. Terahertz-based retrieval of the spectral phase and amplitude of ultrashort laser pulses. *Opt. Lett.* **2018**, *43*, 2783–2786. [CrossRef]
22. Curcio, A.; Petrarca, M. Diagnosing plasmas with wideband terahertz pulses. *Opt. Lett.* **2019**, *44*, 1011–1014. [CrossRef]
23. Kono, S.; Tani, M.; Sakai, K. Ultrabroadband photoconductive detection: Comparison with free-space electro-optic sampling. *Appl. Phys. Lett.* **2001**, *79*, 898–900. [CrossRef]
24. Hebling, J.; Yeh, K.-L.; Hoffmann, M.C.; Bartal, B.; Nelson, K.A. Generation of high-power terahertz pulses by tilted-pulse-front excitation and their application possibilities. *J. Opt. Soc. Am. B* **2008**, *25*, B6–B19. [CrossRef]
25. Yang, Z.; Mutter, L.; Stillhart, M.; Aravazhi, S.; Jazbinšek, M.; Schneider, A.; Gramlich, V.; Günter, P. Large-Size Bulk and Thin-Film Stilbazolium-Salt Single Crystals for Nonlinear Optics and THz Generation. *Adv. Funct. Water* **2007**, *17*, 2018–2023. [CrossRef]
26. Ogawa, J.; Okada, S.; Glavcheva, Z.; Nakanishi, H. Preparation, properties and structures of 1-methyl-4-2-[4-(dimethylamino)phenyl]ethenylpyridinium crystals with various counter anions. *J. Cryst. Growth* **2008**, *310*, 836–842. [CrossRef]
27. Tomasino, A.; Parisi, A.; Stivala, S.; Livreri, P.; Cino, A.C.; Busacca, A.C.; Peccianti, M.; Morandotti, R. Wideband THz Time Domain Spectroscopy based on Optical Rectification and Electro-Optic Sampling. *Sci. Rep.* **2013**, *3*, 3116. [CrossRef]
28. Cunningham, P.D.; Hayden, L.M. Optical properties of DAST in the THz range. *Opt. Express* **2010**, *18*, 23620–23625. [CrossRef]
29. Vicario, C.; Ovchinnikov, A.V.; Ashitokov, S.I.; Agranat, M.B.; Fortov, V.E.; Hauri, C.P. Generation of 0.9-mJ THz pulses in DSTMS pumped by a Cr:Mg_2SiO_4 laser. *Opt. Lett.* **2014**, *39*, 6632–6635. [CrossRef]
30. Brunner, F.D.J.; Kwon, O.P.; Kwon, S.-J.; Jazbinšek, M.; Schneider, A.; Günter, P. A hydrogen-bonded organic nonlinear optical crystal for high-efficiency terahertz generation and detection. *Opt. Express* **2008**, *16*, 16496–16508. [CrossRef]
31. Jeong, J.H.; Kang, B.J.; Kim, J.S.; Jazbinšek, M.; Lee, S.H.; Lee, S.C.; Baek, I.H.; Yun, H.; Kim, J.; Lee, Y.S.; et al. High-power Broadband Organic THz Generator. *Sci. Rep.* **2013**, *3*, 3200–3207. [CrossRef]
32. Hashimoto, H.; Okada, Y.; Fujimura, H.; Moioka, M.; Sugihara, O.; Okamoto, N.; Matsushima, R. Second-Harmonic Generation from Single Crystals of N-Substituted 4-Nitroanilines. *Jpn. J. Appl. Phys.* **1997**, *36*, 6754–6760. [CrossRef]
33. Brunner, F.D.J.; Lee, S.-H.; Kwon, O.-P.; Feurer, T. THz generation by optical rectification of near-infrared laser pulses in the organic nonlinear optical crystal HMQ-TMS. *Opt. Mater. Express* **2014**, *4*, 1586–1592. [CrossRef]
34. Vicario, C.; Monoszlai, B.; Jazbinsek, M.; Lee, S.-H.; Kwon, O.-P.; Hauri, C.P. Intense, carrier frequency and bandwidth tunable quasi single-cycle pulses from an organic emitter covering the Terahertz frequency gap. *Sci. Rep.* **2005**, *5*, 14394. [CrossRef] [PubMed]
35. Schneider, A.; Brunner, F.D.J.; Gunter, P. Determination of the refractive index over a wide wavelength range through time-delay measurements of femtosecond pulses. *Opt. Commun.* **2007**, *275*, 354–458. [CrossRef]

36. Lupi, S.; Nucara, A.; Perucchi, A.; Calvani, P.; Ortolani, M.; Quaroni, L.; Kiskinova, M. Performance of SISSI, the infrared beamline of the ELETTRA storage ring. *J. Opt. Soc. Am. B* **2007**, *24*, 959–964. [CrossRef]
37. Karantzoulis, E.; Penco, G.; Perucchi, A.; Lupi, S. Characterization of coherent THz radiation bursting regime at Elettra. *Infrared Phys. Techn.* **2010**, *53*, 300–303. [CrossRef]
38. Roy, P.; Brubach, J.B.; Calvani, P.; de Marzi, G.; Filabozzi, A.; Gerschel, A.; Giura, P.; Lupi, S.; Marcouillé, O.; Mermet, A.; et al.; Infrared synchrotron radiation: From the production to the spectroscopic and microscopic applications. *Nucl. Instrum. Methods Phys. Res. Sect. A Accel. Spectrom. Detect. Assoc. Equip.* **2001**, *467 Pt 1*, 426–436. [CrossRef]
39. Nichelatti, E. Complex refractive index of a slab from reflectance and transmittance: Analytical solution. *J. Opt. A Pure Appl. Opt.* **2002**, *4*, 300–403. [CrossRef]
40. Kuzmenko, A.B. Kramers–Kronig constrained variational analysis of optical spectra. *Rev. Sci Instrum.* **2005**, *76*, 083108. [CrossRef]
41. Coates, J. Interpretation of Infrared Spectra, A Practical Approach. In *Encyclopedia of Analytical Chemistry*; Meyers, R.A., Ed.; John Wiley & Sons Ltd.: Chichester, UK, 2000; pp. 10815–10837.
42. Buzykin, O.G.; Ivanov, S.V.; Ionin, A.A.; Kotkov, A.A.; Kozlov, A.Y. Spectroscopic detection of sulfur oxides in the aircraft wake. *J. Russ. Laser Res.* **2005**, *26*, 402–426. [CrossRef]
43. Shishlov, N.M.; Khursan, S.L. Effect of ion interactions on the IR spectrum of benzenesulfonate ion. Restoration of sulfonate ion symmetry in sodium benzenesulfonate dimer. *J. Mol. Struct.* **2016**, *1123*, 360–366. [CrossRef]
44. Özel, A.E.; Büyükmurat, Y.; Akyüz, S. Infrared-spectra and normal-coordinate analysis of quinoline and quinoline complexes. *J. Mol. Struct.* **2001**, *565–566*, 455–462. [CrossRef]
45. Lupi, S.; Ortolani, M.; Calvani, P. Optical conductivity of single crystals of Na0.57CoO2. *Phys. Rev. B* **2004**, *69*, 180506. [CrossRef]
46. Phuoc, V.T.; Vaju, C.; Corraze, B.; Sopracase, R.; Perucchi, A.; Marini, C.; Postorino, P.; Chiligui, M.; Lupi, S.; Janod, E.; et al. Optical Conductivity Measurements of GaTa4Se8 Under High Pressure: Evidence of a Bandwidth-Controlled Insulator-to-Metal Mott Transition. *Phys. Rev. Lett.* **2013**, *110*, 037401. [CrossRef] [PubMed]

condensed matter

MDPI

Article

Spatially Resolved Spectral Imaging by A THz-FEL

Akinori Irizawa [1,*], Masaki Fujimoto [1,†], Keigo Kawase [1,‡], Ryukou Kato [1,§], Hidenori Fujiwara [2], Atsushi Higashiya [3], Salvatore Macis [4,5], Luca Tomarchio [4], Stefano Lupi [4,5], Augusto Marcelli [5,6] and Shigemasa Suga [1]

1 The Institute of Scientific and Industrial Research (ISIR), Osaka University, 8–1 Mihogaoaka, Ibaraki, Osaka 5670047, Japan; mfmoto@ims.ac.jp (M.F.); kawase.keigo@qst.go.jp (K.K.); ryukou@post.kek.jp (R.K.); ssmsuga@gmail.com (S.S.)
2 Graduate School of Engineering Science, Osaka University, 1–3 Machikaneyama, Toyonaka, Osaka 5608531, Japan; fujiwara@mp.es.osaka-u.ac.jp
3 Faculty of Science and Engineering, Setsunan University, 17–8 Ikedanaka-Machi, Neyagawa, Osaka 5728508, Japan; higashiya@mpg.setsunan.ac.jp
4 Department of Physics, Sapienza University, P.le Aldo Moro 5, 00185 Rome, Italy; salvatore.macis91@gmail.com (S.M.); luca.tomarchio94@gmail.com (L.T.); stefano.lupi@roma1.infn.it (S.L.)
5 Istituto Nazionale di Fisica Nucleare—Laboratori Nazionali di Frascati (INFN-LNF), Via Enrico Fermi 40, 00044 Frascati, Italy; augusto.marcelli@lnf.infn.it
6 International Centre for Material Science Superstripes, RICMASS, via dei Sabelli 119A, 00185 Rome, Italy
* Correspondence: irizawa@sanken.osaka-u.ac.jp
† Current address: National Institutes of Natural Sciences, Institute for Molecular Science, UVSOR Facility, 38 Nishigo-Naka, Myodaiji, Okazaki, Aichi 4448585, Japan.
‡ Current address: National Institutes for Quantum and Radiological Science and Technology, 2–4 Shirane, Tokai, Ibaraki 3191106, Japan.
§ Current address: High Energy Accelerator Research Organization, KEK, 1–1 Oho, Tsukuba, Ibaraki 3050801, Japan.

Received: 28 April 2020; Accepted: 2 June 2020; Published: 4 June 2020

Abstract: Using the unique characteristics of the free-electron-laser (FEL), we successfully performed high-sensitivity spectral imaging of different materials in the terahertz (THz) and far-infrared (FIR) domain. THz imaging at various wavelengths was achieved using in situ spectroscopy by means of this wavelength tunable and monochromatic source. In particular, owing to its large intensity and directionality, we could collect high-sensitivity transmission imaging of extremely low-transparency materials and three-dimensional objects in the 3–6 THz range. By accurately identifying the intrinsic absorption wavelength of organic and inorganic materials, we succeeded in the mapping of spatial distribution of individual components. This simple imaging technique using a focusing optics and a raster scan modality has made it possible to set up and carry out fast spectral imaging experiments on different materials in this radiation facility.

Keywords: THz; far infrared; FEL; spectroscopy; imaging

1. Introduction

Terahertz (THz)-wave and/or far-infrared photons represent radiation located in the frequency region, which is well known as the "terahertz-gap". This region is a frequency domain that is still particularly interesting in the field of radiation generation and detection technology. In terms of applicability, conventional light sources based on the blackbody emission and well-established spectrometers or interferometers are rather common tools. On the other hand, far-infrared radiation emitted by synchrotron radiation (SR) and free-electron-laser (FEL) sources is powerful for spectroscopy and microspectroscopy. Thanks to the high brilliance and focusing properties, these light sources important results have been obtained with experiments performed with these non-thermal radiation sources. Pump-probe experiments using intense and coherent photon sources to investigate non-linear materials properties can be realized only by using high-intensity FELs. In the recent years, we have performed various linear and non-linear experiments using the terahertz-free-electron-laser (THz-FEL) in the Institute of Scientific and Industrial Research (ISIR) of Osaka University [1–4].

Regarding the electromagnetic waves, THz and far-infrared (FIR) radiation falling in the "terahertz gap" share both the behavior of high-frequency radio waves and low-energy photons. For this reason, in different areas such as physics, chemistry, engineering, and bio-medicine, this radiation can be described by different units such as frequency, wavenumber, wavelength, and energy. Their mutual relationship is shown in Equation (1) with the reference to approximately 3 THz, where only the wavelength is inversely proportional to the others:

$$3\ \text{THz} = 100\ \text{cm}^{-1} = 100\ \mu\text{m} = 12.4\ \text{meV} \tag{1}$$

The terahertz band is conventionally understood as ranging from approximately 0.3 to 3 THz around 1 THz within the International Telecommunication Union (ITU) designated band of frequencies. In addition to the light sources having characteristics of intensity, stability, monochromaticity, or broadbandness, detectors with high sensitivity, high-speed response, and wide (linear) dynamic range are under development. Besides far-infrared photons emitted from a black body in accordance with Planck's law, various other light sources exist, e.g., synchrotron radiation light sources, vacuum tubes including gyrotrons, gas, and solid-state lasers, superconducting devices, non-linear optical devices, and FELs, which are mainly described in this contribution.

The ISIR THz-FEL is a monochromatic, wavelength-tunable and highly coherent pulsed light source with megawatt-class peak intensity. For the first time in the 1970s and in the early phase, FELs produced radiation in the mid-infrared region [5]. The emission wavelength was continuously reduced down to the X-ray range with the goal of studying nano- and sub-nano-size objects and reach the pico- and sub-picosecond time resolution in different research areas [6,7]. On the other hand, although infrared FELs are relatively compact, the operation and the maintenance of these linear accelerators requite important human and/or economic supports from universities and institutions. Mainly for this reason, some FEL facilities in the infrared region were discontinued in the past. However, since the last decade, long wavelengths FELs are again attracting attention as high-intensity coherent radiation sources. Indeed, the increasing number of studies using infrared FEL radiation from the operational facilities is triggering the constructions of new infrared FELs and the plan of new IR and THz beamlines worldwide [8–12].

2. Characteristics of the ISIR THz-FEL

The FEL installed at the quantum beam science research facility in ISIR of the Osaka University can produce high-intensity pulsed light in the THz and FIR domains. The accelerator facility was established in 1957, and after almost 40 years, in 1978, the L-band electron linear accelerator was installed. After the first successful FEL oscillation [13], several upgrades have been realized. The details of the continuous improvement of the electron gun and the accelerator are summarized in Refs. [14–18]. In this contribution, we describe the experimental layout designed and assembled in ISIR and some recent experimental results obtained using this layout. The ISIR THz-FEL can be also used as a pump source thanks to the high intensity and time characteristics or as a probe source if combined with the irradiation of an external laser. In the next section, we will focus on microscopy and spectral-imaging experiments performed with the ISIR THz-FEL as the probe source and will discuss the typical FEL parameters we used for the experiments discussed below.

Figure 1 shows the pulsed time structure of the ISIR THz-FEL radiation reflecting the structure of an electron-bunch train. The FEL pulse train has a two-level structure. The first level is the micro-pulse, and the set of micro-pulses forms the macro-pulse. The long-time structure contains several macro-pulses emitted in sequence every 200 ms (5 Hz repetition). Each macro-pulse contains approximately 100 micro-pulses. The interval among micro-pulses depends on the FEL oscillation mode, i.e., 9.2 ns or 37 ns (= 9.2 ns × 4). The emission is due to ISIR THz-FEL's pre-bunching system, which aims to increase the charge of the electron bunches. The time interval is 12 or 48 times longer than the length of 0.77 ns associated with the RF frequency of 1.3 GHz of the Klystron's. The 9.2 ns interval is associated with the 108 MHz operation of the sub-harmonic buncher, while the 37 ns interval is due to the 27 MHz grid pulser installed on the thermal cathode electron gun.

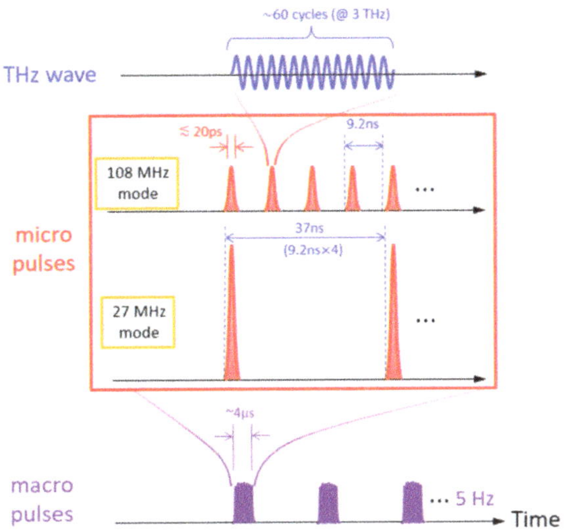

Figure 1. Time structure of the pulsed Institute of Scientific and Industrial Research (ISIR) terahertz-free-electron-laser (THz-FEL). The pulse structure is approximately 4 μs in the macro-pulse and approximately 20 ps in the micro-pulse, respectively.

In the spectroscopy experiments using the FEL as the radiation probe, a stable intensity mode with the micro-pulse interval of 9.2 ns (108 MHz mode) has been usually employed. For irradiation experiments on materials and for non-linear response experiments, the mode with the 37 ns interval (27 MHz mode) has been selected.

In this mode, the light intensity per micro-pulse was greatly increased and used as the pump source. In both modes, beam conditions are searched for the best monochromaticity and stability, and the irradiation energy is up to 10 mJ/macro-pulse in the 27 MHz mode and up to 1 mJ/macro-pulse in the 108 MHz mode. Figure 2 shows the typical wavelength dispersions of the THz-FEL in the (a) 108 MHz mode and the (b) 27 MHz mode. Although the monochromaticity differs much depending on the beam parameters and the FEL mode, the bandwidth is approximately 3% at the 108 MHz mode and approximately 10% at the 27 MHz mode under the best monochromaticity conditions. Actually, deep care must be taken when one adjusts the beam condition to make the energy maximized in the 27 MHz mode, because it often shows a wideband wavelength spectrum that can no longer be considered monochromatic (Appendix A).

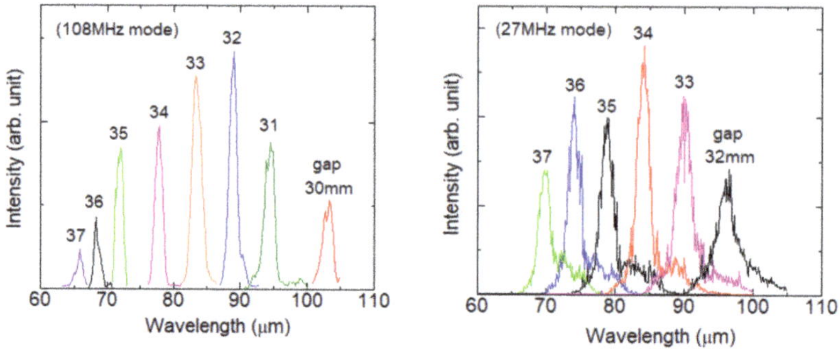

(a)

(b)

Figure 2. FEL spectra for different undulator gap values and modes. (**a**) Spectrum at variable gaps in the range of 30–37 mm at 108 MHz (monochromatic condition); (**b**) Spectrum at variable gaps in the range of 32–37 mm at 27 MHz (monochromatic condition).

3. Spectral Imaging Using THz-FEL

So far, THz imaging using various light sources and detectors has been performed, and technologies have been developed as in other wavelength regions [19–24]. Among them, time domain spectroscopy (TDS), quantum cascade laser (QCL), and terahertz parametric generator (TPG) using $LiNbO_3$ non-linear optical crystals have been employed for the imaging at laboratory level below the 3 THz frequency domain. Moreover, the near field method has enabled the possibility of reaching a spatial resolution below the diffraction limit. However, THz/FIR radiation may not compete with X-rays in terms of transmittance and spatial resolution. That is, the most important motivation for imaging at the THz/FIR domain should be the observation of materials that have a characteristic response in this wavelength range. From this point of view, only a few imaging experiments at arbitrary monochromatic wavelengths, which can be called spectral imaging, are so far reported in the region below 3 THz [25,26]. In other words, conventional THz imaging has been hardly performed by probing the precise wavelength dependence of different materials characterized by specific absorptions, and most researchers have performed imaging by selecting an example that matches the wavelength maximum of each light source.

In the following, we will describe the status of spectroscopy for imaging using long wavelength radiation sources. Such a radiation, when emitted from a storage ring (SR), is competitive because of the brilliance, which is much higher than any standard laboratory light source based on thermal radiation. In this respect, SR is highly effective for any spectroscopic measurement of small areas using condensing optics, e.g., for microspectroscopy [27–29]. However, even in the case of SR sources, the total photon number is not enough for many experiments, especially in the far-infrared region. In general, it is necessary to accumulate spectra for several tens of seconds to several minutes using a Fourier transform infrared (FT-IR) spectrometer even for a single acquisition. Furthermore, it is well known that diffraction grating type spectrometers are more demanding in terms of photon numbers. Therefore, an impractical long-time integration is frequently required in 2D scan spectral imaging. For example, a scan of 50 × 50 points would require approximately 7 h, even if a FT-IR instrument takes only 10 s to measure each point. In contrast, THz-FEL is a quasi-monochromatic light source, and it is several orders of magnitude more intense than SR infrared radiation for both total photon number and illuminance. Thus, the spectral imaging using FEL is certainly realistic even when monochromatized through a diffraction grating-type spectrometer. We will present and discuss high-sensitive spectral imaging experiments performed with arbitrary monochromatic wavelengths in the range of 50–100 μm, i.e., the frequency range of 3–6 THz.

3.1. Optical Scheme and Beam Profile

For spectral imaging, the quasi-monochromatic FEL emission, as shown in Figure 2a, must be further monochromatized. Figure 3 shows the layout of the monochromator, the focusing optics, and the sample stage system to be really employed. FEL radiation transferred through the vacuum tube is first monochromatized by the diffraction grating in the monochromator and propagates from vacuum to atmosphere, as a parallel beam through a diamond window. The monochromatic FEL radiation is focused at the sample position by the lens and refocused on the detector in the transmission arrangement, or in the alternative reflection layout. The lens used is an F = 1.97 (effective aperture Φ = 25.4 mm, focal length f = 50 mm) Tsurupica lens, and the detector used is a COHERENT Energy Max energy sensor calibrated in the THz region. The measurement sample stage is a SIGMAKOKI motorized x-y stage that allows a raster scan in the x-y plane simultaneously with the pulse timing of the FEL.

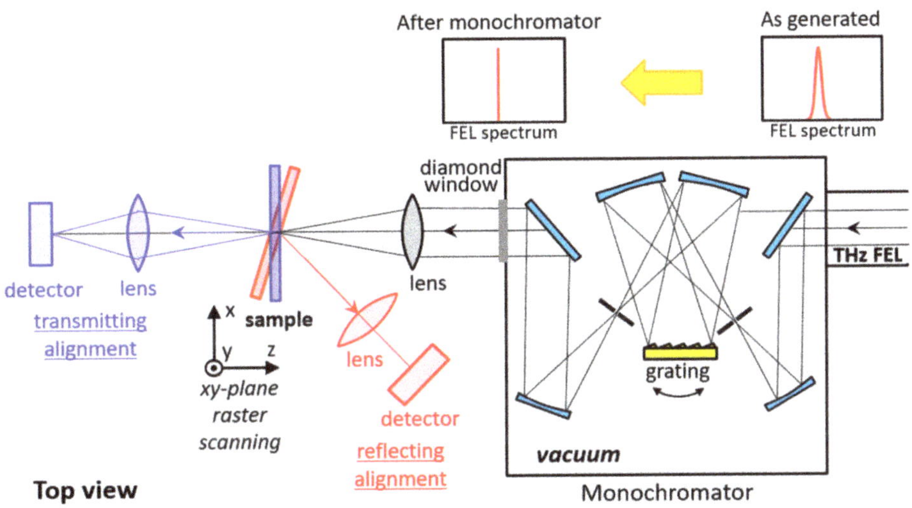

Figure 3. Schematic layout of the monochromator, the focusing optics, and the sample stage system installed downstream the FEL. The blue layout is for transmission imaging, while the red layout refers to reflection imaging.

Figure 4 shows (a) (left hand) the beam profiles of the THz-FEL taken by an uncooled THz imager (NEC Corporation, IR/V–T0831) [30] just after the vacuum window and (right hand) the focal point condensed by the Tsurupica lens and (b) the intensity profiles by a knife-edge scan near the focal point. The parallel beam of the THz-FEL obtained through a 1-inch diameter diamond vacuum window is focused to an area of approximately 350 μm in 2σ at the focal point. The beam profiles observed by the THz imager (Figure 4a), measured with a knife-edge scan (Figure 4b), show an almost Gaussian distribution in both parallel areas just after the vacuum window and the focal point. The beam profile distribution near the focal point was probed by a knife-edge scan at several points in the z-axis direction (optical axis direction). As a result, the Rayleigh length was estimated to be approximately 3.5 mm. As a consequence, we expect that a clear transmitting image can be obtained up to a thickness of the sample of approximately 7 mm or less.

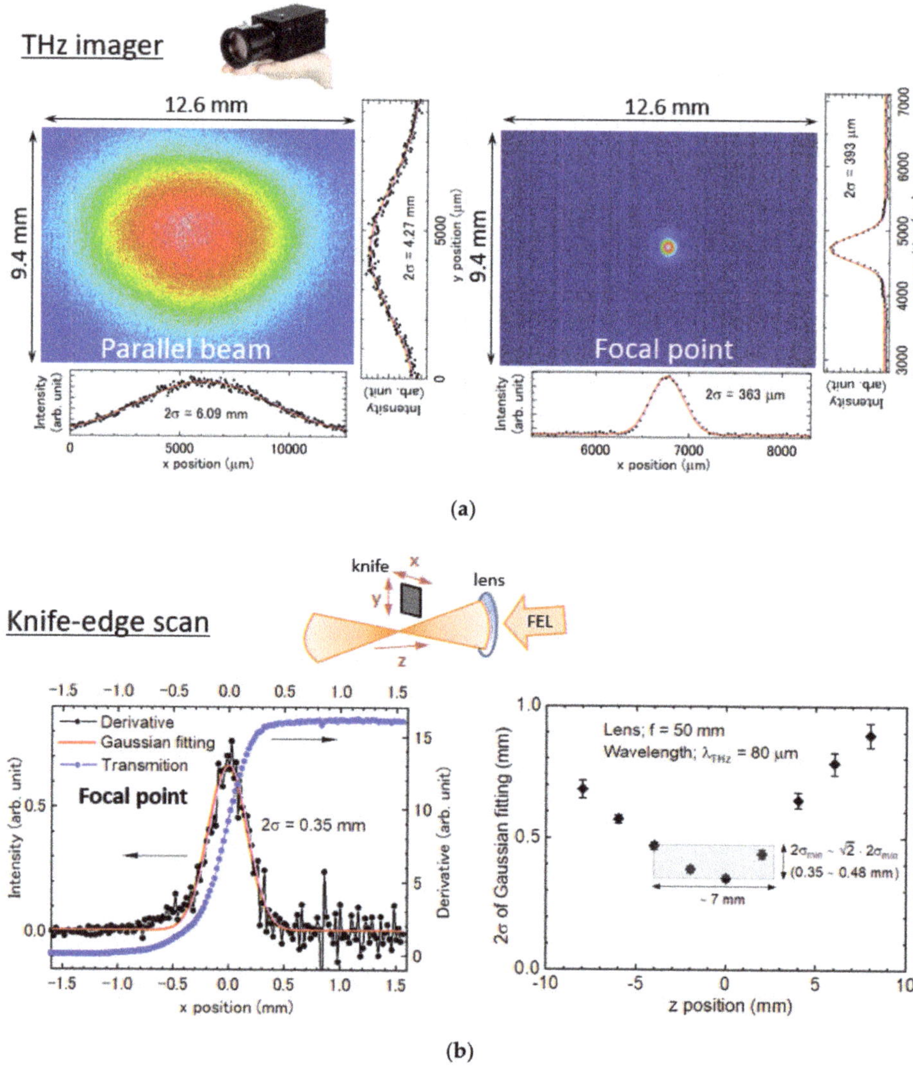

Figure 4. The beam profiles of the ISIR FEL (**a**) taken with a THz camera for the parallel beam (left) and at the focal point (right), and (**b**) the spatial resolution measured with the knife-edge scan around the focal point.

3.2. Spectral Imaging of Solid Samples

In the spectral-imaging experiment, the sample is placed at the FEL focal position. A He-Ne laser, which go through the same FEL optics, is used as a guide to set the sample at the FEL focal spot. In our layout, the FEL radiation after monochromatization by the diffraction grating spectrometer is collected by the lens, and the raster scan of the sample is automated by the stage to obtain a 2D image. The spatial resolution is determined by the degree of FEL focusing and by the step of the translation stage that moves the sample during the scan. In order to ensure that the sample is irradiated with the FEL pulse at each observation point, the stage is moved stepwise at 5 Hz, which is synchronized with the timing system of the linear accelerator [31]. The intensity of the monochromatized FEL radiation is sufficient for transmission and reflection experiments using one macro-pulse per point. When a raster scan of the sample is performed on an area of 2 cm × 2 cm with a scanning step Δ = 500 μm, i.e., 1600 points at 5 Hz, a 2D spectral image can be collected within approximately 6 min.

Figures 5–8 show the results of spectral images collected on pellets, 3D samples made by composite materials, opaque objects, and biological systems. Versatility and short time acquisitions represent the really unique capabilities of this layout. The time required for a single 2D image ranges from 5 to 60 min depending on the sample size and the raster-scan step. The relevance of this experiment resides in the observation of clear changes in the images using high-wavelength resolution. These were the experiments performed for the first time in the THz/FIR domain with such resolution and the first successful imaging of almost opaque objects at these wavelengths. We also successfully collected images showing non-linear behavior due to the high FEL intensity.

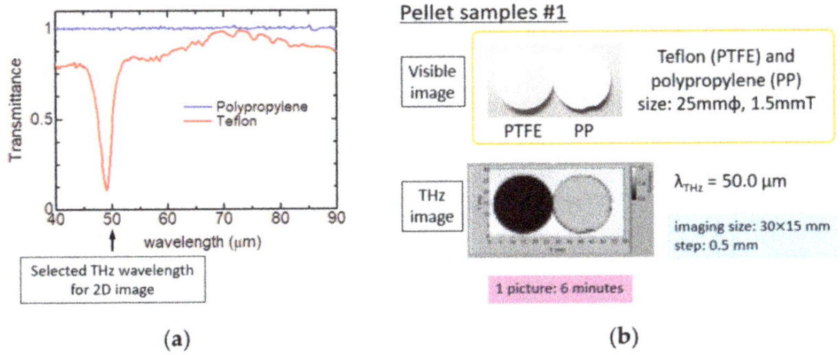

Figure 5. (**a**) Comparison of transmittance spectra of polytetrafluoroethylene (PTFE) and polypropylene (PP) by in situ observation; (**b**) 2D images of PTFE and PP pellets collected in transmission at $\lambda_{THz} = 50$ μm.

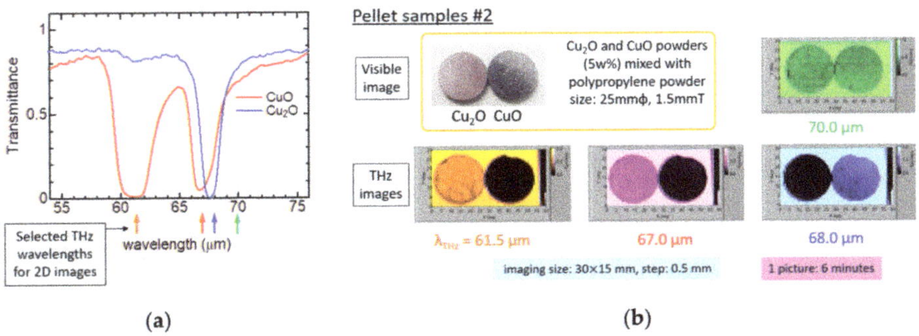

Figure 6. (**a**) Comparison of transmittance spectra of Cu_2O and CuO by in situ observation; (**b**) 2D images of Cu_2O and CuO pellets collected in transmission at several THz wavelengths.

3.2.1. Pellets

Figures 5 and 6 compare the spectral imaging of pellet samples collected in transmission. Figure 5a compares the infrared spectra of the Teflon (polytetrafluoroethylene, PTFE) and of the polypropylene (PP). PP has no intrinsic absorption in the THz/FIR region and is highly transparent. Therefore, as it can be seen in the figure, PP shows only a flat spectrum. On the other hand, PTFE shows a spectrum with an intense absorption peak around 50 μm. Both pellets are made from 100% pure powders of each material using a compression mold. Figure 5b shows the images of these pellets at the wavelength of the absorption peak in PTFE. The upper part is a visible image of the two pellets that can be hardly distinguishable, while collecting images at the absorption wavelength, a clear difference can be observed with a clear contrast. Next, imaging was performed on diluted pellets of CuO and Cu_2O powder. With each powder, we made pellets at the dilution of 5% by weight together with PP powders.

Figure 6a shows the infrared spectra of CuO and Cu_2O, respectively. CuO shows absorption peaks at the wavelengths of 61.5 μm and 67 μm while Cu_2O shows an absorption peak at 68 μm. Figure 6b shows the images at the three absorption wavelengths and at 70 μm that are not absorbed by both CuO and Cu_2O. It takes 6 min to collect each image. In panel 6(b), it can be seen that the image contrast changes with respect to the wavelength. In particular, the contrast between the images of CuO and Cu_2O is clearly reversed with just a difference of 1 μm going from 67 μm up to 68 μm, clearly demonstrating the effectiveness of the monochromatic FEL for long-wavelength high-resolution spectral imaging.

3.2.2. Composite Materials with A Fine Structure

In Figure 7a, transmission imaging of an oily marker having a pen tip of 0.5 mm was performed while keeping the lid so that the pen tip would not dry. The lid is almost transparent, and the metal part and the liquid reservoir inside can be observed as opaque parts. The outer diameter is approximately 1 cm at the maximum. A clear image is obtained with a large depth of focus, which is expected from the Rayleigh length estimated at the focal point.

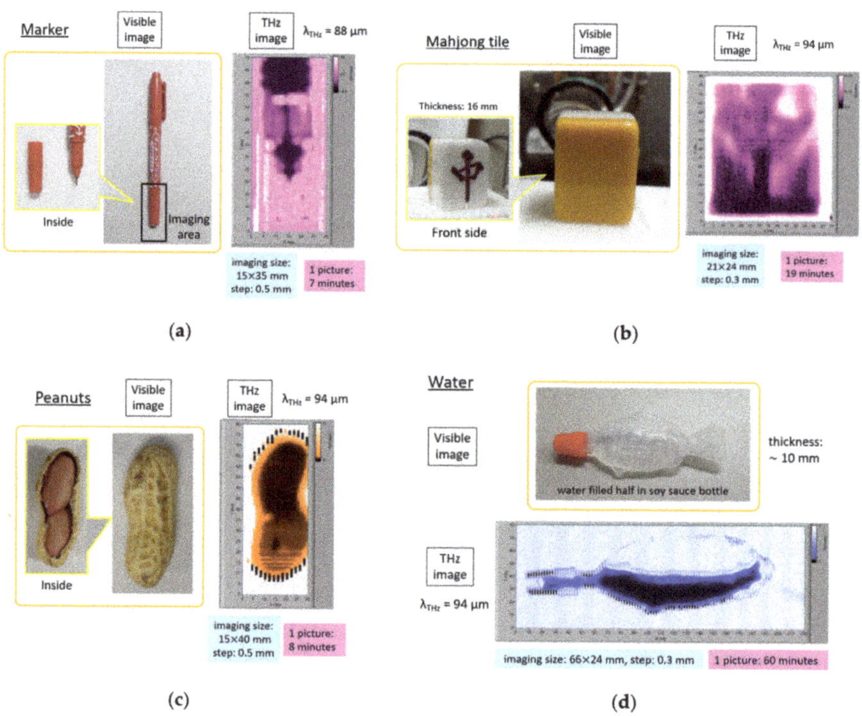

Figure 7. Comparison of 2D images of (**a**) a marker; (**b**) a mahjong tile; (**c**) a peanut (two nuts in a shell); (**d**) water collected in transmission at λ_{THz} approximately 90 μm.

3.2.3. Opaque Systems

Particularly important applications of spectral images refer to hard and soft opaque objects or liquid substances. Mahjong tiles are made of lumps of urea resin and hardly transmit THz/FIR radiation. However, when transmission imaging was performed using FEL with a high-intensity monochromatic beam, we succeeded in observing the engraved character on the surface from behind, as shown in Figure 7b. Since the character is observed from behind, the left and right of the image is reversed. The contrast observed is due to the difference in the reflection and the scattering contributions of the

FEL radiation between the flat part of the tile surface and the regions of the engraved text. There are several successful examples of THz transmission images of nuts at 0.1 THz (λ_{THz} = 3000 μm) [32]. However, as far as we know, Figure 7c is the first in situ observation of dried peanut without peeling at approximately 3 THz (λ_{THz} approximately 100 μm). The outer shell is shown with the orange color, while the two nuts touching each other can be recognized as dark areas in Figure 7c. Between the two nuts inside the shell, the lower nut seems to be drier and the internal embryo is seen through. Since the image including the outer shell is collected at the maximum FEL intensity, the detector was saturated in vacant space outside of the shell, causing a noise pattern due to the incorrect acquisition of the signal.

THz can be used also to collect images of liquids such as water. As an example, an image shown in Figure 7d was taken of a container filled with approximately 1 cm thickness of water. FEL radiation penetrates where the thickness of water is reduced: the neck and edge of the container, and the edge of water surface raised by surface tension. At the neck position with a red cap, the water thickness is approximately 2–3 mm and, to the best of our knowledge, in Figure 7d, we show the first THz transmission image of pure water. Since the image was obtained with a step scan over a wide area with a space resolution of 0.3 mm, it took one hour to collect this high-spatial resolution transmission image. The noise pattern near the edge of container was caused by the same reason of Figure 7c.

3.2.4. Leaves

THz images of leaves showing water distribution are already available in other studies. At ISIR, we tried to investigate their wavelength dependence and time dependence. The wavelength selected for imaging lies in a region where there is no absorption by atmospheric water vapor. At variegation spots highlighted by a dark color in the visible image shown in Figure 8a, the FEL radiation is well transmitted through leaf at any wavelength, even in fresh conditions. It is important to emphasize here, from these experimental results and those of another leaf in Figure 8b, that moisture and nutrients are not homogeneously distributed in leaves.

The image collected at 103 μm is the first image of the present measurement, but it is also measured with the highest transmission compared to the images collected at 89 μm and 83 μm. We would like to point out that the difference between these three images at different wavelengths is not merely due to the amount of the water content through the evaporation occurring with time, but it is due to the different absorbance versus wavelength. The reproducibility of the image at 103 μm was confirmed by measuring it twice, at the beginning and at the end of the experimental series after 30 min.

In order to evaluate how much water remains in a leaf after being detached, several images of a leaf hanged in air for 10 h were collected at several wavelengths, as shown in Figure 8b. Since this measurement is not time-sensitive, four different wavelengths were selected for imaging. Two images at 103 μm, taken twice with different FEL power, are also shown for reference.

Comparisons of the different images in Figure 8b pointed out that at the wavelength of 67 μm, there is still a large absorption in the whole leaf area, while at 103 μm, FEL radiation is almost completely transmitted through, making the pattern of the vein unrecognizable. The strong wavelength dependence in the dried leaf suggests that the decrease of the transmittance in these THz images of leaves is not only due to water, but also due to other substances present in the veins, such as nutrients. Therefore, we assume that water in leaves was almost evaporated after approximately 10 h, while nutrients still remained in the leaf veins, as we can understand from the observed large wavelength dependence. In the freshly detached leaf showed in Figure 8a, nutrients and moisture were homogeneously distributed throughout all veins of the leaf. However, in the leaf that was hanged and dried in air shown in Figure 8b, it is presumed that only the nutrients were concentrated in the lower part of the leaf during water evaporation. In addition, when the power of the FEL is changed as the intensity is halved at the wavelength of 103 μm, the vein pattern comes to be exposed again. One possibility resides in the nutrients in the veins remaining after water evaporation showing non-linear transmittance with respect to FEL intensity. The component analysis and the survey of the distribution of nutrients in leaves are indispensable for future biological science researches. The observed non-linear response can be the first case in THz imaging and could provide very useful information for future studies.

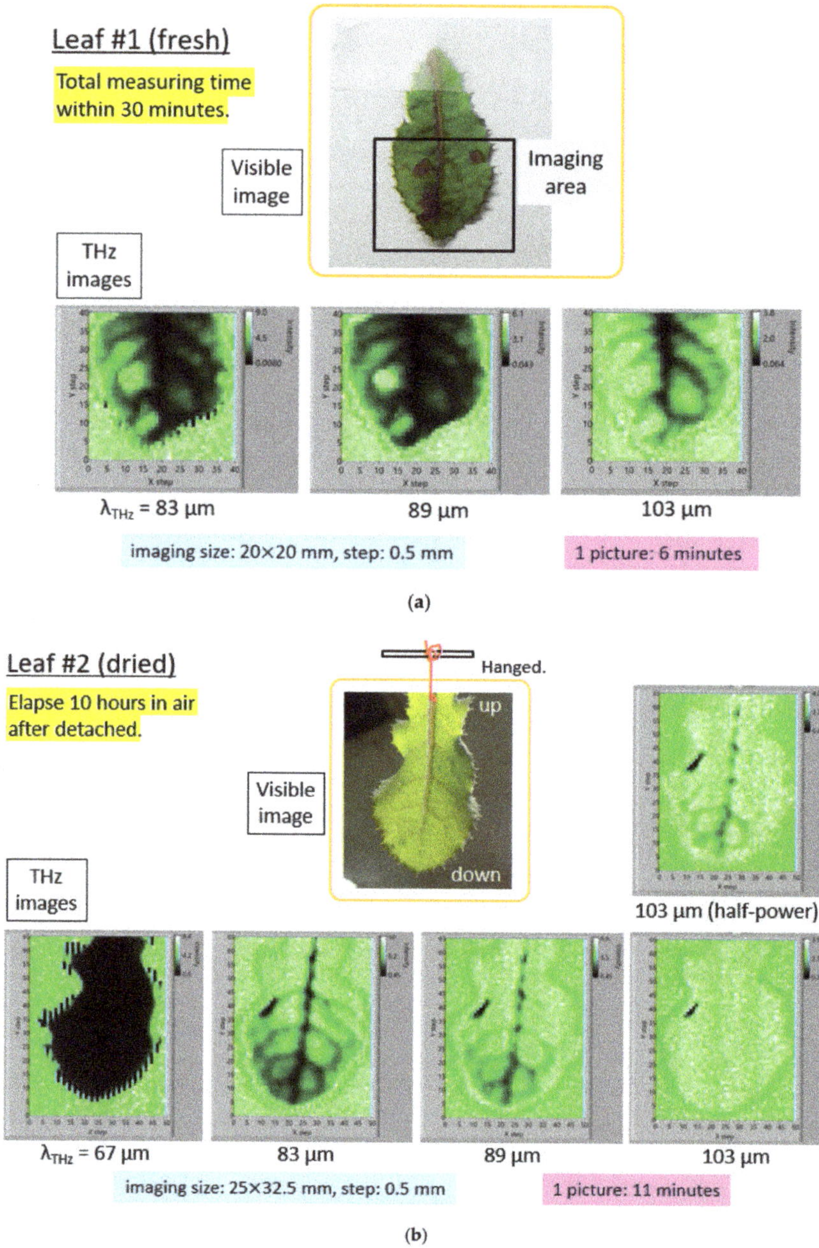

Figure 8. (**a**) shows a comparison of 2D images of a fresh leaf collected in transmission at different THz wavelengths. (**b**). Comparison of 2D images of a dried leaf collected in transmission at different THz wavelengths and different FEL powers.

4. Conclusions

We performed unique spectral-imaging experiments in the THz/FIR range using a grating-type spectrometer, a focusing optics, and a sample stage synchronized with the time structure of the high power ISIR THz-FEL at Osaka University. Successful different images with high wavelength resolution of $\Delta\lambda = 1$ μm have been obtained by tuning the gap of the undulator and the spectrometer. Moreover, thanks to the FEL time structure, fast spectral imaging was also feasible on different samples in various forms. Taking advantage of the high power of the FEL source, for the first time in the THz/FIR domain, we succeeded in the acquisition of transmission images of "opaque" solid materials and liquids. The versatile image acquisition system with raster scanning allowed monitoring changes in the composition of leaves. We also showed for the first time that non-linear phenomenon can be observed in images on changing the FEL intensity.

These results clearly point out the huge opportunities of these coherent high-power, high brilliance sources and the potential of scientific and technological researches in the THz/FIR domain and its interdisciplinary nature. The introduction of new sources, new detectors, control systems, and more advanced analysis techniques will be in great demand in the coming years.

Author Contributions: Conceptualization, project administration, A.I.; data curation, A.I. and M.F.; funding acquisition, A.I., S.L., A.M. and S.S.; investigation, A.I., K.K., M.F., R.K., H.F., A.H., S.M. and L.T.; resources, A.I., M.F., K.K. and R.K.; software, A.I., M.F., K.K. and R.K.; Writing—original draft, A.I; Writing—review and editing, A.M. and S.S. All authors have read and agreed to the published version of the manuscript.

Funding: This work was supported by JSPS KAKENHI Grant No.JP17K18989, by the research program "Dynamic Alliance for Open Innovation Bridging Human, Environment and Materials," and by the Bilateral Cooperation Agreement between Italy and Japan of the *Italian Ministry of Foreign Affairs and of the International Cooperation* (MAECI) in the framework of the project of major relevance N. PGR0072.

Acknowledgments: We acknowledge K. Furukawa, and Y. Okada for their invaluable support during the THz-FEL operation.

Conflicts of Interest: The authors declare no conflict of interest.

Appendix A

The emission characteristics of the ISIR THz-FEL may change greatly depending on the beam setting. When measuring the wavelength spectrum, the bandwidth, the peak intensity, and the total energy have to be changed. As described in the text, when the total energy is set to the maximum at 27 MHz, the bandwidth significantly increases and shows a complex spectral shape (Figure A1).

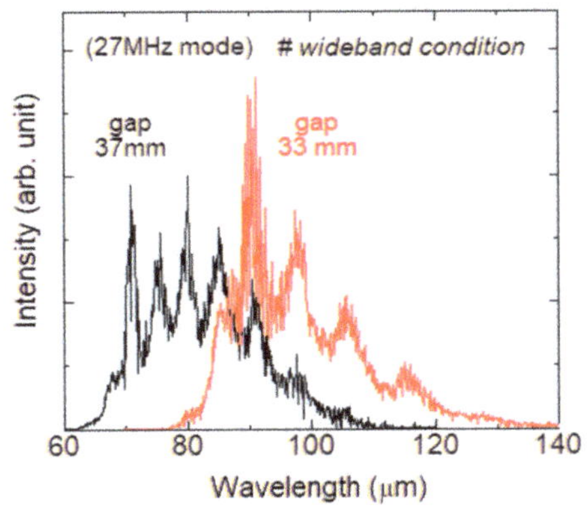

Figure A1. Comparison of two FEL spectra in the 27 MHz mode for the 33 mm (red) and the 37 mm (black) gap (wideband condition).

In this condition, the area of the spectrum, that is proportional to the total energy, increases about three times (approximately 30 mJ/macro-pulse) [33] or more in this wideband scenario if compared with the monochromatic setup (approximately 10 mJ/macro-pulse). However, the maximum intensity at the central wavelength does not increase, as shown in Figure A2. Therefore, such beam conditioning has no advantage for spectral imaging when the FEL is monochromatized by a diffraction grating. At variance, it is detrimental, since it is characterized by higher beam instability. Meanwhile, the pulse width estimated from the autocorrelation of the pulsed FEL is shortened from approximately 20 ps to about several ps by adjusting to the wideband condition, as shown in Figure A3. It must be always clarified whether a short pulse or a monochromatic beam is required for the THz radiation in each experiment, and then it sets the FEL with the necessary parameters.

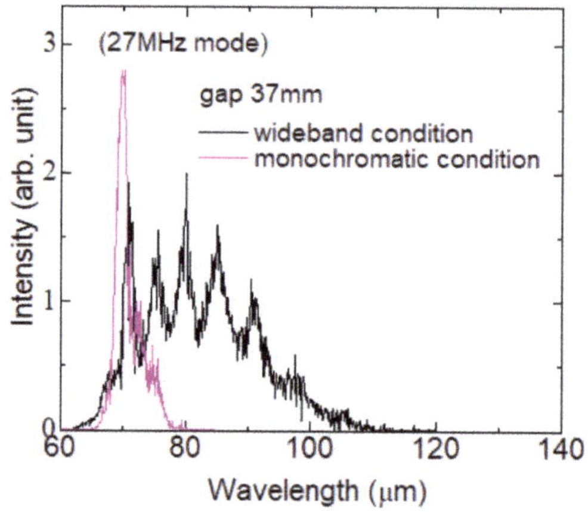

Figure A2. Comparison of two FEL spectra in the wideband (black) and monochromatic (purple) condition in the 27 MHz mode and the undulator gap sets to 37 mm.

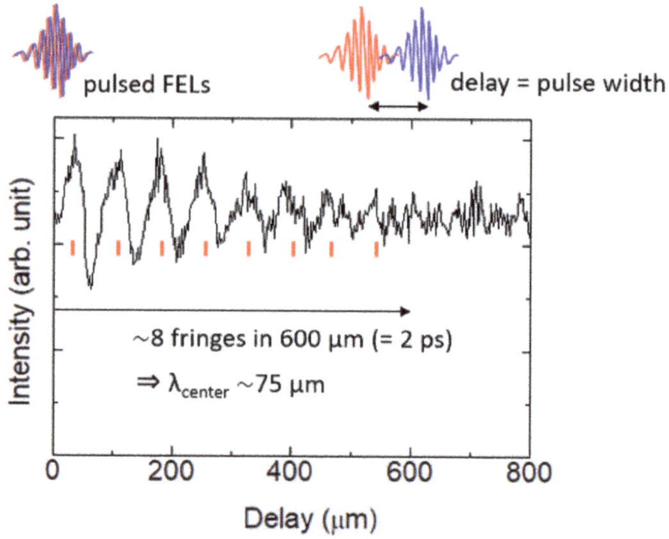

Figure A3. The interference waveform of the pulsed FEL by autocorrelation.

References

1. Hoshina, H.; Suzuki, H.; Otani, C.; Nagai, M.; Kawase, K.; Irizawa, A.; Isoyama, G. Polymer morphological change induced by terahertz irradiation. *Sci. Rep.* **2016**, *6*, 27180. [CrossRef] [PubMed]
2. Nagai, M.; Aono, S.; Ashida, M.; Kawase, K.; Irizawa, A.; Isoyama, G. Luminescence induced by electrons outside zinc oxide nanoparticles driven by intense terahertz pulse trains. *New J. Phys.* **2017**, *19*, 053017. [CrossRef]
3. Irizawa, A.; Suga, S.; Nagashima, T.; Higashiya, A.; Hashida, M.; Sakabe, S. Laser-induced fine structures on silicon exposed to THz-FEL. *Appl. Phys. Lett.* **2017**, *111*, 251602. [CrossRef]
4. Kawasaki, T.; Tsukiyama, K.; Irizawa, A. Dissolution of a fibrous peptide by terahertz free electron laser. *Sci. Rep.* **2019**, *9*, 1–8. [CrossRef]
5. Deacon, D.; Elias, L.R.; Madey, J.-M.J.; Ramian, G.J.; Schwettman, H.A.; Smith, T.I. First operation of a free-electron laser. *Phys. Rev. Lett.* **1977**, *38*, 892. [CrossRef]
6. Shintake, T.; Tanaka, H.; Hara, T.; Tanaka, T.; Togawa, K.; Yabashi, M.; Otake, Y.; Asano, Y.; Bizen, T.; Fukui, T.; et al. A compact free-electron laser for generating coherent radiation in the extreme ultraviolet region. *Nat. Photonics* **2008**, *2*, 555–559. [CrossRef]
7. Ishikawa, T.; Aoyagi, H.; Asaka, T.; Asano, Y.; Azumi, N.; Bizen, T.; Ego, H.; Fukami, K.; Fukui, T.; Furukawa, Y.; et al. A compact X-ray free-electron laser emitting in the sub-angstrom region. *Nat. Photonics* **2012**, *6*, 540–544. [CrossRef]
8. Perucchi, A.; Mitri, S.D.; Penco, G.; Allaria, E.; Lupi, S. The TeraFERMI terahertz source at the seeded FERMI free-electron-laser facility. *Rev. Sci. Instrum.* **2013**, *84*, 022702. [CrossRef]
9. Chiadroni, E.; Bacci, A.; Bellaveglia, M.; Boscolo, M.; Castellano, M.; Cultrera, L.; Di Pirro, G.; Ferrario, M.; Ficcadenti, L.; Filippetto, D.; et al. The SPARC linear accelerator based terahertz source. *Appl. Phys. Lett.* **2013**, *102*, 094101. [CrossRef]
10. Buakor, K.; Chaisueb, N.; Rimjaem, S.; Saisut, J.; Thongbai, C.; Damminsek, K.; Thongpakdi, W. Development of linac-based MIR/THz FEL facility and photocathode RF-gun in Thailand. *Proc. IPAC2017* **2017**, *WEPAB083*, 2763–2766.
11. Miginsky, S.; Jang, K.H.; Gudkov, B.; Bae, S.; Jeong, Y.U.; Mun, J.; Lee, K.; Park, S.H.; Setiniyaz, S. A compact THz FEL at KAERI: The Project and the Status. *Proc. FEL2017* **2017**, *MOP048*, 156–157.
12. Li, H.-T.; Jia, Q.-K.; Zhang, S.-C.; Wang, L.; Yang, Y.-L. Design of FELiChEM: The first infrared free-electron laser user facility in China. *Chinese Phys. C* **2017**, *41*, 018102. [CrossRef]
13. Okuda, S.; Honda, Y.; Kimura, N.; Ohkuma, J.; Yamamoto, T.; Suemine, S.; Okada, T.; Ishida, S.; Takeda, S.; Tsumori, K.; et al. Free-electron laser oscillation with a multibunch electron beam of the 38 MeV L-band linear accelerator at ISIR. *Nucl. Instrum. Meth. A* **1995**, *358*, 244–247. [CrossRef]
14. Kato, R.; Kondo, S.; Igo, T.; Okita, T.; Konishi, T.; Suemine, S.; Okuda, S.; Isoyama, G. Lasing at 150 μm wavelength and measurement of the characteristics of the free-electron laser at ISIR Osaka University. *Nucl. Instrum. Meth. A* **2000**, *445*, 169–172. [CrossRef]
15. Kashiwagi, S.; Kato, R.; Mihara, A.; Noda, T.; Isoyama, G.; Tsuchiya, K.; Shioya, T.; Yamamoto, S. Rigorous evaluation of the edge-focusing wiggler based on the magnetic field measurement. *Phys. Rev. Spec. Top. Accel. Beams* **2009**, *12*, 120703. [CrossRef]
16. Kawase, K.; Kato, R.; Irizawa, A.; Fujimoto, M.; Kashiwagi, S.; Yamamoto, S.; Kamitsukasa, F.; Osumi, H.; Yaguchi, M.; Tokuchi, A.; et al. The high-power operation of a terahertz free-electron laser based on a normal conducting RF linac using beam conditioning. *Nucl. Instrum. Meth. A* **2013**, *726*, 96–103. [CrossRef]
17. Suemine, S.; Kawase, K.; Sugimoto, N.; Kashiwagi, S.; Furukawa, K.; Kato, R.; Irizawa, A.; Fujimoto, M.; Ohsumi, H.; Yaguchi, M.; et al. Grid pulser for an electron gun with a thermionic cathode for the high-power operation of a terahertz free-electron laser. *Nucl. Instrum. Meth. A* **2015**, *773*, 97–103. [CrossRef]
18. Tokuchi, A.; Kamitsukasa, F.; Furukawa, K.; Kawase, K.; Kato, R.; Irizawa, A.; Fujimoto, M.; Osumi, H.; Funakoshi, S.; Tsutsumi, R.; et al. Development of a high-power solid-state switch using static induction thyristors for a klystron modulator. *Nucl. Instrum. Meth. A* **2015**, *769*, 72–78. [CrossRef]
19. Pham, H.H.N.; Hisatake, S.; Minin, O.V.; Nagatsuma, T.; Minin, I.V. Enhancement of spatial resolution of terahertz imaging systems based on terajet generation by dielectric cube. *Appl. Phys. Lett.* **2017**, *2*, 056106.
20. Guerboukha, H.; Nallappan, K.; Skorobogatiy, M. Toward real-time terahertz imaging. *Adv. Opt. Photon.* **2018**, *10*, 843–938. [CrossRef]

21. Smolyanskaya, O.A.; Chernomyrdin, N.V.; Konovko, A.A.; Zaytsev, K.I.; Ozheredov, I.A.; Cherkasova, O.P.; Nazarov, M.M.; Guillet, J.-P.; Kozlov, S.A.; Kistenev, Y.V.; et al. Terahertz biophotonics as a tool for studies of dielectric and spectral properties of biological tissues and liquids. *Prog. Quantum Electron.* **2018**, *62*, 1–77. [CrossRef]
22. Chernomyrdin, N.V.; Kucheryavenko, A.S.; Kolontaeva, G.S.; Katyba, G.M.; Dolganova, I.N.; Karalkin, P.A.; Ponomarev, D.S.; Kurlov, V.N.; Reshetov, I.V.; Skorobogatiy, M.; et al. Reflection-mode continuous-wave 0.15λ-resolution terahertz solid immersion microscopy of soft biological tissues. *Appl. Phys. Lett.* **2018**, *113*, 111102. [CrossRef]
23. Dolganova, I.N.; Zaytsev, K.I.; Yurchenko, S.O.; Karasik, V.E.; Tuchin, V.V. The Role of Scattering in Quasi-Ordered Structures for Terahertz Imaging: Local Order Can Increase an Image Quality. *IEEE Trans. Terahertz Sci. Technol.* **2018**, *8*, 403–409. [CrossRef]
24. Zaytsev, K.I.; Dolganova, I.N.; Chernomyrdin, N.V.; Katyba, G.M.; Gavdush, A.A.; Cherkasova, O.P.; Komandin, G.A.; Shchedrina, M.A.; Khodan, A.N.; Ponomarev, D.S.; et al. The progress and perspectives of terahertz technology for diagnosis of neoplasms: A review. *J. Opt.* **2020**, *22*, 013001. [CrossRef]
25. Angeluts, A.A.; Balakin, A.V.; Evdokimov, M.G.; Esaulkov, M.N.; Nazarov, M.M.; Ozheredov, I.A.; Sapozhnikov, D.A.; Solyankin, P.M.; Cherkasova, O.P.; Shkurinov, A.P. Characteristic responses of biological and nanoscale systems in the terahertz frequency range. *Quantum Electron.* **2014**, *44*, 614. [CrossRef]
26. Murate, K.; Kawase, K. Perspective: Terahertz wave parametric generator and its applications. *J. Appl. Phys.* **2018**, *124*, 160901. [CrossRef]
27. Kimura, S.; Nakamura, E.; Yamazaki, J.; Katoh, M.; Nishi, T.; Okamura, H.; Matsunami, M.; Chen, L.; Nanba, T. New infrared and terahertz beam Line BL6B at UVSOR. *AIP Conf. Proc.* **2004**, *CP705*, 416–419.
28. Irizawa, A.; Suga, S.; Isoyama, G.; Shimai, K.; Sato, K.; Iizuka, K.; Nanba, T.; Higashiya, A.; Niitaka, S.; Takagi, H. Direct observation of a pressure-induced metal-insulator transition in LiV2O4 by optical studies. *Phys. Rev. B* **2011**, *84*, 235116. [CrossRef]
29. Marcelli, A.; Cinque, G. Infrared synchrotron radiation beamlines: High brilliance tools for IR spectromicroscopy. A practical guide to the characteristics of the broadband and brilliant non-thermal sources. In *Biomedical Applications of Synchrotron Infrared Microspectroscopy*; Moss, D., Ed.; Royal Society of Chemistry: London, UK, 2011; Chapter 3; pp. 67–104.
30. Oda, N.; Sudou, T.; Ishi, T.; Okubo, S.; Isoyama, G.; Irizawa, A.; Kawase, K.; Kato, R. Externally triggered imaging technique for microbolometer-type terahertz imager. In *Proc. SPIE 9856, Terahertz Physics, Devices, and Systems X: Advanced Applications in Industry and Defense*; SPIE: Bellingham, WA, USA, 2016; p. 98560I.
31. Kashiwagi, S.; Isoyama, G.; Kato, R.; Suemine, S. Development of a precise timing system for the ISIR L-band linac at Osaka University. *Proc. FEL 2007* **2007**, *WEPPH030*, 413–416.
32. Terahertz Inspection of Agricultural Products. Available online: https://terasense.com/applications/terahertz-agriculture/ (accessed on 24 April 2020).
33. Kawase, K.; Nagai, M.; Furukawa, K.; Fujimoto, M.; Kato, R.; Honda, Y.; Isoyama, G. Extremely high-intensity operation of a THz free-electron laser using an electron beam with a higher bunch charge. *Nucl. Instrum. Meth. A* **2020**, *960*, 163582. [CrossRef]

condensed matter

MDPI

Article

Evaluation of Bunch Length by Measuring Coherent Synchrotron Radiation with a Narrow-Band Detector at LEBRA

Takeshi Sakai [1,*], Ken Hayakawa [1], Toshinari Tanaka [1], Yasushi Hayakawa [1], Kyoko Nogami [1] and Norihiro Sei [2]

1 Laboratory for Electron Beam Research and Application, Nihon University, 7-24-1 Narashinodai, Funabashi 274-8501, Japan; hayakawa@lebra.nihon-u.ac.jp (K.H.); tanaka@lebra.nihon-u.ac.jp (T.T.); yahayak@lebra.nihon-u.ac.jp (Y.H.); nogami@lebra.nihon-u.ac.jp (K.N.)

2 Research Institute of Instrumentation Frontier, National Institute of Advanced Industrial Science and Technology, 1-1-1 Umezono, Tsukuba, Ibaraki 305-8568, Japan; sei.n@aist.go.jp

* Correspondence: sakai@lebra.nihon-u.ac.jp; Tel.: +81-47-469-5489

Received: 7 April 2020; Accepted: 7 May 2020; Published: 9 May 2020

Abstract: This study presents a novel technology to measure electron bunch length with a high time resolution by measuring coherent synchrotron radiation using a narrow-band detector at Laboratory for Electron Beam Research and Application (LEBRA)—an S-band linear accelerator facility for free-electron lasers. The form factor was observed to decrease exponentially with charge—in concordance with the relationship between the intensity of the coherent synchrotron radiation and the magnitude of electron bunch charge—in the region in which the effect of electron bunch charge on bunch length is negligible. The calculated root-mean-square bunch length was observed to agree well with the value determined from the spectral shape obtained. The aforementioned results are expected to be useful in real-time observation of small changes in electron bunches in advanced accelerators.

Keywords: coherent synchrotron radiation; bunch length; FEL; linac; terahertz

1. Introduction

The recent development of compact light sources in the terahertz band has led to the frontier research field of terahertz wave spectroscopy. Several vibration modes exist in the terahertz region characteristic of molecules, and measurement of terahertz wave spectra is widely used to identify materials. A major advantage of a terahertz light source using an accelerator is its high peak power, leading to the development of various light sources such as free-electron lasers (FELs) and coherent radiation [1,2]. With the development of accelerator technology, various techniques have been developed to produce short-pulse electron bunches. Consequently, active research and development of terahertz light sources using accelerators have been conducted [3–11]. In particular, at Nihon University, various types of terahertz light sources have been developed over the past decade. At the straight sections of FELs, coherent synchrotron radiation (CSR) and coherent edge radiation (CER) have been developed and implemented, and the produced light has been transmitted to laboratories through FEL beamlines. As these terahertz light sources can be used simultaneously with infrared FELs, they are used for experiments including those involving biologic tissue imaging. Further, at the straight sections of parametric X-ray radiation (PXR), coherent transition radiation (CTR) with a power output of one millijoule per macro-pulse has been developed and implemented, which can be used in laboratories via PXR beamlines. Additionally, prospective application of CTR in product management of tablets is being researched.

A terahertz light source produced in an accelerator also serves as an excellent tool for evaluation of the accelerator. In advanced accelerators like X-ray FELs, the size of an electron bunch is comparable to that of the wavelength of terahertz light. Therefore, it is possible to measure the shape of a bunch by measuring the spectrum of coherent radiation produced by that electron bunch [12–16]. To date, streak cameras and RF deflectors are the only tools that have been developed to measure bunch shapes. However, the time resolution of conventional streak cameras is insufficient to measure sub-ps bunch lengths. Further, electron bunches are often distorted by conventional RF deflectors, making it impossible to non-destructively measure the states of electron bunches prior to FEL oscillation. Additionally, in the case of electron beams with high repetition frequency, such as X-ray FELs using superconducting linear accelerators that are currently under development, it is difficult to measure the bunch length for each micro-pulse via the existing methods.

Given these circumstances, we conducted measurements of bunch length using CSR that was capable of non-destructively observing electron beams. As the light source, we utilized the CSR source in the terahertz band that had been developed at the LEBRA accelerator facility at Nihon University [17–21]. Such bunch length measurements obtained using coherent radiation produced by electron bunches have also been studied and used in several accelerator facilities [12–16]. However, unlike other studies, we succeeded in measuring the root-mean-square (RMS) bunch length with high precision by strictly considering the detector's characteristic sensitivity during the observation of the coherent radiation. In the present study, we first outline the CSR light source from Nihon University that was used. Following this, we discuss the proposed measurement technology that is capable of detecting small changes in bunch length using the narrow-band characteristic of the detector. On the basis of the results obtained, we conclude that a correlation exists between electron bunch charge and bunch length for high-charge bunches.

2. Coherent Synchrotron Radiation Source at LEBRA

The LEBRA Linac at Nihon University consists of a DC electron gun of 100 kV, one buncher, three S-Band 4 m acceleration tubes and two high-frequency source klystrons, as depicted in Figure 1. Its operating acceleration frequency is 2856 MHz, the acceleration energy is 40–100 MeV, the repetition rate is 2–10 Hz and the macro-pulse width of the associated beam is 5–20 μs. Besides the normal full-bunch mode operation, the electron gun system is capable of operating in a high-charge burst mode at intervals of 1/64 or 1/128 using a high-speed grid pulser of 600 ps full width at half maximum (FWHM). During full-bunch mode operation, bunches are aligned to produce a peak current of 200 mA and bunch intervals of 350 ps (Figure 2a). During burst-mode operation, it is possible to produce one (single-bunch mode) or two micro-pulses (two-bunch mode) within an interval of 600 ps at a peak current of about 2 A, depending on the operation acceleration frequency and high-speed grid pulse timing (Figures 2b and 3) [19–22]. The experiments conducted in this study carry out the burst-mode operation in two-bunch mode.

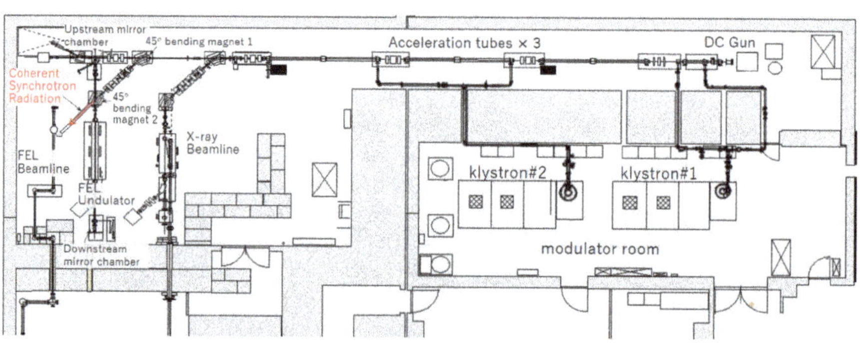

Figure 1. Layout of the LEBRA linac at Nihon University.

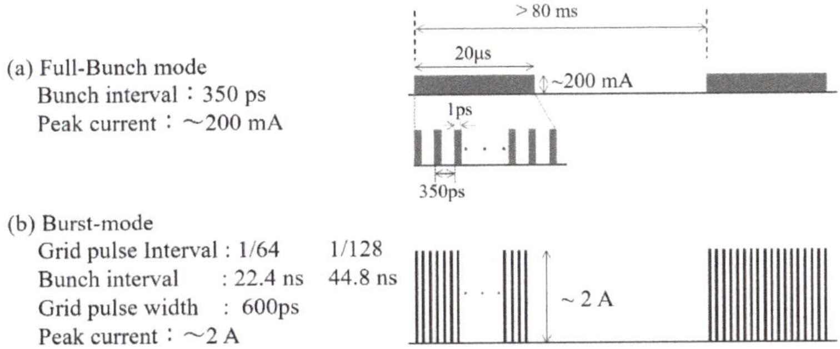

Figure 2. Pattern diagrams of full-bunch mode and burst-mode.

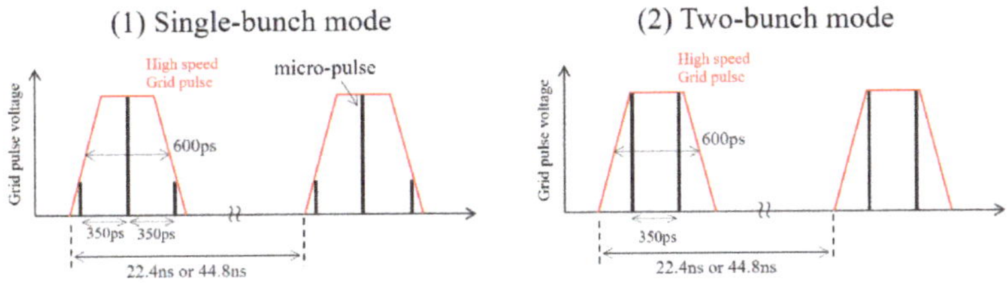

Figure 3. High-speed grid pulse timing of burst-mode operations for single and two-bunch modes.

The electron beam produced by the electron gun is accelerated by the linear accelerator, and then transmitted to the FEL 2.4-m undulator after being bent by 90° using two 45-degree bending magnets. The electron beam is bent by a 45-degree bending magnet after passing through the undulator, and then it loses its energy due to the beam dump. When the electron beam passes through the two bending magnets, its bunch length w observed to shorten from 3 ps to less than 1 ps due to pulse compression. The bunch length of the electron beam is monitored using the CSR generated at the entrance of the downstream bending magnet during the compression process. The design bunch length is observed to be about 2 ps [23]. The CSR passes through a straight pipe with a diameter of 20 mm and is ejected from the vacuum chamber through a quartz window. Then, the CSR is reflected by a plane mirror towards a detector set that is at a distance of 1.3–1.5 m from the source point. A schematic of the CSR light source is presented in

For the detection of CSR, the D-band diode detector (Millitech, Inc. USA, DXP-06) with a square antenna and an aperture of H17 mm × V11 mm was used. This detector is a commercially available Schottky barrier zero bias diode type, which is characterized by small, robust and thermally stable [24]. Its sensitivity to CW light source was determined to be approximately 5 mV/10 μW, and the linearity was calibrated by varying the thickness of the BK7 plate. The relative sensitivity spectrum $B(f)$ of a D-band diode detector with a pyramidal horn antenna was measured using a terahertz wave beamline in the L-band linac at the Institute for Integrated Radiation and Nuclear Science at Kyoto University [25]. As depicted in Figure 5, the diode detector used exhibited a sensitivity in the range of 0.08–0.17 THz, making it appropriate for use in spectroscopy measurements. The output of the D-band diode detector was measured using an oscilloscope located at an approximate distance of 30 m in the laboratory. The rise time of the measurement system was observed to be approximately 1.3 ns, making it difficult to distinguish the micro-pulses from each other during operation in the full-bunch mode. However, during operation in the burst mode, the time change of the THz wave could be measured because the interval time of the micro-pulse was sufficiently wider than that in the full-bunch mode. The results also confirmed that the CSR beam reflected in the vacuum chamber of the bending magnet was emitted into the atmosphere through a vacuum window [17]. The frequency of the obtained CSR was measured to be between approximately 0.1 to 0.2 THz and its power per macro-pulse was measured to be approximately 0.4 μJ in this frequency region. The absorption at the quartz window during CSR extraction was ascertained to be negligible at frequencies below 0.2 THz [26,27]. Further, based on the HITRAN database (high resolution transmission molecular absorption database) [28], the influence of the air absorption was expected to be negligible in the frequency range between 0.1 and 0.2 THz. Figure 4.

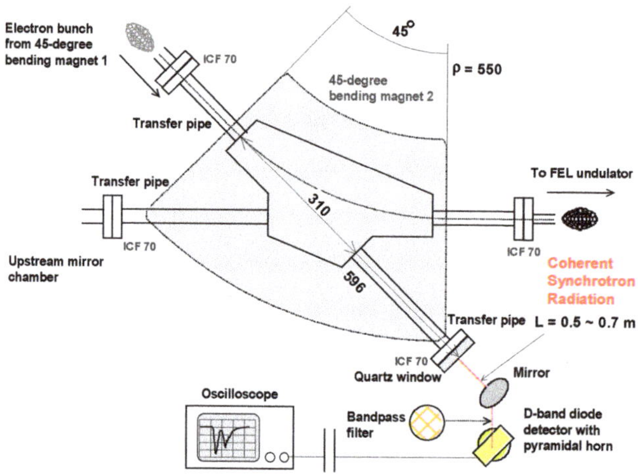

Figure 4. Outline of CSR measurement equipment at LEBRA.

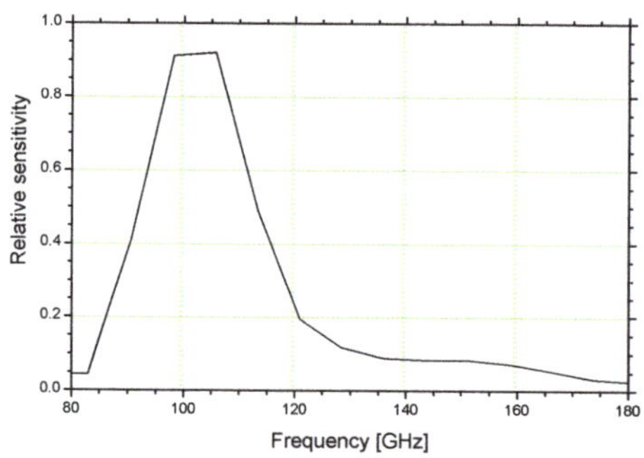

Figure 5. Measurement results of sensitivity characteristics for the D-band diode detector.

3. Bunch Length Measurement of CSR Power with Narrow-Band Detector

Assuming that there are N electrons in a bunch, the intensity spectrum of the CSR is given by the following equation using the intensity spectrum $I_e(f)$ of the CSR generated by one electron [29].

$$I(f) = NI_e(f) + N(N-1)F(f)I_e(f), \tag{1}$$

where $F(f)$ denotes the form factor and f denotes a frequency of the CSR. The form factor is a dimensionless number, not exceeding 1, defined by the following equation:

$$F(f) = \left| \int dz S(z) exp\left(-i\frac{2\pi f z}{c}\right)\right|^2, \tag{2}$$

where $S(z)$ denotes a normalized electron density function. If the number of electrons is sufficiently large and $1/N << F(f)$, then Equation (1) can be approximated by the following equation:

$$I(f) \cong N^2 F(f) I_e(f) \tag{3}$$

If the bunch shape is known, $F(f)$ can be calculated as a function of the bunch length. A method of evaluation of the bunch length of an electron beam by measuring the powers of associated wake fields or the transition radiation at multiple frequencies has been developed based on a simple method of measurement of bunch length [15,16]. Although this technique is easily affected by the bunch shape, it can be applied in the real-time measurement of the bunch length. Due to this advantage, we evaluated the bunch lengths corresponding to the observations of the CSR at LEBRA via this measurement technique. Based on previous measurements, it was known beforehand that the distribution of the electron bunch shapes corresponding to LEBRA was very close to a Gaussian one [23]. Therefore, as radiation intensity is a function of wavelength, the contributions of the wavelengths to bunch length are observed to gradually accumulate during the measurement of the intensity using the detector. Although the sensitivity range of the D-band diode detector is relatively narrow, we used the band-pass filter with a transmission center frequency of 0.09 THz to further narrow it down. Figure 6 depicts the transmission characteristic $T(f)$ of the band-pass filter over an extended range of frequencies. As is evident, it was only possible to extract the component of CSR corresponding to 0.08–0.10 THz by combining the D-band diode detector and a band-pass filter. As mentioned above, the diode detector can, in principle, measure the bunch length per micro-pulse of the electron beam because of its extremely short response time (about 1.3 ns).

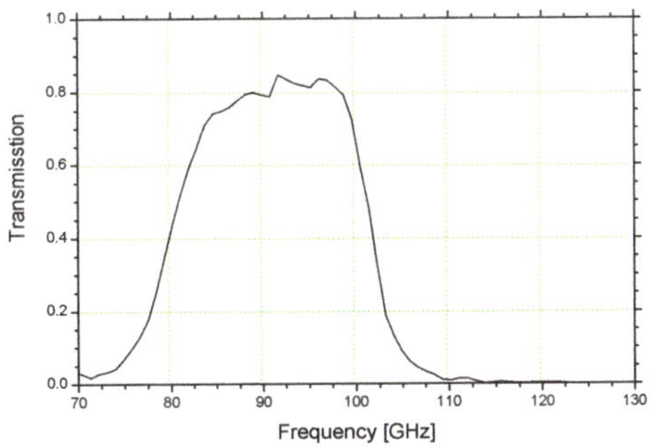

Figure 6. Measurements of the transmission characteristics corresponding to the band-pass filter.

The intensity of the CSR measured using the diode detector is given by

$$P_{\mathrm{BPF}} \cong \int_{f_{\mathrm{L}}}^{f_{\mathrm{H}}} df N^2 B(f) F(f) I_e(f), \tag{4}$$

where f_{L} and f_{H} denote the lower and upper limits of transmission frequency corresponding to the band-pass filter, respectively. In this case, if the width of $B(f)$ is sufficiently narrow compared to the variation in $F(f)I_e(f)$, the intensity of the CSR can be approximately expressed by

$$P_{\mathrm{BPF}} \cong N^2 \overline{B}(f_c) \overline{F}(f_c) \overline{I_e}(f_c) \Delta f, \tag{5}$$

where f_c denotes the frequency at which the measured CSR intensity is the maximum and Δf denotes the full width at half maximum of $B(f)$. The bar above each symbol indicates its average value within the band of $B(f)$. In the case in which the bandwidth of the diode detector is constant, the values of $\overline{B}$ and $\overline{I_e}$ are also constant. Therefore, the value obtained by dividing P_{BPF} by N^2, that is, the value obtained by dividing the CSR intensity by the square of the charge, is equal to the variation in $\overline{F}$ with respect to the magnitude of the charge.

In the case in which the electron-bunch charge is low, as in the case of the full-bunch mode, it is expected that the bunch length will not change even if the magnitude of charge is varied. During curve fitting using the square of the magnitude of the charge in the region in which the electron bunch charge is less than 30 ps in the full-bunch mode, the fitting coefficient of determination, R^2, was ascertained to be 0.9999, as depicted in Figure 7. This demonstrates very close agreement.

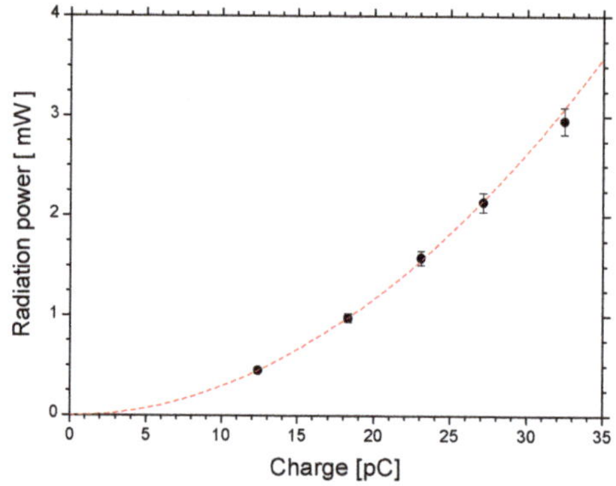

Figure 7. Relation between electron charge and radiation intensity.

The datum at 32 pC was observed to deviate from this fitting curve by more than one standard deviation, leading to the expectation that the value of $\overline{F}$ is prone to change in the region above 30 pC. We further measured the relationship between the magnitude of the charge and the CSR intensity P_{BPF} during burst mode corresponding to large magnitudes of charge. Charge saturation was observed at around 0.65 nC/2 bunches, as depicted in Figure 8. In other words, the CSR power is not proportional to the square of the charge in the burst mode where the amount of charge per micro-pulse is high. This can be attributed to the increase in bunch length due to a space-charge effect as the electron-bunch charge increases. The bunch lengthening caused by the space-charge effect is often observed for short electron bunches whose bunch lengths are less than 1 ps [30].

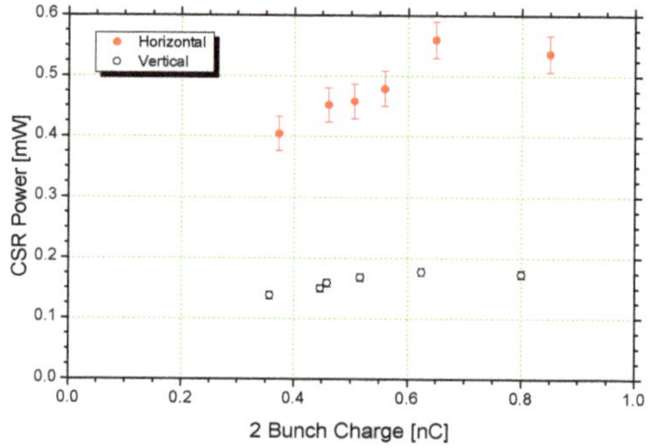

Figure 8. Relationship between the magnitude of electron charge and CSR intensity P_{BPF} during burst mode.

According to the literature, when the magnitude of electron bunch charge exceeds a threshold, q_{th}, the bunch length σ increases following the equation [30]:

$$\sigma^2 = \sigma_{th}^2 + \sigma_{th}^2\left(\frac{q - q_{th}}{q_{th}}\right)^n, \; q > q_{th}, \tag{6}$$

where σ_{th} denotes the bunch length at the threshold magnitude of the electron bunch charge. If the distribution of electron bunches is Gaussian, as is the case with LEBRA at Nihon University, the relationship between the form factor F at a certain wavelength and the bunch length is expressed by [6].

$$F(f,\sigma) = exp\left(-\frac{4\pi^2 f^2}{c^2}\sigma^2\right). \tag{7}$$

By substituting Equation (6) into Equation (7), the mean form factor over the band range of the diode detector, when $q >> q_{th}$, can be expressed by

$$\overline{F}(f_c,\sigma) \cong \overline{F}(f_c,\sigma_{th})exp\left[-\frac{4\pi^2 f_c^2 \sigma_{th}^2}{c^2}\left(\frac{q}{q_{th}}\right)^n\right]. \tag{8}$$

During the performance of measurement in the region where the magnitude of the electron bunch charge exceeds the threshold charge amount, the argument of the exponential function of Equation (8) can be calculated. The charge dependence of the form factor then becomes possible to be evaluated.

Hence, based on Figure 8, the dependences of the horizontal and vertical polarization components of P_{BPF}/q^2 on q^n were investigated. As depicted in Figure 9, assuming n = 2/3, the value of P_{BPF}/q^2 was observed to decrease exponentially.

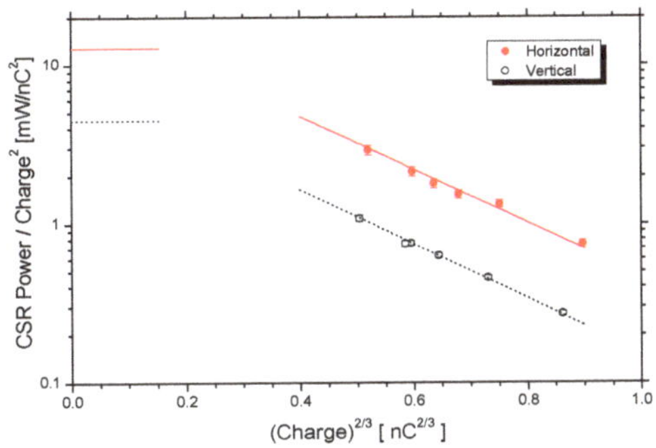

Figure 9. Dependence of P_{BPF}/q^2 on $q^{2/3}$ corresponding to the horizontal and vertical polarization components.

During curve fitting using Equation (8), n = 0.64 ± 0.04 was obtained for the horizontal polarization component and n = 0.68 ± 0.02 was obtained for the vertical component. This elucidated the following approximate relationship between bunch length and magnitude of charge:

$$\sigma \propto q^{\frac{1}{3}}. \tag{9}$$

The difficulty in accurately measuring the value of radiation intensity makes it difficult to evaluate the form factor based solely on the measurement of radiation intensity. Generally, RMS bunch length is evaluated by measuring the spectra of coherent radiation and conducting curve fitting for the spectra based on Equations (1) and (7). For LEBRA, during preliminary measurement using the spectrum, the RMS bunch length was observed to be 1.9 ps in the full-bunch mode during which the magnitude of the electron-bunch charge was observed to be 30 pC or less. Further, the bunch length was ascertained to be 2.9 and 3.1 ps corresponding to charge magnitudes of 0.49 nC/2 bunches and 0.60 nC/2 bunches in burst mode, respectively. If the standard value of RMS bunch length (i.e., the bunch length at a certain electron-bunch charge magnitude) is known in advance, the RMS bunch length at other charge amounts can be estimated using Equation (8). Figure 10 depicts the RMS bunch length evaluated for the data of vertical polarization with smaller errors than those presented in Figure 9, using $q = 0.49$ nC/2 bunches at the bunch length of 2.9 ps. As we have indicated above, the accuracy of the data used in the evaluation is $n = 0.68 \pm 0.02$. As depicted in Figure 10, the bunch length corresponding to the charge of about 0.62 nC/2 bunches was 3.1 ps. This verifies that the value is consistent with the bunch length corresponding to the charge of about 0.60 nC/2 bunches evaluated using the spectra.

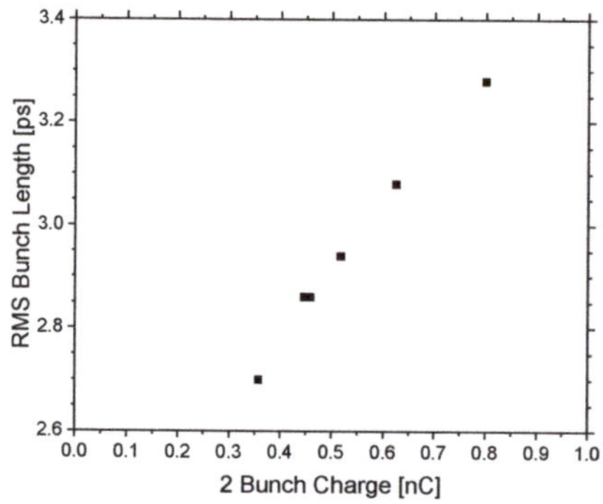

Figure 10. RMS bunch length evaluated for vertical polarization data in Figure 9.

Generally, the band region of a diode detector is sufficiently narrow compared to the variation in $F(f)I_e(f)$. However, as depicted in Figure 11, when the diode detector used in LEBRA was evaluated, $B(f)F(f)I_e(f)$ was observed to attain its maximum value at approximately 0.1 THz, and the FWHM was measured to be approximately 23 GHz, i.e., the peak positions and the spectrum shapes are nearly unchanged. As a result, it did not differ much corresponding to bunch lengths in the range of 2.6–3.3 ps. Therefore, it was deemed appropriate to use Equation (8) during the form factor evaluation corresponding to this measurement. Additionally, the data presented in Figure 11 suggests that variations in the bunch length could be measured with high sensitivity by using a narrow-band diode detector corresponding to each bunch length.

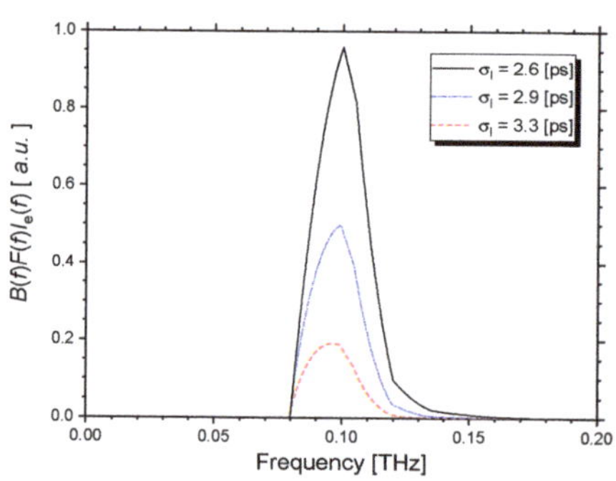

Figure 11. Relation between frequency and $B(f)F(f)I_e(f)$.

4. Conclusions

In this study, we evaluated RMS bunch lengths in full-bunch mode and in burst mode using the CSR at LEBRA at Nihon University. Corresponding to regions in which CSR intensity was observed to not be proportional to the square of the magnitude of the electron bunch charge, the form factor was observed to decrease exponentially with the charge magnitude. Further, using the fact that the form factor was observed to decrease exponentially with the square of the bunch length, we performed precise evaluations of the variations in bunch length. In this method, the commercially available product-based detector can be used to easily evaluate the change in bunch length for each micro-pulse. Currently, the development of self-amplified spontaneous emission X-ray FELs are being actively promoted. Further, the development of X-ray FEL oscillators using energy-recovery linacs are also under popular consideration. Our results are expected to be applicable to the precise evaluation of FEL gain with respect to changes in the magnitude of the charge.

At LEBRA at Nihon University, the development of high-power terahertz light sources using various coherent radiations is currently being carried out. For CER light sources in the straight section of FEL beamlines, it is possible to observe CER simultaneously with FELs by inserting a hole-coupled mirror that separates CER and FELs. In the future, we plan to elucidate the effect of FELs on the electron bunch by using the bunch length measurement method proposed in this study.

Author Contributions: Methodology, N.S.; software, K.H.; investigation, T.S., N.S., K.H., T.T., Y.H. and K.N.; writing—original draft preparation, T.S. and N.S.; writing—review and editing, T.S. and N.S.; funding acquisition, N.S. All authors have read and agreed to the published version of the manuscript.

Funding: This work was supported by JSPS KAKENHI JP19H04406.

Acknowledgments: We would like to express our gratitude for the help of Takahashi of the Research Reactor Institute, Kyoto University for measuring the sensitivity curve of the D-band diode detector.

Conflicts of Interest: The authors declare no conflict of interest.

References

1. Elias, L.R. Free-electron laser research at the University of California, Santa Barbara. *IEEE J. Quantum Electron.* **1987**, *23*, 1470–1475. [CrossRef]
2. Nakazato, T.; Oyamada, M.; Niimura, N.; Urasawa, S.; Konno, O.; Kagaya, A.; Kato, R.; Kamiyama, T.; Torizuka, Y.; Nanba, T.; et al. Observation of coherent synchrotron radiation. *Phys. Rev. Lett.* **1989**, *63*, 1245. [CrossRef] [PubMed]
3. Abo-Bakr, M.; Feikes, J.; Holldack, K.; Kuske, P.; Peatman, W.B.; Schade, U.; Wüstefeld, G.; Hübers, H.-W. Brilliant, Coherent Far-Infrared (THz) Synchrotron Radiation. *Phys. Rev. Lett.* **2003**, *90*, 094801. [CrossRef] [PubMed]
4. Gensch, M.; Bittner, L.; Chesnov, A.; Delsim-Hashemi, H.; Drescher, M.; Faatz, B.; Feldhaus, J.; Fruehling, U.; Geloni, G.; Gerth, C.; et al. New infrared undulator beamline at FLASH. *Infrared Phys. Technol.* **2008**, *51*, 423. [CrossRef]
5. Sei, N.; Kuroda, R.; Yasumoto, M.; Toyokawa, H.; Ogawa, H.; Koike, M.; Yamada, K. Measurement of intense coherent synchrotron radiation at frequencies around 0.1 THz using the compact S-band linac. *J. Appl. Phys.* **2008**, *104*, 114908. [CrossRef]
6. Carr, G.L.; Martin, M.C.; McKinney, W.R.; Jordan, K.; Neil, G.R.; Williams, G.P. High-power terahertz radiation from relativistic electrons. *Nature* **2002**, *420*, 153. [CrossRef]
7. Kim, K.J.; Shvyd'ko, Y.; Reiche, S. A Proposal for an X-Ray Free-Electron Laser Oscillator with an Energy-Recovery Linac. *Phys. Rev. Lett.* **2008**, *100*, 244802. [CrossRef]
8. Emma, P.; Akre, R.; Arthur, J.; Bionta, R.; Bostedt, C.; Bożek, J.; Brachmann, A.; Bucksbaum, P.; Coffee, R.; Decker, F.-J.; et al. First lasing and operation of an ångstrom-wavelength free-electron laser. *Nat. Photonics* **2010**, *4*, 641. [CrossRef]
9. Pile, D. First light from SACLA. *Nat. Photonics* **2011**, *5*, 456. [CrossRef]
10. Hajima, R. Energy Recovery Linacs for Light Sources. *Rev. Accel. Sci. Technol.* **2010**, *3*, 121. [CrossRef]
11. Bilderback, D.H.; Sinclair, C.; Gruner, S.M. Technical Report: The Status of the Energy Recovery Linac Source of Coherent Hard X-rays at Cornell University. *Synchrotron Radiat. News* **2006**, *19*, 30–35. [CrossRef]
12. Roux, R.; Couprie, M.-E.; Hara, T.; Bakker, R.; Visentin, B.; Billardon, M.; Roux, J. The Super-ACO FEL dynamics measured with a streak camera. *J. Nucl. Instrum. Methods Phys. Res. Sect. A* **1997**, *393*, 33–37. [CrossRef]

13. Sei, N.; Yamada, K.; Ogawa, H.; Yasumoto, M.; Mikado, T. Electron Beam Qualities with and without Free Electron Laser Oscillations in the Compact Storage Ring NIJI-IV. *Jpn. J. Appl. Phys.* **2003**, *42*, 5848. [CrossRef]
14. Lihn, H.; Kung, P.; Settakom, C.; Wiedemann, H.; Bocek, D. Measurement of subpicosecond electron pulses. *Phys. Rev. E* **1996**, *53*, 6413. [CrossRef]
15. Ieiri, T. A real time bunch-length monitor using the beam spectrum and measurements of bunch lengthening. *Nucl. Instrum. Methods Phys. Res. Sect. A* **1993**, *329*, 371. [CrossRef]
16. Otake, Y.; Maesaka, H.; Matsubara, S.; Inoue, S.; Yanagida, K.; Ego, H.; Kondo, C.; Sakurai, T.; Matsumoto, T.; Tomizawaakurai, H.; et al. Beam monitor system for an x-ray free electron laser and compact laser. *Phys. Rev. Spéc. Top. Accel. Beams* **2013**, *16*, 042802. [CrossRef]
17. Sei, N.; Ogawa, H.; Hayakawa, K.; Tanaka, T.; Hayakawa, Y.; Nakao, K.; Sakai, T.; Nogami, K.; Inagaki, M. Observation of intense terahertz-wave coherent synchrotron radiation at LEBRA. *J. Phys. D Appl. Phys.* **2013**, *46*, 045104. [CrossRef]
18. Sei, N.; Ogawa, H.; Hayakawa, K.; Tanaka, T.; Hayakawa, Y.; Nakao, K.; Sakai, T. Measurement of a bunch length of a relativistic electron beam using a narrow-band diode detector in a terahertz region. *J. Jpn. Soc. Infrared Sci. Technol.* **2015**, *25*, 97. (In Japanese)
19. Hayakawa, Y.; Sato, I.; Hayakawa, K.; Tanaka, T.; Nakazawa, H.; Yokoyama, K.; Kanno, K.; Sakai, T.; Ishiwata, K.; Enomoto, A.; et al. First lasing of LEBRA FEL at Nihon University at a wavelength of 1.5 μm. *Nucl. Instrum. Methods Phys. Res. Sect. A* **2002**, *483*, 29. [CrossRef]
20. Nakao, K.; Hayakawa, K.; Tanaka, T.; Hayakawa, Y.; Sakai, T.; Sato, I.; Nogami, K.; Inagaki, M. Analysis on variation factors of optical powerat the LEBRA FEL. In Proceedings of the FEL, Liverpool, UK, 23–28 August 2009; Volume 675. Available online: http://accelconf.web.cern.ch/AccelConf/FEL2009/papers/wepc75.pdf (accessed on 6 April 2020).
21. Hayakawa, Y.; Sato, I.; Hayakawa, K.; Tanaka, T.; Yokoyama, K.; Kanno, K.; Sakai, T.; Ishiwata, K.; Nakao, K.; Hashimoto, E. Characteristics of the fundamental FEL and the higher harmonic generation at LEBRA. *Nucl. Instrum. Methods Phys. Res. Sect. A* **2003**, *507*, 404. [CrossRef]
22. Nakao, K.; Tanaka, T.; Hayakawa, K.; Hayakawa, Y.; Nogami, K.; Inagaki, M. Lasing of FEL with the burst mode beam at LEBRA Nihon University. In Proceedings of the 8th Annual Meeting of Particle Accelerator Society of Japan, Tsukuba, Japan, 1–3 August 2011; Volume 1051. Available online: https://www.pasj.jp/web_publish/pasj8/proceedings/poster/TUPS072.pdf (accessed on 6 April 2020). (In Japanese).
23. Yokoyama, K.; Sato, I.; Hayakawa, K.; Tanaka, T.; Hayakawa, Y.; Nakao, K. Measurement of electron bunch length at LEBRA. In Proceedings of the 1st Annual Meeting of Particle Accelerator Society of Japan and 29th Linear Accelerator Meeting in Japan, Funabashi, Japan, 4–6 August 2004; Volume 602. Available online: https://www.pasj.jp/web_publish/pasj1_lam29/WebPublish/5P58.pdf (accessed on 6 April 2020). (In Japanese).
24. Smiths Interconnect on the Web (The Current Model Number is DET-06). Available online: https://www.smithsinterconnect.com/products/multi-function-rf-systems/mixers-and-detectors/general-purpose-detector-(det)/ (accessed on 24 April 2020).
25. Takahashi, T.; Takami, K. Observation of THz coherent transition radiation from single-bunch beam at KURRI-LINAC as an intense pulsed light source. *Infrared Phys. Technol.* **2008**, *51*, 363. [CrossRef]
26. Grischkowsky, D.; Keiding, S.; Van Exter, M.P.; Fattinger, C. Far-infrared time-domain spectroscopy with terahertz beams of dielectrics and semiconductors. *J. Opt. Soc. Am. B* **1990**, *7*, 2006. [CrossRef]
27. Mira, N.; Robert, E.M. Terahertz Time-Domain Spectroscopy for Material Characterization. *Proc. IEEE* **2007**, *95*, 1658. [CrossRef]
28. HITRAN on the Web. Available online: http://hitran.iao.ru/ (accessed on 29 February 2020).
29. Jackson, J.D. Longitudinal plasma oscillations. *J. Nucl. Energy* **1960**, *1*, 171. [CrossRef]
30. Kan, K.; Yang, J.; Kondoh, T.; Norizawa, K.; Yoshida, Y. Effects of emittance and space-charge in femtosecond bunch compression. *Nucl. Instrum. Methods Phys. Res. Sect. A* **2008**, *597*, 126. [CrossRef]

condensed matter

MDPI

Article

Proximity Array Device: A Novel Photon Detector Working in Long Wavelengths

S. Javad Rezvani [1,2,*], Daniele Di Gioacchino [1], Claudio Gatti [1], Carlo Ligi [1], Mariangela Cestelli Guidi [1], Sara Cibella [3], Matteo Fretto [2], Nicola Poccia [4], Stefano Lupi [5] and Augusto Marcelli [1,6]

1 Istituto Nazionale di Fisica Nucleare, Laboratori Nazionali di Fisica Nucleare, Via Enrico Fermi 54, 00044 Frascati, Italy; daniele.digioacchino@lnf.infn.it (D.D.G.); claudio.gatti@lnf.infn.it (C.G.); carlo.ligi@lnf.infn.it (C.L.); mariangela.cestelliguidi@lnf.infn.it (M.C.G.); augusto.marcelli@lnf.infn.it (A.M.)
2 Istituto Nazionale di Ricerca Metrologica (INRiM), Strada delle Cacce 91, 10135 Torino, Italy; m.fretto@inrim.it
3 Consiglio Nazionale delle Ricerche, Istituto di Fotonica e Nanotecnologie, Rome 00156, Italy; sara.cibella@ifn.cnr.it
4 Institute for Metallic Materials, IFW-Dresden, Dresden 01069, Germany; n.poccia@ifw-dresden.de
5 INFN and Department of Physics, Università degli Studi di Sapienza, 00185 Rome, Italy; stefano.lupi@roma1.infn.it
6 International Centre for Material Science Superstripes, RICMASS, via dei Sabelli 119A, 00185 Rome, Italy
* Correspondence: rezvani@lnf.infn.it

Received: 7 April 2020; Accepted: 23 April 2020; Published: 1 May 2020

Abstract: We present here an innovative photon detector based on the proximity junction array device (PAD) working at long wavelengths. We show that the vortex dynamics in PAD undergoes a transition from a Mott insulator to a vortex metal state by application of an external magnetic field. The PAD also evidences a Josephson I-V characteristic with the external field dependent tunneling current. At high applied currents, we observe a dissipative regime in which the vortex dynamics is dominated by the quasi-particle contribution from the normal metal. The PAD has a relatively high photo-response even at frequencies below the expected characteristic frequency while, its superconducting properties such as the order parameter and the Josephson characteristic frequency can be modulated via external fields to widen the detection band. This device represents a promising and reliable candidate for new high-sensitivity long-wavelength detectors.

Keywords: niobium; proximity effects; superconductivity; detectors; terahertz; Vortex dynamics

1. Introduction

Long-wavelength radiation has recently become one of the most significant regions of the electromagnetic spectrum in terms of multi-disciplinary use not only in basic science research, but also in different technologies [1–3]. This wavelength range λ = 250–2500 µm (energy range 0.5–4 meV) allows the investigation of many fundamental physical phenomena, e.g., phonon and plasmon dynamics, elementary particle physics and probably in the near future even the cold dark matter [4–9]. Furthermore, with its high transmission through a wide range of non-conducting materials, long-wavelength radiation and in particular terahertz (THz) radiation holds a significant potential in many applications such as non-destructive imaging, security screening and material characterizations. However, in comparison with adjacent energy ranges, e.g., IR and UV, in this wavelength domain the instrumentation and in particular the detectors are still limited, expensive or difficult to use. For instance semiconducting bolometers are widely used as cryogenic and non-cryogenic THz detectors [10], but their sensitivity to temperature and mechanical fluctuations and their limitations in the high frequency band operation, renders its use mainly confined to laboratories. Recent studies have shown that also superconducting devices can be employed for long wavelengths and in particular for THz generation and detection [11–14]. These devices exhibit extreme low noise compared to their semiconducting counterparts, while having response time orders of magnitude lower and a higher frequency range of operation. However, the frequency range of the operation is set by the superconducting gap of the material (usually in the range 300 GHz to 1 THz in conventional superconductors and up to 30 THz in cuprates). Furthermore, there are studies showing the possibility to control semiconducting and superconducting properties in materials with physico-chemical treatments [15,16] and systems with low dimensionality [17–20]. Meanwhile, in recent years several distinctive mechanisms are being explored as new detection methods in this energy domain such as tunnel junctions, high gap superconducting junctions, transition edge sensors, bi-crystal junctions, etc. [21–23].

In this work we present and characterize a novel device based on the superconducting proximity junction array that can be used for detection of photons at long wavelengths. The superconducting characteristics of this device based on non-localized vortex dynamics will be discussed. The effect of the quasi-particle currents and the subsequent limitations of this detector will be also reviewed. The response of the device in the radio-frequency domain well below the expected characteristic frequency of the proximity junctions will be presented and further advancements based on non-equilibrium phenomena due to the external perturbations will be discussed. The proposed device represents a step forward toward the design of a robust, low-noise, broadband, high-sensitivity, and long-wavelength radiation detector operating in a wide energy range covering the THz domain.

2. Experiment

The device we designed and realized consists of an array of about 90000 Nb superconducting islands regularly deposited on a non-superconducting substrate of 80 × 80 μm^2 in size and with a period of ∼270 nm, with an island diameter of ∼220 nm and a thickness of 45 nm (see Figure 1). Standard photo-lithographic tools have been employed to obtain a 40 nm thick Au template on the non-superconducting Si/SiO_2 substrate (1 × 1 cm^2). The template consists of a central square of 82 × 82 μm^2, with the corners connected to four terminals used for electric measurements. The size of these terminals is 200 × 200 μm^2, large enough for micro bonding. The Nb pattern is then created on top of the central Au square employing electron beam lithography and DC sputtering. To ensure a uniform current injection inside the array, two 45 nm Nb bus bars were deposited along two opposite sides of the array. The IV measurements were carried out in a shielded cryostat at 4.2 K. A current bias was applied using the signal generator at 107 Hz. Two measurements mode were used. In one the I-V curves were measured amplifying the output voltage via a low-noise amplifier and digitized by the National instruments sampling card (USB-6363). In the second mode, the voltage sinusoidal waveform was measured via a lock-in amplifier locked to the applied current signal, resulting in the direct measurement of the dynamic resistance. The magnetic field was applied by placing a solenoid around the sample with the field perpendicular to the plane of the device. The current through the solenoid increased in a step-wise fashion and separate I-V traces were recorded at each field step.

For the radio-frequency response of the device, a custom-made diamagnetic dipole antenna with length L = $\lambda/2$, centered at 8 GHz was inserted inside the cryostat in front of the magnetic coils and the device. The highest transmission line (resonant frequency) using a vector analyzer connected to the antenna was found at f = 7.78 GHz at which the measurements were done. The response measurements were carried out via a continuous measurement of the electrical response of the device with and without the radio-frequency (RF) signal.

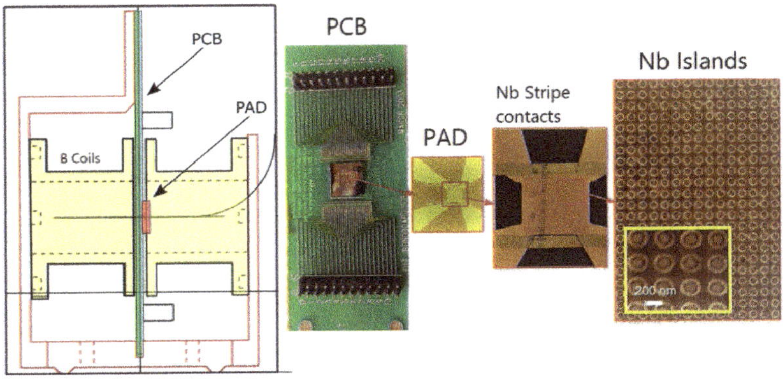

Figure 1. The sample positioning within the cryostat between the magnetic coils. The device was mounted on a PCB (green) connected to the electrical measurement apparatus. The enlarged images of the PCB and the mounted PADs show the presence of the uniform injecting Nb stripes while the SEM images show the Nb islands of the device.

3. Results and Discussion

The transition behavior of the PAD was investigated while cooling to the liquid helium temperature and is shown in Figure 2. The experimental curve points out a percolative transition around the critical temperature T_c. It occurs via a multistage coherence development by decreasing the temperature. At $T \geq T_1$, grains of each islands are in an incoherent non-superconducting phase (i.e., the normal state). As they are poly-crystalline islands, initially ($T \leq T_1$) a coherent phase is established within island grains, after which the system resistance decreases. Lowering the temperature, the intra-island phase coherence continuously strengthens, and the resistance of the system similarly decreases rather than steeply dropping at T_1. Within this temperature domain the normal-metal coherence length increases until it becomes comparable to the spacing among islands. Then, inter-island phase coherence begins to emerge and finally, reaching T_2, the system undergoes the transition to a fully superconducting state [24–26]. In our device we could observe a high temperature limit T_1 between 15–8 K with T_C = 7.5 K and a low fully coherent limit at T_2 = 4.2 K.

The dynamics resistance measurement of the device was performed at 4.2 K in the fully intra-islands coherent phase as a function of the applied magnetics and of the electric field (see Figure 3). The dynamic resistance exhibits a clear evolution from the minima to maxima at the specific applied magnetic fields (frustration fields). The conversion occurs specifically at integer and fractional frustration fields f being the B/B_0 with $B_0 = 27.8$ mT, the ratio of the magnetic quantum flux on the area of the junction (Φ_0/a^2). The presence of minima in the dynamic resistance (and its analogous counterpart in the resistance) and singularities in the magnetization, reflects the modulation of the ground state energy due to the formation of a periodic vortex pattern, which have been previously observed [27]. However, in our experiment the conversion from minima to maxima in the dynamic resistance increasing the electric field intensity is the direct consequence of the phase transformation from a vortex Mott insulator into a metallic state, a mechanism previously observed in similar systems [28]. These results indicate the possibility to tune the vortex dynamic and to modulate the energy states with an external fields and with an enhanced response at integer and half-integer frustration factors ($f = 1/2$ and $f = 1$) [29].

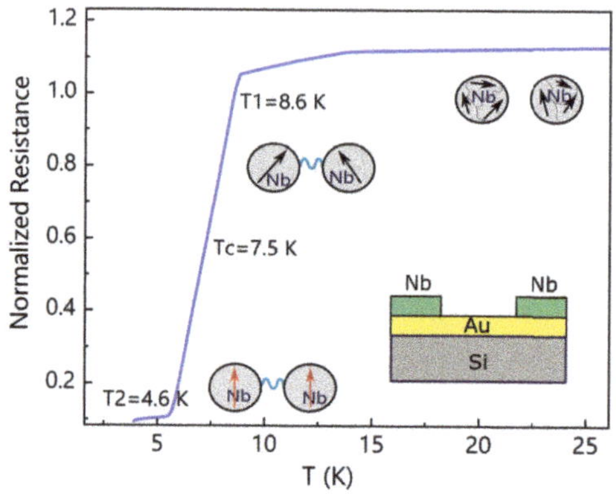

Figure 2. Percolative superconducting transition of the Nb islands on the Au layer. The island transition occurs at T_1 while the transition of the film is at T_2. Above T_1, the Nb islands behave as normal metals and the arrows indicate the coherence in grains and islands. Below T_1 grains start to become coherent throughout each island and at T_1, Cooper pairs start to diffuse from Nb into Au, and the resistance decreases. Gradually the coherence length of the film becomes comparable to the spacing among islands, and the entire system crosses toward a global phase coherence. As the temperature decreases (T_2), the film undergoes a transition to a superconducting state. (Inset): The schematics of the islands on the gold surface deposited on the silicon substrate.

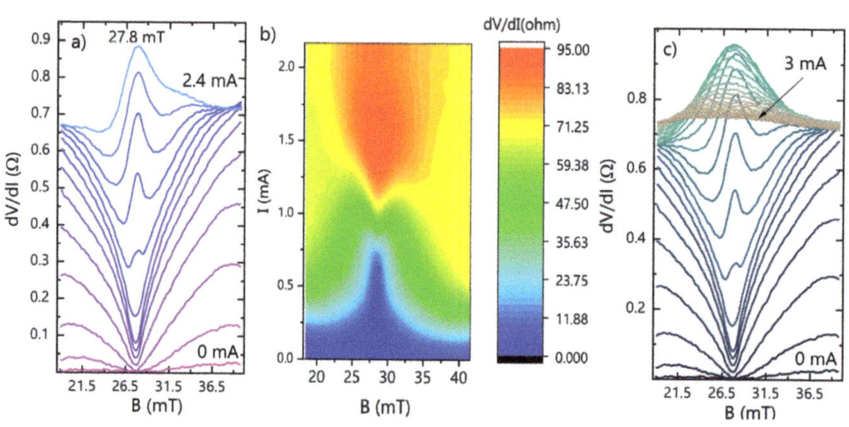

Figure 3. (**a**) Representative dV/dI versus magnetic field curves at different bias currents. At low current bias, dV/dI minima at different B values indicate the formation of the vortex Mott insulator. Increasing the current, minima reverse into maxima according to an insulator-to-metal transition. The transitions are most pronounced at integer and half-integer frustration factors ($f = 1/2$ and $f = 1$); (**b**) dynamic resistance as a function of applied current and magnetic field and (**c**) evolution of the dynamic resistance as a function of the applied magnetic field at high applied currents; inset) the layout of the junctions formed via Nb islands and the Au normal layer.

The intensity increase at the integer transition points continues with the increase of the applied electric field up to a critical point of the applied current of I = 2.4 mA at which it is severely hindered. In this region the maxima is smeared by the increase of the applied current reaching lower values if compared with the lower applied current, while the maximum also shifts toward a lower frustration value. This behavior corresponds to the emergence of a dissipative regime in which the singularities tends to be wiped out. However, since the resistance of our device shows pronounced dips even at large currents (not shown), indicating strong pinning, neither the minima-to-maxima transitions nor the dissipative behavior can be correlated with a vortex depinning or to the overcome of the critical current [29,30]. Consequently, it can be deduced that at such high current values, the metallic state features are reduced by the contribution of the quasi-particles current, occurring in the normal-metal layer [31].

The I-V measurement of the proximity junction arrays exhibits the typical Josephson behavior (see Figure 4) with the maximum tunneling current (i.e., scape current) I_t = 1.2 mA, over which the increase of current is accompanied by the increase of the array's potential. Above the tunneling current, the device shows a shunted behavior that might be due to either the resistive component of the device or to pairs hopping between the potential wells in a medium electric field range in which the field is not sufficient to force the junctions to the normal state. In the measurement range carried out in our experiment we did not observe the transition to the resistive normal state. Hence, the absolute value of the critical current (I_C), being relatively higher than the I_t, could not be calculated. The application of the external magnetic field significantly modifies the I-V characteristics of the junction array, towards the hindrance of the junction properties (i.e., lower I_t) due to the superimposition of the screening current generated at the edges of the superconducting contacts, to the vortex current.

On the other hand, the I-V measurements at different applied magnetic fields (see Figure 4) gradually tend to exhibit a S shape behavior. This may indicate a super-relativistic motion of non-localized vortices, in which they can overcome the characteristic electromagnetic wave velocity [32]. This motion should be accompanied by Cherenkov radiation of Josephson plasma waves with a dispersion relation of the eigenmodes given by $\omega_m^2 = \omega_p^2 + c_m^2 k^2$ using the Sine-Gordon equations where c_m are the characteristic velocities [32–34]. However, in this type of proximity array the Cherenkov oscillations induced by moving fluxons are damped by the increased quasi-particle conductance. As a consequence, the decay length of the oscillating tail associated with the fluxon decreases and the resonances on the fluxon step disappear, in agreement with the observed S shape behavior. The observation of the resonance steps can appear either increasing the magnetic field (below H_{c2}) or by decreasing the measurement temperature to values well below T_2.

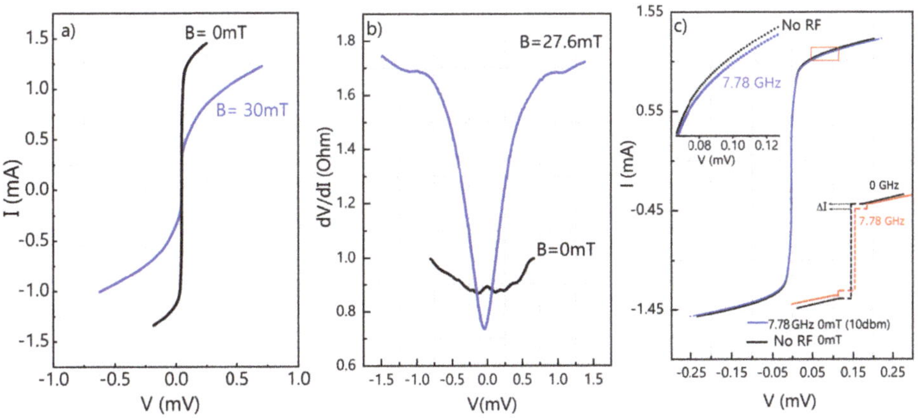

Figure 4. (**a**) I-V curves measured with and without the applied external magnetic field of B = 27.6 mT; (**b**) Andreev spectra obtained from the numerical derivative of the I-V curves. The reflection peaks observable at zero field are shifted towards the higher potentials; (**c**) I-V response of the PAD at 7.78 GHz. The red rectangular range above I_t is expanded in the plot. The lower inset schematically shows the expected formation of the Shapiro steps and the voltage response of the device.

The representative of the Andreev spectra (shown in Figure 4b) of our device was achieved by the numerical derivative of the I-V curves. The reflection peaks at zero field spectra are clearly visible. From them, the gap of the PAD is $\Delta \sim$1.3 meV. At the magnetic field corresponding to the frustration $f = 1$, the gap peculiarities in the spectra are not only shifted towards lower energies as is expected in the classical case, but they move in the inverse direction and the intensity increases. This anomalous behavior implies that in the non-localized vortex state a triplet Cooper pairing may occur, with the total spin moment of the superconducting configuration $S = 1$. In this case, in agreement with the observed vortex dynamics reported in Figure 3, a moderate magnetic field can contribute to the stabilization or even to the enhancement of the superconducting state [35] and hence to the superconducting order parameter.

The response of the device to a radio-frequency (RF) signal was measured in a non-bolometric regime in absence of applied magnetic field ($f = 0$). The response of the device to the signals with frequency lower than that of characteristic frequency (RF signal in our case) can be characterized as a perturbation to the system at non-equilibrium that results in the hindrance of the scape current (I_t) and the increase of the scape rate. This modulation directly corresponds to the RF intensity power and frequency [36]. However, the response of the device at frequencies close to the characteristic frequency (THz, in our case), should result in formation of the Shapiro steps, corresponding to the incident frequency [37].

The RF frequency was set to 7.78 GHz, the resonance frequency of the dipole antenna, though far lower than the expected characteristic frequency of the device calculated using the I_t and the measured normal resistance of the device. The characteristic frequency was calculated considering the tunneling current maximum (I_t) and the normal resistance of the device ($R_n = 1.1\Omega$) as $\omega_c = I_t R_n / \Phi_0$ in which $\Phi_0 = h/2e$ is the magnetic flux quantum with the value of $\omega_c = 300$ GHz [38]. Based on the sample dimensions and the distance from the RF source, the overall RF power collected by the sample was estimated to be of the order of 1×10^{-6} times the incident power sent to the antenna. The I-V characteristic of the device was measured with and without the RF signal superimposition and it is shown in Figure 4c. The reduction of the I_t, indicating the increase of the escaping rate is clearly visible. Though, as discussed above, because of the difference between the RF frequency and the characteristic frequency and the low RF power, the formation of Shapiro steps could not be observed. Nonetheless, the I_t contraction, usually associated with the scape rate, can marks the relative response of the device (see Figure 4c-inset).

From the I-V characteristics and the power coupled into the junction we were able to estimate the device voltage sensitivity. Considering the dynamic resistance of the $R_d = dV/dI \sim 0.9\ \Omega$ (either via the direct ac dynamic resistance, comparable to the numerically derived dc resistance) at the device operation point, slightly above the tunneling current limit I_t, we calculated the voltage modulation amplitude, $\Delta V = R_d \Delta I$ where ΔI is the induced current change at the operation point. Accordingly, the voltage sensitivity $\eta = \Delta V / P_{rf}$ of a non-magnetically modulated device was calculated to be of the order of $\eta \sim 7000\ VW^{-1}$ [39], which is relatively high compared with the semiconductor detector counterpart, with sensitivities of the order of $\eta \sim 3000\ VW^{-1}$ [40]. However, considering the possible impedance mismatch between the antenna and the junction, further reflections could be present and an optimized signal-coupling could increase the precision of the measurements. Since the dynamic resistance of the device is subject to the modulation via the external magnetic field, it is expected that at a certain operation bias, the device response can be significantly enhanced with the application of the magnetic field [41]. Furthermore, because of its characteristic device frequency, an enhanced response in the THz domain is expected.

4. Conclusions

In this report, we present a proximity junction arrays device (PAD), as a novel long-wavelength radiation detector. We characterized the vortex dynamics of this proximity array device (PAD) vs. external parameters, e.g., magnetic and electric field, showing the occurrence of the transition from a

Mott insulating to a metallic state. We also demonstrated that the application of the magnetic field can modulate and significantly enhance the superconducting order parameter. The device shows a shunted collective Josephson characteristics with the characteristic frequency of $\omega = 300$ GHz. The application of the magnetic field also introduces a superimposition of the screening current to the tunneling current, which determines its reduction. The response of the device was experimentally measured at the frequency of $f = 7.78$ GHz immersed in liquid He at the temperature of T = 4.2 K. As a detector, this device shows a relatively high voltage sensitivity of 7000 VW^{-1}, in absence of magnetic field modulations. Our results suggest that photo-response and voltage sensitivity of the proximity junctions array, can be additionally enhanced via external parameters, i.e., geometry and magnetic field at specific operation bias.

Author Contributions: S.J.R., D.D.G., C.G., N.P., S.L., M.C.G. and A.M. designed the experiment. S.J.R., D.D.G., C.L., M.F., S.C. and C.G. performed the experiment. S.J.R., D.D.G. and C.G. analyzed the data. All authors have read and agreed to the published version of the manuscript.

Funding: This project was financially supported by project TERA, an internal call of group V^{th} of INFN at LNF.

Acknowledgments: S.J.R. would like to thank Antonio Grilli and Agostino Raco for their valuable technical support. We acknowledge the financial support of the Bilateral Cooperation Agreement between Italy and Japan of the Italian Ministry of Foreign Affairs and of the International Cooperation (MAECI) in the framework of the project of major relevance N. PGR0072.

Conflicts of Interest: The authors declare no conflict of interest.

References

1. Dhillon, S.S.; Vitiello, M.S.; Linfield, E.H.; Davies, A.G.; Hoffmann, M.C.; Booske, J.; Paoloni, C.; Gensch, M.; Weightman, P.; Williams, G.P.; et al. The 2017 terahertz science and technology roadmap. *J. Phys. D. Appl. Phys.* **2017**, *50*. [CrossRef]
2. Ortolani, M.; Lupi, S.; Baldassarre, L.; Schade, U.; Calvani, P.; Takano, Y.; Nagao, M.; Takenouchi, T.; Kawarada, H. Low-Energy Electrodynamics of Superconducting Diamond. *Phys. Rev. Lett.* **2006**, *97*, 097002. [CrossRef] [PubMed]
3. Nanni, E.A.; Huang, W.R.; Hong, K.H.; Ravi, K.; Fallahi, A.; Moriena, G.; Dwayne Miller, R.J.; Kärtner, F.X. Terahertz-driven linear electron acceleration. *Nat. Commun.* **2015**, *6*, 8486, [CrossRef] [PubMed]
4. Jepsen, P.; Cooke, D.; Koch, M. Terahertz spectroscopy and imaging - Modern techniques and applications. *Laser Photon. Rev.* **2011**, *5*, 124–166. [CrossRef]
5. Saeedkia, D. *Handbook of Terahertz Technology for Imaging, Sensing and Communications*; Woodhead Publishing: Sawston, UK, 2013.
6. Karasik, B.S.; Sergeev, A.V.; Prober, D.E. Nanobolometers for THz Photon Detection. *IEEE Trans. Terahertz Sci. Technol.* **2011**, *1*, 97–111. [CrossRef]
7. Rezvani, S.; Perali, A.; Fretto, M.; De Leo, N.; Flammia, L.; Milošević, M.; Nannarone, S.; Pinto, N. Substrate-Induced Proximity Effect in Superconducting Niobium Nanofilms. *Conden. Matter* **2018**, *4*, 1. [CrossRef]
8. Toma, A.; Tuccio, S.; Prato, M.; De Donato, F.; Perucchi, A.; Di Pietro, P.; Marras, S.; Liberale, C.; Proietti Zaccaria, R.; De Angelis, F.; et al. Squeezing Terahertz Light into Nanovolumes: Nanoantenna Enhanced Terahertz Spectroscopy (NETS) of Semiconductor Quantum Dots. *Nano Lett.* **2015**, *15*, 386–391. [CrossRef]
9. D'Apuzzo, F.; Piacenti, A.R.; Giorgianni, F.; Autore, M.; Guidi, M.C.; Marcelli, A.; Schade, U.; Ito, Y.; Chen, M.; Lupi, S. Terahertz and mid-infrared plasmons in three-dimensional nanoporous graphene. *Nat. Commun.* **2017**, *8*, 14885. [CrossRef]
10. Richards, P.L. Bolometers for infrared and millimeter waves. *J. Appl. Phys.* **1994**, *76*, 1–24, [CrossRef]
11. Ozyuzer, L.; Koshelev, A.E.; Kurter, C.; Gopalsami, N.; Li, Q.; Tachiki, M.; Kadowaki, K.; Yamamoto, T.; Minami, H.; Yamaguchi, H.; et al. Emission of Coherent THz Radiation from Superconductors. *Science* **2007**, *318*, 1291–1293. [CrossRef]
12. Du, J.; Smart, K.; Li, L.; Leslie, K.E.; Hanham, S.M.; Wang, D.H.C.; Foley, C.P.; Ji, F.; Li, X.D.; Zeng, D.Z. A cryogen-free HTS Josephson junction detector for terahertz imaging. *Supercond. Sci. Technol.* **2015**, *28*, 084001. [CrossRef]

13. Hammar, A.; Cherednichenko, S.; Bevilacqua, S.; Drakinskiy, V.; Stake, J. Terahertz Direct Detection in YBa2Cu3O Microbolometers. *IEEE Trans. Terahertz Sci. Technol.* **2011**, *1*, 390–394. [CrossRef]
14. Seliverstov, S.; Maslennikov, S.; Ryabchun, S.; Finkel, M.; Klapwijk, T.M.; Kaurova, N.; Vachtomin, Y.; Smirnov, K.; Voronov, B.; Goltsman, G. Fast and sensitive terahertz direct detector based on superconducting antenna-coupled hot electron bolometer. *IEEE Trans. Appl. Supercond.* **2015**. [CrossRef]
15. Carturan, S.; Maggioni, G.; Rezvani, S.; Gunnella, R.; Pinto, N.; Gelain, M.; Napoli, D. Wet chemical treatments of high purity Ge crystals for γ-ray detectors: Surface structure, passivation capabilities and air stability. *Mater. Chem. Phys.* **2015**, *161*, 116–122. [CrossRef]
16. Sartale, S.; Ansari, A.; Rezvani, S.J. Influence of Ti film thickness and oxidation temperature on TiO2 thin film formation via thermal oxidation of sputtered Ti film. *Mater. Sci. Semicond. Proc.* **2013**, *16*, 2005–2012. [CrossRef]
17. Rezvani, S.J.; Pinto, N.; Boarino, L.; Celegato, F.; Favre, L.; Berbezier, I. Diffusion induced effects on geometry of Ge nanowires. *Nanoscale* **2014**, *6*, 7469–7473. [CrossRef]
18. Rezvani, S.J.; Pinto, N.; Enrico, E.; D'Ortenzi, L.; Chiodoni, A.; Boarino, L. Thermally activated tunneling in porous silicon nanowires with embedded Si quantum dots. *J. Phys. D Appl. Phys.* **2016**, *49*, 105104. [CrossRef]
19. Pinto, N.; Rezvani, S.J.; Favre, L.; Berbezier, I.; Fretto, M.; Boarino, L. Geometrically induced electron-electron interaction in semiconductor nanowires. *Appl. Phys. Lett.* **2016**, *109*, 123101, [CrossRef]
20. Rezvani, S.; Pasqualini, M.; Witkowska, A.; Gunnella, R.; Birrozzi, A.; Minicucci, M.; Rajantie, H.; Copley, M.; Nobili, F.; Di Cicco, A. Binder-induced surface structure evolution effects on Li-ion battery performance. *Appl. Surface Sci.* **2018**, *435*, 1029–1036. [CrossRef]
21. Luukanen, A.; Hadfield, R.H.; Miller, A.J.; Grossman, E.N. A superconducting antenna-coupled microbolometer for THz applications. In Proceedings of the Terahertz for Military and Security Applications II, Orlando, FL, USA, 8 September 2004.
22. Ariyoshi, S.; Otani, C.; Dobroiu, A.; Sato, H.; Kawase, K.; Shimizu, H.M.; Taino, T.; Matsuo, H. Terahertz imaging with a direct detector based on superconducting tunnel junctions. *Appl. Phys. Lett.* **2006**, *88*, 203503, [CrossRef]
23. Kislinskii, Y.; Constantinian, K.; Borisenko, I.; Ovsyannikov, G.; Yagoubov, P. Submillimeter wave signal detection by bicrystal YBCO Josephson junctions at liquid nitrogen temperatures. *Physica C* **2002**, *372–376*, 436–439. [CrossRef]
24. Resnick, D.J.; Garland, J.C.; Boyd, J.T.; Shoemaker, S.; Newrock, R.S. Kosterlitz-Thouless Transition in Proximity-Coupled Superconducting Arrays. *Phys. Rev. Lett.* **1981**, *47*, 1542–1545. [CrossRef]
25. Abraham, D.W.; Lobb, C.J.; Tinkham, M.; Klapwijk, T.M. Resistive transition in two-dimensional arrays of superconducting weak links. *Phys. Rev. B* **1982**, *26*, 5268–5271. [CrossRef]
26. Pinto, N.; Rezvani, S.J.; Perali, A.; Flammia, L.; Milošević, M.V.; Fretto, M.; Cassiago, C.; De Leo, N. Dimensional crossover and incipient quantum size effects in superconducting niobium nanofilms. *Sci. Rep.* **2018**, *8*, 4710. [CrossRef]
27. Baturina, T.I.; Vinokur, V.M.; Mironov, A.Y.; Chtchelkatchev, N.M.; Nasimov, D.A.; Latyshev, A.V. Nanopattern-stimulated superconductor-insulator transition in thin TiN films. *EPL* **2011**, *93*, 47002. [CrossRef]
28. Rezvani, J.; Di Gioacchino, D.; Gatti, C.; Poccia, N.; Ligi, C.; Tocci, S.; Cestelli Guidi, M.; Cibella, S.; Lupi, S.; Marcelli, A. Tunable Vortex Dynamics in Proximity Junction Arrays: A Possible Accurate and Sensitive 2D THz Detector. *Acta Phys. Polonica A* **2020**, *137*, 17–20, [CrossRef]
29. Poccia, N.; Baturina, T.I.; Coneri, F.; Molenaar, C.G.; Wang, X.R.; Bianconi, G.; Brinkman, A.; Hilgenkamp, H.; Golubov, A.A.; Vinokur, V.M. Critical behavior at a dynamic vortex insulator-to-metal transition. *Science* **2015**, *349*, 1202–1205. [CrossRef]
30. Benz, S.P.; Rzchowski, M.S.; Tinkham, M.; Lobb, C.J. Critical currents in frustrated two-dimensional Josephson arrays. *Phys. Rev. B* **1990**, *42*, 6165–6171. [CrossRef]
31. Argaman, N. Nonequilibrium Josephson-like effects in wide mesoscopic SNS junctions. *Superlattices Microstruct.* **1999**, *25*, 861 – 875. [CrossRef]
32. Zitzmann, J.; Ustinov, A.V.; Levitchev, M.; Sakai, S. Super-relativistic fluxon in a Josephson multilayer: Experiment and simulation. *Phys. Rev. B* **2002**, *66*, 064527. [CrossRef]

33. Sakai, S.; Bodin, P.; Pedersen, N.F. Fluxons in thin-film superconductor-insulator superlattices. *J. Appl. Phys.* **1993**, *73*, 2411–2418. [CrossRef]
34. Goldobin, E.; Wallraff, A.; Thyssen, N.; Ustinov, A. Cherenkov radiation in coupled long Josephson junctions. *Phys. Rev. B* **1998**, *57*. [CrossRef]
35. Tran, V.H.; Zaleski, T.A.; Bukowski, Z.; Tran, L.M.; Zaleski, A.J. Tuning superconductivity in $Eu(Fe_{0.81}Co_{0.19})_2As_2$ with magnetic fields. *Phys. Rev. B* **2012**, *85*, 052502. [CrossRef]
36. Schöndorf, M.; Govia, L.C.G.; Vavilov, M.G.; McDermott, R.; Wilhelm, F.K. Optimizing microwave photodetection: input–output theory. *Quantum Sci. Technol.* **2018**, *3*, 024009, [CrossRef]
37. Shapiro, S. Josephson Currents in Superconducting Tunneling: The Effect of Microwaves and Other Observations. *Phys. Rev. Lett.* **1963**, *11*, 80–82. [CrossRef]
38. Tachiki, M.; Koyama, T.; Takahashi, S. Electromagnetic phenomena related to a low-frequency plasma in cuprate superconductors. *Phys. Rev. B* **1994**, *50*, 7065–7084. [CrossRef] [PubMed]
39. Barone, A.; Paternò, G. *Physics and Applications of the Josephson Effect*; Wiley: New York, NY, USA, 1982. [CrossRef]
40. Jeon, Y.; Jung, S.; Jin, H.; Mo, K.; Kim, K.R.; Park, W.K.; Han, S.T.; Park, K. Highly-Sensitive Thin Film THz Detector Based on Edge Metal-Semiconductor-Metal Junction. *Sci. Rep.* **2017**, *7*, 16830. [CrossRef] [PubMed]
41. Du, J.; Hellicar, A.D.; Li, L.; Hanham, S.M.; Nikolic, N.; Macfarlane, J.C.; Leslie, K.E. Terahertz imaging using a high-Tcsuperconducting Josephson junction detector. *Supercond. Sci. Technol.* **2008**, *21*, 125025. [CrossRef]

condensed matter

MDPI

Article

Infrared Synchrotron Radiation and Its Application to the Analysis of Cultural Heritage

Yuka Ikemoto [1,*], Manako Tanaka [2], Tomohiro Higuchi [3], Toshirou Semba [4], Taro Moriwaki [1], Emi Kawasaki [5] and Masayoshi Okuyama [5]

1 Spectroscopy and Imaging Division, Japan Synchrotron Radiation Institute, Hyogo 679-5187, Japan; moriwaki@spring8.or.jp
2 Department of History and Culture, Showa Women's University, Tokyo 154-8533, Japan; ma-tanaka@swu.ac.jp
3 Tokyo Metropolitan Industrial Technology Research Institute, Tokyo 135-0064, Japan; higuchi.tomohiro@iri-tokyo.jp
4 Department of Home Economics, Iida Women's Junior College, Nagano 395-8567, Japan; senba@iidawjc.ac.jp
5 Archaeological Institute of Kashihara, Nara 634-0065, Japan; capeli@hotmail.co.jp (E.K.); m-okuyama05@kashikoken.jp (M.O.)
* Correspondence: ikemoto@spring8.or.jp

Received: 29 February 2020; Accepted: 6 April 2020; Published: 9 April 2020

Abstract: Infrared synchrotron radiation (IR-SR) is a broad-band light source. Its brilliance is the main advantage for microspectroscopy experiments, when the limited size of the sample often prevents the use of conventional thermal radiation sources. Cultural heritage materials are delicate and valuable; therefore, nondestructive experiments are usually preferred. Nevertheless, sometimes, small pieces can be acquired in the process of preservation and conservation. These samples are analyzed by various experimental techniques and give information about the original material and current condition. In this paper, four attempts to analyze cultural heritage materials are introduced. All these experiments are performed at the microspectroscopy station of IR beamline BL43IR in SPring-8.

Keywords: infrared analysis; microspectroscopy; cultural heritage; synchrotron radiation

1. Introduction

Recently, many infrared (IR) beamlines have been in operation in synchrotron facilities around the world. The IR synchrotron radiation (SR) is emitted from a bending magnet, and it covers a broad spectral range, from near-IR to THz waves. Owing to the high brilliance of the IR-SR, broad-band microspectroscopy is the most frequently used technique at the IR beamlines, and scientific studies in various fields have been conducted. In particular, the low-wavenumber region is important, because commercial IR microspectroscopy equipment covers down to about 700 cm^{-1} and does not cover the lower range in many cases.

IR-SR has been applied to many fields of study, such as physics, chemistry, biology and others. Cultural heritage is one of the important fields among them. The materials are unique because they differ depending on the fabricated age, place, creator, material and storage conditions. Therefore, a nondestructive measurement is required in many cases. Nevertheless, sometimes, small pieces can be obtained from the cultural heritage for the investigation or in the process of preservation and conservation. The pieces are analyzed by many experimental techniques, to characterize the materials and investigate the conditions. IR analysis is often used for the analysis both in the mid-IR and THz region, using conventional source [1–6]. IR-SR provides high spatial resolutions, especially in the mid-IR region, which has played critical roles for the analysis of small fragments of cultural heritage materials [7–9]. IR-SR is a broad band light source. The low-wavenumber region is also important, since an absorption band of a metal complex is often observed and gives valuable information.

In this paper, the properties of infrared beamline BL43IR in SPring-8 are described in Section 2. In Section 3, four attempts to analyze cultural heritage materials at BL43IR are presented.

2. Infrared Beamline BL43IR at SPring-8

One of the most commonly used IR broad-band light sources for spectroscopy is a thermal radiation source, such as a globar lamp. IR-SR is another broad-band source for spectroscopy, with a high-brilliance feature. Figure 1 shows the brilliance of IR-SR estimated using BL43IR parameters, and that of the thermal radiation calculated using Planck's formula at 1500 °C. The definition of brilliance is the number of photons emitted from electrons in a storage ring per second, per angular divergence of the photons, per cross-sectional area of the beam and per bandwidth of 0.1% of the central photon energy. In Figure 1, the brilliance of IR-SR is shown to be more than two orders higher than that of the thermal radiation in the whole spectral range, from near-IR to THz waves [10]. Various kinds of IR sources, other than IR-SR and the thermal radiation source, have been developed recently. The IR free electron laser is one of the high-power sources, and it is often used as an excitation source [11]. Recently, broad-band IR lasers have become popular, such as quantum cascade lasers and IR fiber lasers. They have strong power, and the bandwidth is large enough to be used as sources of spectroscopy [12]. The spectral range, however, is still limited in the mid-IR region. The lowest wavenumber is about 1000 cm^{-1}. In the THz region, time-domain spectroscopy (TDS) is often used. The highest wavenumber of TDS is about 200 cm^{-1}, and there is a gap between the broad-band IR laser. The IR-SR is an important source that covers the gap and also has a high-brilliance feature.

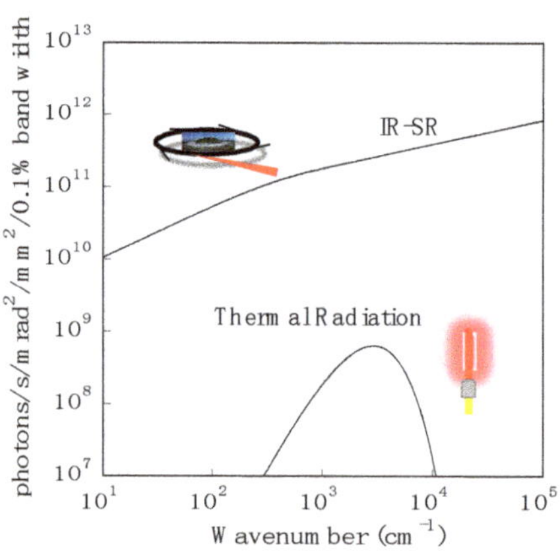

Figure 1. Brilliance of IR-SR and thermal radiation sources.

The light at BL43IR covers 100 to 10,000 cm^{-1}, which includes part of THz waves. The lowest wavenumber is limited by the structure of the beam extraction port of BL43IR [13]. Commercial IR microspectroscopy equipment covers down to about 700 cm^{-1} in many cases. We describe wavenumber region between 700 and 100 cm^{-1} as low-wavenumber region in this paper. There are three microscope stations; a long-working-distance microscope station, a magneto-optical microscope station and a high-spatial-resolution microscope station [14]. The first and second stations use the same Fourier transform infrared (FTIR) spectrometer, Bruker IFS 120, and the microscopes are custom made by Bunko–Keiki. The working distance of the first station's microscope is as long as 50 mm. Attachments that need a long working distance, such as a diamond anvil cell for high-pressure experiments, are used at this station. The microscope at the second station is assembled in a superconducting magnet. The maximum magnetic field is ± 14 T. Low-dimensional materials, as well as materials that undergo a phase transition in a magnetic field, are investigated at this station. The importance of the low-wavenumber region for physical research is that the electronic state close to the Fermi level of metallic materials, as well as low-energy-gap materials, can be investigated [15,16]. The magnifications of the objective mirrors in the first and second microscopes are 10 and 8, respectively. The FTIR spectrometer and the microscope at the third station are Bruker VERTEX70 and HYPERION2000 with objective mirror magnifications of 15× and 36×, respectively. All the cultural heritage experiments in this paper were performed at the third station. The beamsplitter in VERTEX70 was KBr and the detector was HYPERION2000 equipped with HgCdTe (MCT) for the mid-IR experiments. For experiments in the low-wavenumber-region, the beamsplitter was Si, and the detector was a Si bolometer.

3. Cultural Heritage Experiments

Many kinds of cultural heritage experiments have been conducted at BL43IR. Among them, four attempts are introduced in this section.

3.1. Iron Oxide in Bengala Pigment

This study was conducted by Dr. Manako Tanaka. Bengala is a kind of red pigment that has been used for arts, crafts and buildings since ancient times. The main components are iron oxides that show different colors, depending on chemical composition: the red α-Fe_2O_3, blown γ-Fe_2O_3 and black Fe_3O_4. There are two shapes of bengala: pipe-shaped and non-pipe-shaped [17]. The former has been mainly found from Jomon and Kofun period archaeological sites in Japan. The pipe-shaped bengala consists of small particles, and the size of the pipe is about 1 μm in diameter and 10–20 m in length. They are considered to be derived from the iron-oxidizing bacterium, *Leptothrix ochracea*, but details of raw materials and making methods have not been revealed. In this study, the effectiveness of spectra in low-wavenumber region at BL43IR was investigated, to identify the kinds of iron oxide included in pipe-shaped bengala. Spectroscopy in this region is a powerful tool for analyzing metal oxide materials. In this study, powder samples of α-Fe_2O_3 (Nilaco Corp., Tokyo, Japan), γ-Fe_2O_3 (JFE Chemical Corporation, Tokyo, Japan) and Fe_3O_4 (Rare Metallic Co. LTD., Tokyo, Japan) were used as standard iron oxides. Note that α- and γ-Fe_2O_3 are called hematite, and the crystal structures are rhombohedral and cubic, respectively. Fe_3O_4 is called magnetite, and it has a cubic inverse spinel structure. The quality of the IR spectrum strongly depends on the external form of the sample. A flat and uniform shape is suitable, and IR-SR is advantageous because the small sample size, which is about 200 μm in diameter, is enough to be measured even in the low-wavenumber region. Moreover, because our final target is minute pipe-shaped bengala sample excavated from an archaeological site, microspectroscopy is indispensable. The powder iron oxide samples were pressed gently, using a hand press (Diamond EX'Press, S.T. Japan, Tokyo, Japan), to form a flat and uniform sample. Figure 2a–c shows absorption spectra of α-Fe_2O_3, Fe_3O_4 and γ-Fe_2O_3, respectively. The wavenumber resolution was 4 cm^{-1}, and the number of scans was 256. In the shaded area, there is absorption from the beamsplitter of the FTIR spectrometer. Optical densities are shown by the bars in Figure 2. Fe-O stretching modes are observed at 302, 450 and 530 cm^{-1} in (a), 360 and 567cm^{-1} in (b), and 384 and 557 cm^{-1} in (c). These three iron oxides can be clearly identified from the absorption spectra. The low-wavenumber-region spectroscopy with IR-SR at BL43IR is expected to be a powerful tool for the analysis of pipe-shaped bengala samples.

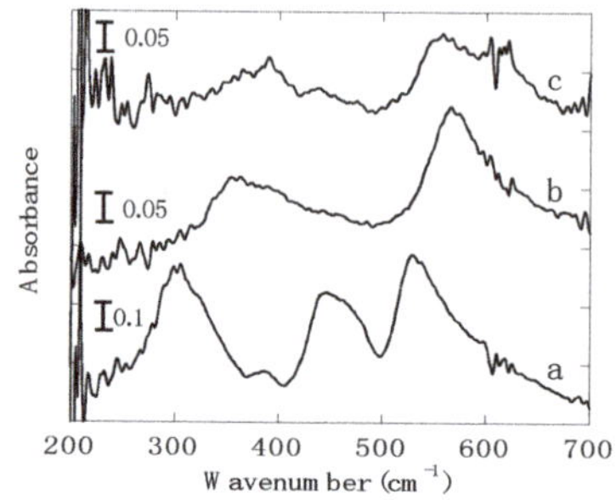

Figure 2. THz absorption spectra of (**a**) α-Fe_2O_3, (**b**) Fe_3O_4 and (**c**) γ-Fe_2O_3.

3.2. Biomineral Distribution Generated by a Lichen Growing on a Stone Cultural Heritage

This attempt is to analyze biominerals generated by a lichen growing on a stone cultural heritage. This study was conducted by Dr. Emi Kawasaki. In order to evaluate the risk of stone heritage deterioration, the distribution of biomineral components at the interface between stone and lichen is analyzed. Lichen samples were collected from the Bayon temple in the Angkor archaeological complex in Cambodia. In this study, two kinds of lichen are used, *Althoniales* sp. (International Nucleotide Sequence Databases (INSD): AB764061) and *Lepraria* sp. (INSD: AB764076) [18]. The forms of *Althoniales* sp. and *Lepraria* sp. are crustose and granular, respectively, and the microscope photographs are shown in Figure 3. The tip of a needle indicates the algal layer in *Althoniales* sp. and the medulla in *Lepraria* sp. The illustration in Figure 3 shows the structure of lichen on stone. The algal layer was wrapped in fungal hyphae. The surface and interface between lichen and stone correspond to the top and bottom of the sample, respectively. The lichens were cut in the direction perpendicular to the interface, and a small sample was collected from the cut surface, using a manipulator needle or a knife. The size of the collected sample was as small as about 50 μm to identify each layer, and microspectroscopy using IR-SR played a critical role in this study. The pieces were made thin and flat, using a hand press (Diamond EX'Press, S.T. Japan, Tokyo, Japan). Sample preparation is an important process in the study of cultural heritage materials. Handling a small piece using a micro-sampling manipulator is a typical technique. The thin, flat shape is also important for the reason explained in Section 3.1.

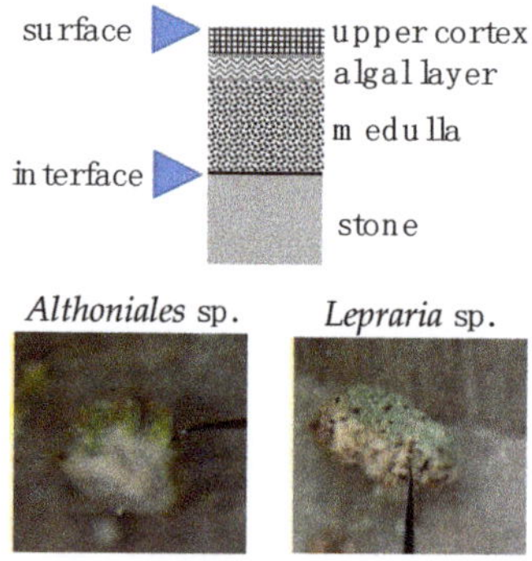

Figure 3. Schematic illustration of the structure of lichen on stone (top). Microscope photographs of *Althoniales* sp. and *Lepraria* sp. (bottom).

Figure 4a,b shows the absorption spectra of *Althoniales* sp. and *Lepraria* sp. The samples are set on a diamond substrate. The wavenumber resolution was 4 cm^{-1}, and the number of scans was 64. An aperture with the size of 10×10 μm^2 was used in the HYPERION 2000 microscope. Calcium oxalate is one of the typical biomineral components, and it is known to exhibit infrared absorption at about 1610 and 1315 cm^{-1} [19]. These peaks are observed in the medulla and are shown by the arrows in Figure 4a. In Figure 4b, the peaks are observed at the interface. The result indicates that the distribution of the biomineral component differs depending on the kind of lichen. Calcium oxalate also has a strong band at 274 cm^{-1}, in the low-wavenumber region [20]. Further revelations are expected from the results of low-wavenumber-region experiments.

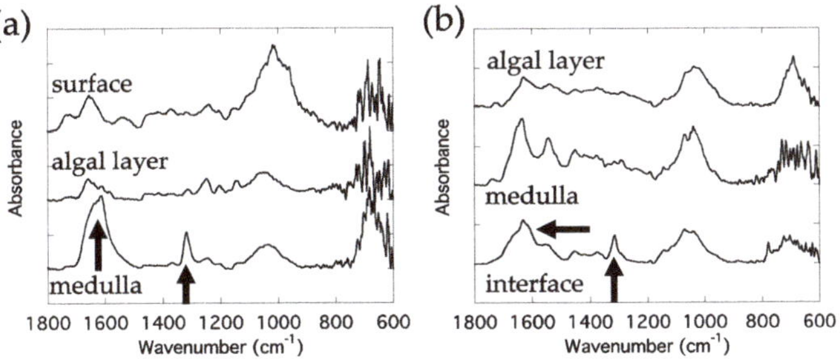

Figure 4. Absorption spectra of (**a**) *Althoniales* sp. and (**b**) *Lepraria* sp.

3.3. Adhesives on Fiber of Traditional Snow Festival Costume in Niino

This study was conducted by Mr. Toshiro Semba. There is a project to restore a traditional costume worn in the snow festival in Niino, Nagano prefecture. The festival has a history of more than 500 years. The material is known to be made of *Wisteria floribunda* fibers. When the fibers are observed under an optical microscope, adhesives can be seen on them. The aim of this study is to determine the adhesive materials. There are several possible candidates, with lacquer being one of them. The other candidates are the materials used when the *Wisteria floribunda* fibers were colored, such as *suou* (a kind of dye made from *Casaplania sappan*) and soy bean. Figure 5 shows the mid-IR absorption spectra. The sample image under the HYPERION 2000 microscope is shown in the inset. The bar in the image corresponds to 300 μm. Only a small piece of sample was obtained from the costume, and microspectroscopy by IR-SR was indispensable. The measurement position was specified by the microscope. Spectra for adhesives at two different positions are shown by curves a and b. Curves c to f show the absorption spectra of lacquer, *suou*, *Wisteria floribunda* and soy bean. The bar in the image corresponds to 200 μm. The wavenumber resolution was 4 cm^{-1}, and the number of scans was 500. An aperture with the size of 10×10 μm^2 was used in the HYPERION 2000 microscope.

From Figure 5, the absorption spectra of the adhesives are seen to be different depending on the position. This means that the adhesives are composed of multiple substances. Some of the absorption peaks in Figure 5 can be assigned as follows. The peak at about 3300–3400 cm^{-1} corresponds to OH, double bands at 2926 and 2847 cm^{-1} correspond to CH_3 and CH_2 symmetric stretching modes, and the bands at 1660 and 1546 cm^{-1} correspond to amide I and II. For the assignments of other absorption bands, composition analysis may be useful. The lower wavenumber region experiments are also expected to yield information because the main components, such as protein and textile fibers, sometimes have peaks in that region [21]. The overall shape of spectrum a is similar to curve f of soy bean, and curve b is similar to the lacquer curve c, although there are some differences. Comparing spectra b and c, we see that the double band at about 2800 cm^{-1} is small in c and broad in b. The band at 2600 cm^{-1} in c splits in b. Spectra a and f are similar, but the band at 1740 cm^{-1} in f is not present in a. One possible reason for the differences may be the deterioration of samples a and b.

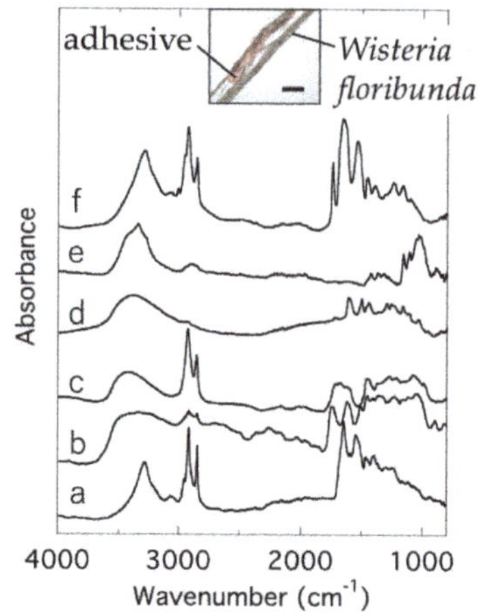

Figure 5. Absorption spectra of the adhesives and the candidate materials. Spectra of a and b are of adhesive samples. Spectra c to f show those of lacquer, *suou* (a kind of dye), *Wisteria floribunda* fibers and soy bean. Inset is the sample image under HYPERION 2000 microscope.

3.4. Organic Materials in Overglaze Layers on Ceramics

Dr. Tomohiro Higuchi conducted a study to analyze organic residuals in the overglaze layer of Japanese ceramics. When a Japanese ceramic is painted with overglaze, some kind of glue is used before it is fired at a high temperature [22]. After firing, a small amount of organic materials has been found to remain in the overglaze layer. The residual is thought to originate from the glue [23]. It is also predicted that the organic materials form compounds with metallic elements included in glass frit. These remains are expected to be observed in mid-IR and the low-wavenumber region. The final purpose of this study is to identify the glue from the residual organic materials observed in the fired ceramics. Information about the kind of glue is expected to provide knowledge about the place where the ceramic vessels were made. In this study, simulated samples were prepared, to observe the change in the state of the glue during the firing process. Two kinds of commonly used glue, nikawa (Kissho, granular animal glue) and funori (Isekyu, red seaweed glue) were used. They were dissolved in water at 10% m/m and 2% m/m, in mass concentration, and 1 mL of solution was mixed with 1 g of colored overglaze that contains Pb-included frits and iron oxides. The mixed colored overglaze was applied to a CaF_2 substrate. Absorption spectra of the sample were measured during heating to 600 °C, on a temperature-controlled stage (LINKAM 10036L). The heating rate was 100 °C/h, and the atmosphere was air. The wavenumber resolution was 4 cm^{-1}, and the number of scans was 64. Figure 6a,b shows the absorption spectra of the nikawa- and funori-containing samples, respectively. The observed temperatures were room temperature, 200, 240, 300 and 600 °C in Figure 6a; and room temperature, 200, 230, 350 and 600 °C in Figure 6b. The optical densities (OD) are shown in the figures by the bars. The ODs are different between Figure 6a,b because the sample thicknesses where the spectra were measured were different. At a high temperature, such as 600 °C, the thermal radiation from the sample reaches the detector and increases the background noise. The spectrum at 600 °C in Figure 6b was measured under such conditions.

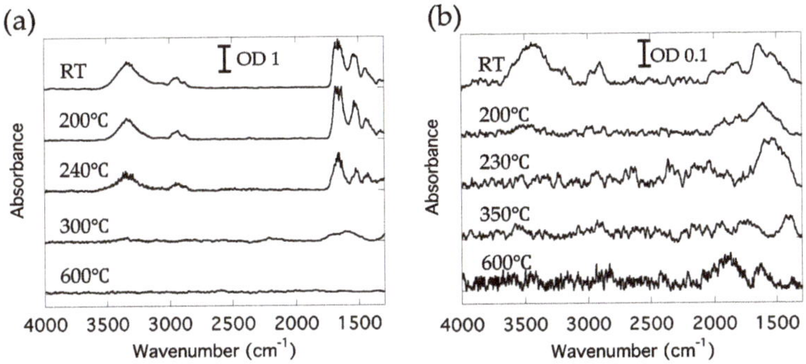

Figure 6. Absorption spectra from samples containing (**a**) nikawa and (**b**) funori.

The absorption spectrum observed at room temperature (RT) in Figure 6a has bands at 3300, 2900 and 1800–1500 cm^{-1}. These bands are assigned to a protein included in nikawa. The structures at 1800–1500 cm^{-1} are amide I and II bands. The RT spectrum in Figure 6b has absorption bands at 3300, 2900 and 1750–1400 cm^{-1}, and they are assigned to a polysaccharide included in funori. In Figure 6a, the shape of the bands observed at RT does not change much, and only the intensities decrease up to 300 °C. At 600 °C, there is no obvious band. In Figure 6b, the absorption band intensity decreases and vanishes once at 230 °C. At 600 °C, a different absorption band appears at 2100–1000 cm^{-1}. From Figure 6, the temperature dependence is found to differ depending on the kind of glue. The original organic component of the glue may be burned out at between 200 and 300 °C. Then, in the sample of Figure 6b, the transformed organic materials appear on the substrate at 600 °C, but not in the case of Figure 6a. In the overglaze material, metal elements were included, and the IR experiments in the low-wavenumber region are expected to yield information about the metal composition.

4. Summary

IR-SR has two important features, a high brilliance and a broad-band coverage. The brilliance was calculated by using the parameters of infrared beamline BL43IR in SPring-8 and was shown to be more than two orders of magnitude higher than that of a thermal radiation source. Compared with other IR light sources, low-wavenumber region between 700 and 200 cm^{-1} is a valuable region for IR-SR. BL43IR is dedicated to IR microspectroscopy and has three stations with different characteristics. Four attempts to analyze cultural heritage materials were introduced: iron oxides in bengala red pigment, biomineral distribution caused by lichen growing on a stone cultural heritage, adhesives of fibers of a traditional snow festival costume and organic materials in overglaze layers on ceramics. All these experiments were performed at the high-spatial-resolution microspectroscopy station, at BL43IR. In the analysis of cultural heritage materials, IR-SR can play an important role, especially in the low-wavenumber region.

Author Contributions: Y.I. wrote the manuscript; M.T., T.M., T.H., M.T., T.S., E.K. and M.O. conducted the experiments. All authors have read and agreed to the published version of the manuscript.

Funding: This research was partially supported by JPSJ KAKENHI grant numbers JP16K16339 and JP16K01188.

Acknowledgments: The authors thank Kazami Mizumoto (Tokyo University of the Arts), Toshiyasu Shinmen and Shuji Ninomiya (Tokyo Gakugei University) for their collaboration. Synchrotron radiation measurements were performed at SPring-8, with the approval of JASRI (2016B1795, 2017B1758, 2016A1696, 2016B1809, 2018A1330, 2017B1768, 2013B1232, 2016B1804, 2016B1808, 2014A1203, 2014A1827, 2017B1765, 2016B1803, 2016B1190, 2016A1380).

Conflicts of Interest: The authors declare no conflict of interest.

References

1. Fukunaga, K.; Sekine, N.; Hosako, I.; Oda, N.; Yoneyama, H.; Sudou, T. Real0time terahertz imaging for art conservation science. *J. Eur. Opt. Soc. Rapid* **2008**, *3*, 08027. [CrossRef]
2. Gallerano, G.P.; Doria, A.; Giovenale, E.; Messina, G.; Petralia, A.; Spassovsky, I.; Fukunaga, K.; Hosako, I. THz-ARTE: Non-Invasive Diagnostics for Art Conservation. In Proceedings of the 2008 33rd International Conference on Infrared, Millimeter and Terahertz Waves, Pasadena, CA, USA, 15–19 September 2008; pp. 1–2.
3. Manceau, J.-M.; Nevin, A.; Fotakis, C.; Tzortzakis, S. Terahertz time domain spectroscopy for the analysis of cultural heritage related materials. *Appl. Phys. B* **2008**, *90*, 365–368. [CrossRef]
4. Jackson, J.B.; Mourou, M.; Whitaker, J.F.; Duing, I.N.; Williamson, S.L.; Menu, M.; Mourou, G.A. Terahertz imaging for non-destructive evaluation of mural paitings. *Opt. Commun.* **2008**, *281*, 527–532. [CrossRef]
5. Fukunaga, K. Terahertz spectroscopy for non-invasive analysis of cultural properties. *J. Natl. Inst. Inf. Commun. Technol.* **2009**, *55*, 67–71.
6. van Loon, A.; Gambardella, A.A.; Gonalez, V.; Cotte, M.; Nolf, W.; Keune, K.; Leonhardt, E.; Groot, S.; Gaibor, A.N.P.; Vandivere, A. Out of the blue: Vermeer's use of ultramarine in girl with a pearl earring. *Herit. Sci.* **2020**, *8*, 25. [CrossRef]
7. Pronti, L.; Romani, M.; Vivian, G.; Stani, C.; Gioia, P.; Cestelli, M. Advanced methods for the analysis of Roman wall paintings: Elemental and molecular detection by means of synchrotron RT-IR and SEM icro-imaging spectroscopy. *Rend. Lincei. Sci. Fis. Naurali* **2020**, 1–9. [CrossRef]
8. Salvado, N.; Buti, S.; Tobin, M.J.; Pantos, E.; John, A.; Prag, N.W.; Pradell, T. Advantages of the use of SR-FT-IR microspectroscopy: Applications to cultural heritage. *Anal. Chem.* **2005**, *77*, 3444–3451. [CrossRef] [PubMed]
9. Cinque, G.; Marcelli, A. Synchrotron radiation inraRed microspectroscopy and imaging in the characterization of archaeological materials and cultural heritage artefacts. *EMU Notes Mineral.* **2019**, *20*, 411–444.
10. Ikemoto, Y.; Ishikawa, M.; Nakashima, S.; Okamura, H.; Haruyama, Y.; Matsui, S.; Moriwaki, T.; Kinoshita, T. Development of scattering near-field optical microspectroscopy apparatus using an infrared synchrotron radiation sources. *Opt. Commun.* **2012**, *285*, 2212–2217. [CrossRef]
11. Torgasin, K.; Morita, K.; Zne, H.; Masuda, K.; Bakr, M.; Nagasaki, K.; Kii, T.; Ohgaki, H. Study on anormalous photoemission of LaB6 at high temperatures. *Phys. Scr.* **2019**, *94*, 7. [CrossRef]
12. Bechtel, H.A.; Muller, E.A.; Olmon, R.L.; Martin, M.C.; Raschke, M.B. Ultrabroadband infrared nanospectroscopic imaging. *PNAS* **2014**, *111*, 7191–7196. [CrossRef] [PubMed]
13. Kimura, H.; Moriwaki, T.; Takahashi, S.; Aoyagi, H.; Matsushita, T.; Ishizawa, Y.; Masaki, M.; Oishi, M.; Ohkuma, H.; Namba, T.; et al. Infrared beamline BL43IR at SPring-8:: design and commissioning. *Nucl. Instrum. Methods Phys. Res. A* **2001**, *467–468*, 441–444. [CrossRef]
14. Moriwaki, T.; Ikemoto, Y. BL43IR at SPring-8. *Infrared Phys. Technol.* **2008**, *51*, 400–403. [CrossRef]
15. Kobayashi, R.; Hashimoto, K.; Yoneyama, N.; Yoshimi, K.; Motoyama, Y.; Iguchi, S.; Ikemoto, Y.; Moriwaki, T.; Taniguchi, H.; Sasaki, T. Dimer-Mott and charge-ordered insulating states in the quasi-one-dimensional organic conducting δ'_{P}- and δ'_{C}-(BPDT-TTF)$_2$ICl$_2$. *Phys. Rev. B* **2017**, *96*, 115112. [CrossRef]
16. Okamura, H.; Shoji, K.; Miyata, K.; Sugawara, H.; Moriwaki, T.; Ikemoto, Y. Pressure suppression of spin-density wave gap in the optical conductivity of $SrFe_2As_2$. *J. Phys. Soc. Jpn.* **2013**, *82*, 074720. [CrossRef]
17. Kamijoh, T. Red Pigments used from Jomon Period to Kofun Period. *J. Jpn. Soc. Colour Mater.* **2004**, *77*, 86–90. [CrossRef]
18. Kawasaki, E.; Matsui, T.; Yamamoto, Y.; Hara, K. The documentation method using lichens growing on stoneworks in order to protect cultural heritage stone monuments—The case of Angkor monuments of Cambodia. *Lichenology* **2013**, *11*, 39–52.
19. Parsons, J.G.; Dokken, K.M.; McClure, J.; Gardea-Torresdey, J.L. FTIR, XA, and XRD study of cadmium complexes with l-crysteine. *Phlyhedron* **2013**, *56*, 237–242. [CrossRef]
20. Pucetaite, M.; Tamosaityte, S.; Engdahl, A.; Ceponkus, J.; Sablinskas, V.; Uvdal, P. Microspectroscopic infrared specular reflection chemical imaging of multi-component urinary stones: MIR vs. FIR. *Cent. Eur. J. Chem.* **2014**, *12*, 44–52. [CrossRef]
21. Yanchen, H.; Shengjie, L.; Zeming, Q.; Xin, C. Application of far-infrared spectroscopy to the structural identification of protein materials. *Phys. Chem. Chem. Phys.* **2018**, *18*, 11643–11648.

22. Ohnishi, S. Chapter 6 Techniques of Overglaze. In *Traditional Techniques of Ceramics Art*, 1st ed.; Rikogakusya Publishing Co., Ltd.: Tokyo, Japan, 2001; pp. 231–278.
23. Higuchi, T.; Ninomiya, S. Attempt to Analyze the Organic Residuals in Overglaze Layers on Ceramics. In Proceedings of the 31st Annual Meeting of the Japan Society for Scientific Studies on Cultural Property, Nara, Japan, 5 July 2014; pp. 258–259.

condensed matter

MDPI

Article

Cross-Correlation of THz Pulses from the Electron Storage Ring BESSY II

Ulrich Schade [1,*], Peter Kuske [1], Jongseok Lee [2], Barbara Marchetti [3] and Michele Ortolani [4]

1 Helmholtz-Zentrum Berlin für Materialien und Energie GmbH, Albert-Einstein-Strasse 15, 12489 Berlin, Germany; peter.kuske@helmholtz-berlin.de
2 Department of Physics and Photon Science, Gwangju Institute of Science and Technology, Gwangju 61005, Korea; jsl@gist.ac.kr
3 Deutsches Elektronen-Synchrotron DESY, Notkestrasse 85, 22607 Hamburg, Germany; barbara.marchetti@desy.de
4 Department of Physics, Sapienza University of Rome, Piazzale Aldo Moro 2, 00185 Roma, Italy; michele.ortolani@roma1.infn.it
* Correspondence: ulrich.schade@helmholtz-berlin.de

Received: 17 February 2020; Accepted: 25 March 2020; Published: 27 March 2020

Abstract: Coherent synchrotron radiation from an electron storage ring is observed in the THz spectral range when the bunch length is shortened down to the sub-mm-range. With increasing stored current, the bunch becomes longitudinally unstable and modulates the THz emission in the time domain. These micro-instabilities are investigated at the electron storage ring BESSY II by means of cross-correlation of the THz fields from successive bunches. The investigations allow deriving the longitudinal length scale of the micro bunch fluctuations and show that it grows faster than the current-dependent bunch length. Our findings will help to set the limits for the possible time resolution for pump-probe experiments achieved with coherent THz synchrotron radiation from a storage ring.

Keywords: THz; coherent synchrotron radiation; low alpha; cross-correlation; pump-probe

1. Introduction

Coherent synchrotron radiation (CSR) emission from a relativistic electron bunch in a storage ring is a natural process of collective in-phase emission of all electrons and occurs for wavelengths λ longer than the electron bunch length [1]. The total power $P(\lambda)$ emitted by N electrons can be calculated from the power $p(\lambda)$ emitted by one electron [2]:

$$P(\lambda) = [N + N(N-1)\, f(\lambda)]\, p(\lambda), \tag{1}$$

where $f(\lambda)$ is the form factor given by the square of the Fourier transform of the normalized longitudinal electron distribution $n(z)$ of the bunch:

$$f(\lambda) = \left| \int_{-\infty}^{\infty} e^{2\pi i \frac{z}{\lambda}}\, n(z)\, dz \right|^2. \tag{2}$$

Since the propagation of long wavelengths radiation is suppressed by the wave-guiding properties of the finite dimensions of the metallic walls of the vacuum chambers [3], bunch lengths have to be shortened below the cut-off wavelength in order to become detectable at the end of a beamline. This becomes valid for most of the electron storage rings for sub-mm bunches in the THz spectral range.

The longitudinal bunch length of the electrons in a storage ring can be controlled by the momentum compaction factor α—the ratio between the deviation of an electron from the reference orbit and its momentum deviation—and the temporal gradient of the *rf* (radiofrequency) cavity voltage. Short bunch lengths emitting coherently in the THz spectral range are obtained by reducing the momentum compaction factor from the value normally applied by tuning the storage ring optics [4] into the so-called "*low* α" mode. Such shorter bunches cause a coherent enhancement for a given number of electrons in the bunch which can be expressed by the amplification factor AF, the ratio between the coherent and incoherent emission spectrum, e.g., of the spectrum from the short electron bunch and from the bunch of nominal length. The longitudinal electron distribution in a storage ring in equilibrium can be treated as Gaussian for low bunch charges. For a Gaussian electron distribution, the coherent enhancement factor becomes Gaussian itself with the mean σ_z as the *rms* (root mean square) bunch length [2] having a maximum of N for infinite wavelength for a large ensemble of electrons:

$$AF(\lambda) = N\, e^{-(\frac{4\pi\sigma_z}{\lambda})2}. \tag{3}$$

Typically, 10^7–10^9 electrons per bunch are stored. According to Equation (1) the total power emitted from the bunch would then be proportional to the square of these values yielding an increase of seven to nine orders of magnitude compared to the incoherent bunch emission. Such an enormous power increase influences the electron bunch and produces bunch instabilities. The own CSR field of the bunch and its space charges on the conductive chamber walls affect the bunch shape in such a way that the bunch becomes longer and distorted from the Gaussian form. These longitudinal distortions contain higher Fourier frequency components responsible for a broader CSR power spectrum. Numerical calculations show a steepening of the leading edge of the bunch due to its own CSR wake field [5]. Further, the electron distribution becomes unstable above a certain stored charge threshold and initial small density perturbations can then cause micro-bunching due to the intense wake field finally emitting stochastically THz bursts [6,7]. All these longitudinal density fluctuations in a scale shorter than the nominal bunch length are associated with a coherent increase of emitted THz power at higher photon energies than expected for the nominal Gaussian bunch shape.

More than fifteen years have passed since it was proven that CSR is a suitable source for diffraction-limited Fourier transform spectroscopy [8] and the first successful scientific experiment using THz radiation from short electron bunches was performed [9]. The *low* α technology is now becoming available at several storage rings worldwide (e.g., [10–15]) and this powerful broadband source allows for many scientific experiments [16] including THz- electron paramagnetic resonance (EPR) [17] and high-resolution spectroscopy with THz frequency combs [18]. Currently, storage rings offer to users the *low* α mode at reduced currents stored. However, strong longitudinal bunch focusing by superconducting *rf* cavities in a storage ring will extent the bursting threshold to higher currents and shortens CSR pulses to a couple of THz [19] making this short-bunch scheme not only more attractive to far infrared/THz but also for time-resolved X-ray experiments [20].

Even though a significant progress has been achieved over the last years of offering short-bunch operation, a consistent picture of bunch deformation and bunch instabilities in *low* α operational mode of an electron storage ring is still under discussion. Especially, the transition from stable CSR emission via quasi harmonic micro-bunching to the chaotic bursting is very different among the various storage ring facilities and call for sophisticated diagnostics. The electron bunch spatio-temporal dynamics have been studied by the experimentally obtained spectral content of THz pulses in comparison with numerical simulations [21–23]. A straightforward approach to study the emitted CSR pulses is not available due to the lack of suitable THz recording electronics, but the quasi-direct observation of sub-picosecond microstructures within an electron bunch became recently possible by a time-stretching scheme [24–26].

This paper reports on cross-correlation measurements from subsequent bunches emitting THz pulses at the storage ring BESSY II. By applying a simple model for the projection of the longitudinal phase space quantitative information on the dynamics of bunch instabilities, which are derived and compared to experimental findings.

2. Experimental Section

The measurements presented were conducted at the IRIS beamline [27] at the electron storage ring BESSY II during several dedicated *low* α shifts. The beamline accepts THz and infrared dipole radiation 60 mrad horizontally and 40 mrad vertically from the storage ring. The fundamental *rf* of BESSY II is 500 MHz with a harmonic number of 400 in a circumference of 240 m. Thus, the revolution frequency is 1.25 MHz with a minimum bunch spacing of 2 ns. A train of 350 consecutive bunches build up the fill pattern. For our investigation, the storage ring operated with a *low-*α*-*optics and a synchrotron frequency of f_s = 1.75 kHz providing an *rms* bunch length of 1 mm in the limit of zero current [28]. The same synchrotron frequency is applied for normal *low* α operation. This operational mode is offered to the storage ring users for pump-probe and THz spectroscopic experiments two times a year.

The IRIS beamline is equipped with a commercial Fourier transform (FT) vacuum spectrometer which operates down to 2 cm^{-1} and which utilizes several bolometers as detectors suitable for diverse applications. The spectra shown in this study were obtained by a 1.6 K Si composite bolometer. The bolometer has a response time of about 2 ms and a high-frequency cut-off at about 2 THz.

A free optical port of the beamline, only separated from the vacuum of the beamline to the ambient by a Picarin™ window, delivers a quasi-collimated beam. This beam feeds a home-built Martin–Puplett type interferometer (MPI) [29]. The MPI allows for the measurement of the auto and cross-correlation function of the CRS. While for the auto-correlation mode of the MPI the two mirror arms are set to equal lengths for the cross-correlation mode one mirror arm length is extended by 0.3 m in order to set the delay to the temporal distance between two bunches of 2 ns.

The correlation function measured was a mean correlation over the integration time of the detection and which was much longer than the revolution time of the storage ring. In the case of the auto-correlation the mean of all individual bunch correlations was measured. The cross-correlation yields the mean correlation of the fields from all pairs of successive bunches. In this context, it should be mentioned that the inverse Fourier transformation of the auto-correlation function results in the spectrum of the CRS emission.

In this study, we performed cross-correlation measurements for bunch charges between 5 and 50 pC with the MPI schematically shown in Figure 1. A mechanical chopper (65 Hz) modulated the THz radiation. The aforementioned 1.6 K Si composite bolometer was used as a detector and a lock-in amplifier recovered the modulated THz signal. A room-temperature band-pass filter was placed in front of the bolometer to produce a transform-limited Gaussian pulse. The filter was chosen so that the center frequency is in the range of maximum of the recorded CSR spectrum.

For time-domain analyses the power of the CSR and incoherent infrared synchrotron radiation were recorded by directly focusing the collimated beam from the optical port into an appropriate detector. A magnetically enhanced 4.2 K InSb hot-electron bolometer was used for the THz range. The detector has a response time of 2 µs (-3 dB) and is sensitive between 0.06 and 1.5 THz. In addition, a mercury-cadmium-telluride (MCT) detector was employed. The MCT detector has a response time of 5 µs and is sensitive down to 650 cm^{-1} (about 20 THz) and therefore only recording the incoherent spectral part of the synchrotron radiation. A fast Fourier transform (FFT) spectrum analyzer provided the time-domain power spectra of the signal from both detectors.

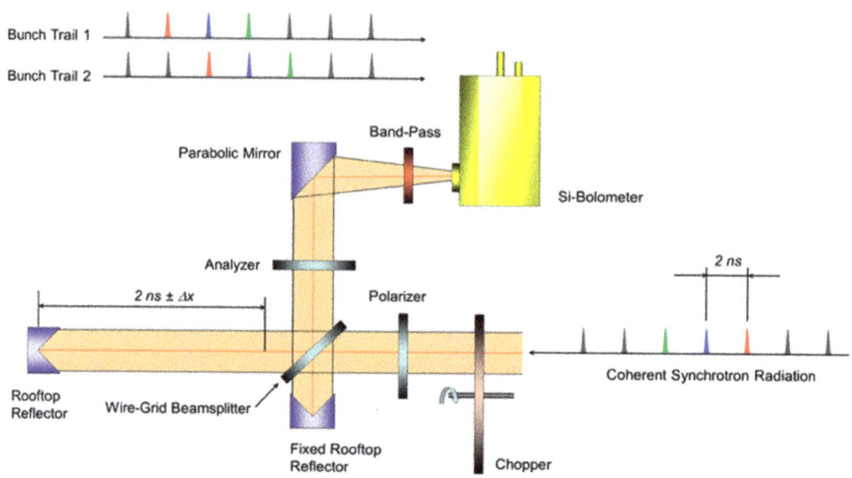

Figure 1. Optical scheme of the Martin–Puplett type interferometer (MPI) for cross-correlation measurement of subsequent bunches. The collimated coherent synchrotron radiation (CSR) comes from the exit port of the beamline and is modulated by a chopper before entering the MPI. The radiation is split into the two arms by the beamsplitter and is reflected back from the rooftop mirrors via the beamsplitter to the exit of the interferometer. Both beams pass a band-pass filter before being focused and recombined at the Si-detector.

3. Results and Discussion

3.1. Coherent Synchrotron Emission

Figure 2 shows the coherent enhancement above the bursting threshold for a 46 pC bunch (3×10^8 electrons) and a synchrotron frequency of $f_s = 1.75$ kHz measured with the vacuum FT spectrometer. Assuming a Gaussian bunch form, the experimental data fit neither the number of electrons nor the nominal *rms* bunch length of 1 mm. A pure Gaussian electron distribution with the above parameters would produce a spectral power. Below 60 GHz the finite dimensions of the dipole chamber of a free height of 35 mm at BESSY II almost shield the radiation. Additional diffraction losses from the beamline limit a realistic detection of the CSR even below 0.1 THz. Figure 2 cartoons this spectral region as well. If one uses the Equations (1) and (2) with the nominal *rms* bunch length to describe the CSR process, bunch shapes different from Gaussian have to be assumed (e.g., saw tooth shape). Such bunch shapes would contain higher frequency components explaining the broad experimentally observed amplification factor. However, the measured coherent enhancement fits perfectly a Gaussian over four orders of magnitudes indicating that the CSR is emitted from a longitudinal micro bunch structure with an *rms* micro bunch length of 0.17 mm. Moreover, only less than 0.1‰ of the electrons from the nominal bunch charge are involved in the detected coherent process.

Below a charge threshold of about 30 pC, the bunch distortion is stable in time and so the CSR power does. Above this threshold, the bunch distortion becomes longitudinally unstable and modulates the THz emission in the time-domain. Figure 3 shows the power spectrum of the temporal behavior of the CSR for two different bunch charges, below and above the threshold. The synchrotron frequency f_s is clearly seen in the power spectra for the coherent and incoherent part of the synchrotron radiation. For the coherent radiation below the bursting threshold, a longitudinal instability develops in a very regular repetition of a fundamental frequency f_f at about 3.75 kHz. This frequency is about twice the synchrotron frequency f_s. As it is also observed at other storage ring facilities [24,30], the fundamental frequency of instability shifts to higher frequencies and gains in strength with increasing bunch charge and even mix with different other modes. A higher bunch charge produces saw-tooth type instability and the duration of a saw-tooth determines the side-band structure. A further increase of bunch charge leads to a turbulent longitudinal bunch distortion causing CSR bursts emitting randomly shown in Figure 3 for the CSR power spectrum for a bunch charge of 46 pC. A more detailed picture of the temporal structure of CSR bursts at BESSY II is published elsewhere [31,32].

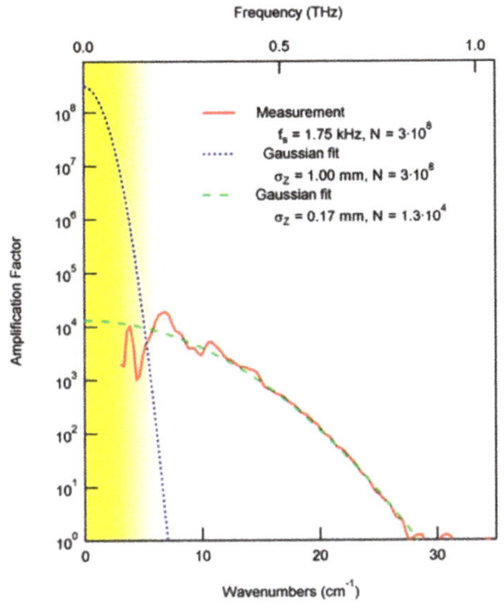

Figure 2. Measured amplification factor for the nominal BESSY *low* α operation at 1.75 kHz synchrotron frequency (red curve) and the corresponding Gaussian fit (dashed green curve). The dotted blue curve shows the expected amplification factor assuming a Gaussian bunch of 46 pC (3×10^8 electrons) with nominal *rms* bunch length of 1 mm. The yellow-shaded region below 5 cm^{-1} indicates the cut-off for CSR propagation.

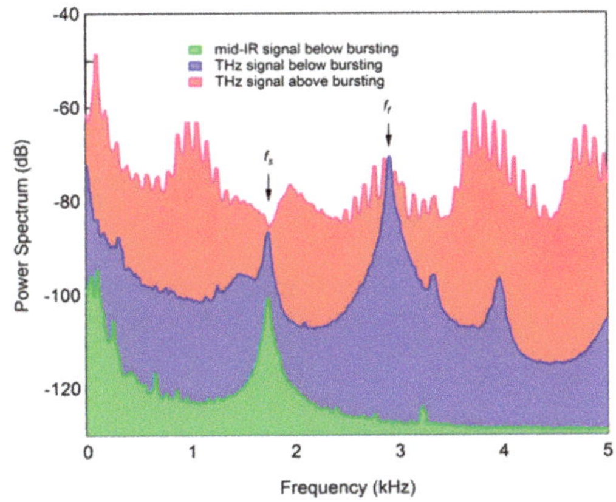

Figure 3. Power spectrum for the time-domain of the integral emitted power for a bunch charge of 26 pC below the bursting threshold (green and blue curve) and for 46 pC well above the bursting threshold (red curve). The green curve was taken with a mercury-cadmium-telluride (MCT) detector only recording the incoherent synchrotron radiation while the blue and red curve were obtained from an InSb-bolometer with an entrance filter only passing through radiation below 100 1.5 THz.

The existence of two thresholds in the electron bunch spatio-temporal dynamics while increasing the total charge of the bunch has been pointed out in [21]. The first threshold found indicates the presence of micro-structures drifting in the bunch profile and is signaled by the appearance of a resonance in the power spectral density at high frequencies (above 30 kHz) which is outside the range of our measurement. The second threshold indicates that the micro-structures are strong enough to persist after about half a revolution period of the electron bunch in the longitudinal phase space. It is linked with the appearance of another resonance in the power spectral density at about twice the f_s, in perfect agreement with our observation. An analytical link between the structure of the longitudinal phase space of the electrons and the frequencies seen in the power spectrum of the CSR emission during instabilities is derived in [33].

3.2. Cross-Correlation of Subsequent Bunches

Next, we like to draw a simple picture of the mechanism of the longitudinal instability by assuming a longitudinal oscillation with frequency f_f of the emitting micro bunch. Despite this scheme is extremely simple, it reproduces well the mechanism described in [21], where the longitudinal phase space of the electron bunch, after the first threshold, presents some sub-structures that evolve in time with a rotating movement due to the synchrotron rotation. Those sub-structures translate into a micro-bunching in the associated bunch profile. Each micro-bunch oscillates harmonically in a region whose length depends on how long the sub-structure persists in the bunch.

In contrast to established temporal approaches we follow a statistical approach to describe the bunch dynamics. Our simple model assumes that the micro bunch has an oscillating phase against the nominal bunch position and tries to describe the amplitude of this phase oscillation in respect to the bunch charge.

The oscillating phase shift $\tau\ (t)$ of the individual micro bunch in time units can be expressed by

$$\tau(t) = \tau_o(q)\cos(2\pi\ f_f\ t), \tag{4}$$

with τ_o as the maximum oscillation phase which is a function of the bunch charge q. With the assumption that the instability is a single bunch effect each micro bunch behaves independently and experiences all possible phase shifts over the integration time of the detector. Therefore, a measured cross-correlation of CSR pulses from pairs of successive micro-bunches can mathematically be treated as the auto-correlation of the mean from all CSR pulses. The mean electrical field of the pulse is given by the convolution integral of the single CSR pulse shape with the probability $P(\tau, d\tau) = p(\tau)\ d\tau$ that the individual pulse shift can be found in the interval $[\tau, \tau + d\tau]$:

$$\overline{E(t)} = E(t) \otimes P(\tau(t), d\tau) = \int_{-\infty}^{\infty} E(\xi) \cdot P(\tau(t-\xi), d\tau)\ d\xi. \tag{5}$$

In analogy to the harmonic oscillator model (e.g., [34]) we assume that the probability density function $p(\tau)$ is proportional to the time dt in which the CSR pulse phase changes by $d\tau$ and can be written in its normalized form to:

$$p(\tau) = \frac{1}{\pi\ \tau_o} \frac{1}{\sqrt{1-\left(\frac{\tau}{\tau_o}\right)^2}}, -\tau_o \le \tau \le \tau_o. \tag{6}$$

In Figure 4 the probability density function multiplied by the maximum of the oscillating phase is presented. It is obvious, that the turning points at $\pm\tau_o$ yield the highest probability for the phase shift. The inset of this figure shows a cartoon of the longitudinal projection of the resulting mean pulse trace under investigation. As stated above the theoretical cross-correlation of the successive CSR pulses can be obtained by the correlation of the mean CSR pulse from Equation (5):

$$\overline{E(t)} * \overline{E(t)} = \int_{-\infty}^{\infty} \overline{E(\xi)} \cdot \overline{E(t+\xi)}\, d\xi \tag{7}$$

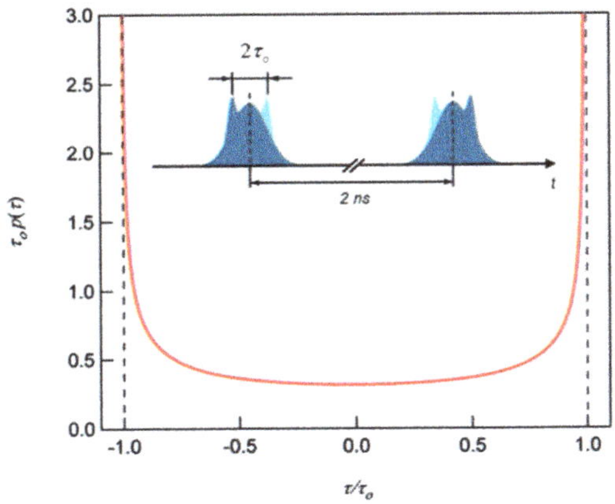

Figure 4. Probability density function for the normalized phase shift τ/τ_o. Dashed lines indicate the turning points of the sinusoidal oscillation. The inset shows schematically the longitudinal projection of the trail of mean bunches in the storage ring with a nominal bunch separation of 2 ns. $2\tau_o$ translates into a characteristic length where the micro-bunch is found in the longitudinal phase space over time.

The experimental cross-correlation was measured utilizing the Martin–Puplett interferometer as shown in Figure 1 with a nominal optical pass difference between the two interferometer arms equal to the distance between subsequent bunches in the bunch trail. For experimental and mathematical simplicity, we restrict our consideration not to the entire CSR spectrum of the micro bunch but to a limited bandwidth achieved by passing the radiation through a band-pass filter.

Figure 5 shows the power spectrum of the electrical field of the new created CSR pulse when passing the filter and which is according to the Wiener–Khinchin theorem [35] the squared Fourier transform of the measured field auto-correlation. A Gaussian fit to the spectrum yields an *rms* spectral width of 8.5 GHz (0.29 cm^{-1}) and a center frequency of 0.22 THz (7.4 cm^{-1}). These values are then used to construct an analytical transform-limited Gaussian pulse *E(t)* by using the time-bandwidth product [36]. Figure 5 also shows this pulse with a *rms* temporal width for the Gaussian envelope of 9.4 ps.

The amplitude of the cross-correlation as the function of the time delay between the two spectrometer arms is exemplarily shown in Figure 5. For the calculation τ_o was set to zero and the result compares well with the measured cross-correlation for a pulse emitted from a bunch below the bursting threshold. Note that the experimentally obtained cross-correlation function is superimposed by a constant *dc* voltage signal. This *dc* component corresponds to the mean of the power of the two pulses (e.g., [37]) and therefore is used for power normalization of the experimental cross-correlation function.

From both the theoretical and measured cross-correlation functions, shown in Figure 6, a contrast calculates as the difference between the maximum and the minimum of the modulation amplitudes. This contrast, shown in Figure 7, is independent of the bunch charge in the stable mode below 28.6 pC (vertical dashed line in Figure 7) where no oscillating instabilities appear. Above that charge, instabilities start to appear in the nominal bunch and the contrast drops rapidly to zero and develops with further increasing charge into an expiring oscillation. The oscillation period in the contrast function relates to the center frequency and the shape of the band path filter applied. In order to fit the theoretical contrast given as a function of the phase shift amplitude to the ring current (and accordingly to bunch charge units), distinctive data points were chosen from the theoretical and measured contrast to scale the abscissa. The inset of Figure 7 shows the polynomial fit function to this scaling procedure.

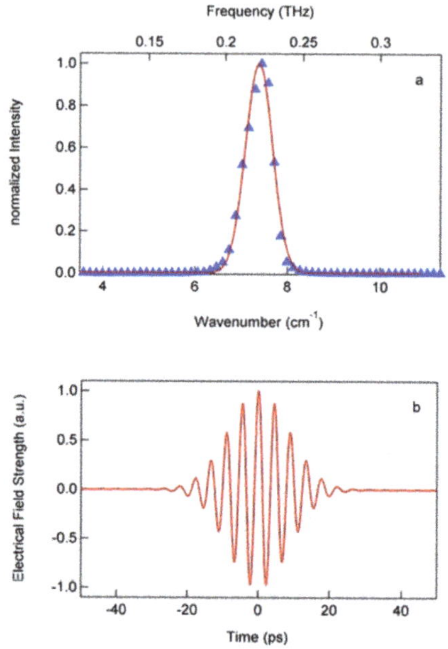

Figure 5. (**a**) measured transmittance of the band-pass filter (blue) and the corresponding Gaussian fit (red) with a center at 7.4 cm^{-1} and an *rms* spectral width of 0.29 cm^{-1}. (**b**) temporal structure of a constructed Fourier-transform limited pulse when passed through the band-pass filter with the fitted characteristics shown in (a).

The theoretical contrast as a function of ring current describes remarkably well the measured data as shown in Figure 7. This holds, even the experimental data were obtained at several temporal experiment shifts over one year of observation at the storage ring and proofs the stability and reproducibility of the BESSY II ring and the *low α* set up offered to THz users. Note, that for a given center frequency of the band-pass filter the relation between the contrast of the cross-correlation function and the phase shift of the pulse is completely determined and the only assumption made is that the micro pulse is oscillating in longitudinal direction. With the scaling relation shown in Figure 7 one has a ruler in hands to measure the longitudinal phase shift amplitude of the micro-bunch directly from the ring current stored. It is obvious that the model discussed above is not only valid for one oscillating micro bunch but also applies to higher instability modes where more than one micro bunch is involved. The projection of higher azimuthal modes, for example quadrupole or sextupole modes, are very similar to two or three harmonically oscillating micro-bunches also described by Equation (6).

In [28] an empirical relation is given to describe the measured *rms* length of the nominal bunch at BESSY II for *low α* operation and different synchrotron frequencies as a function of the bunch charge. Here, the nominal lengths were measured by means of a streak camera for bunches longer than 1.5 ps. As an example, this empirical relation results an *rms* length of about 3.9 ps for a bunch of 28.6 pC and a synchrotron frequency of 1.75 kHz just when the oscillating instability becomes observable by our measurements. As introduced before, $2\tau_o$ describes a longitudinal region in the nominal bunch profile of the oscillating instability whose length depends on how long the sub-structure persists in the bunch.

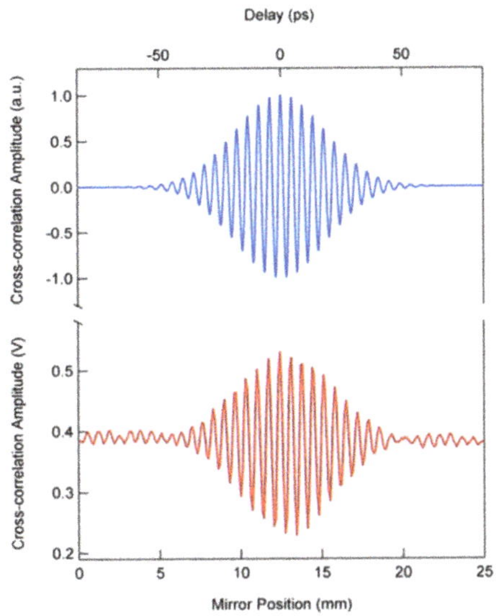

Figure 6. Comparison between the measured cross-correlation (lower red curve) of a CSR pulse from a bunch of 20 pC (below bursting threshold) and the calculated cross-correlation (blue curve above, normalized and *dc* component removed) using the transform-limited pulse in Figure 5. While the overall envelope of the measured cross-correlation agrees very well with the calculated one additional side wings appear caused by the asymmetric filter edges.

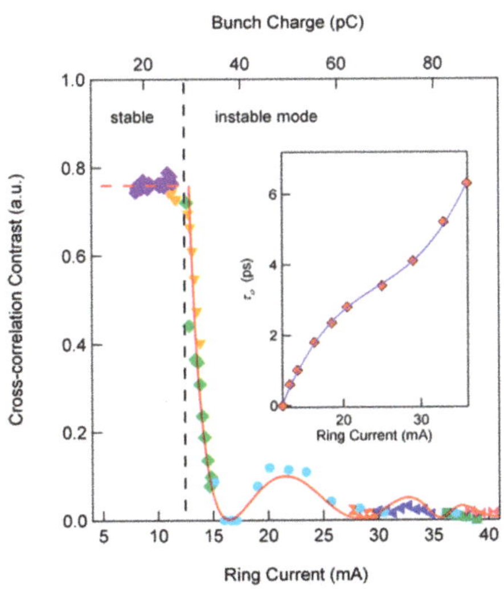

Figure 7. Theoretical (red curve) and measured contrast (different markers and colors represent the different experimental runs) of the cross-correlation as a function of the ring current stored. The abscissa for the theoretical contrast is scaled according to the fit for the phase shift amplitude shown in the inset (blue curve).

Figure 8 shows this characteristic length for given bunch charges in units of the *rms* length of the nominal bunch taken from [28]. The characteristic micro bunch length increases stronger than the nominal bunch length. It reaches about the 2.5-fold of the nominal *rms* length for the limit of the highest bunch charge analyzed by the cross-correlation method proposed. Further higher bunch charges define the limit of our model when the bunch instabilities become chaotic and a cross-correlation signal is not anymore detectable.

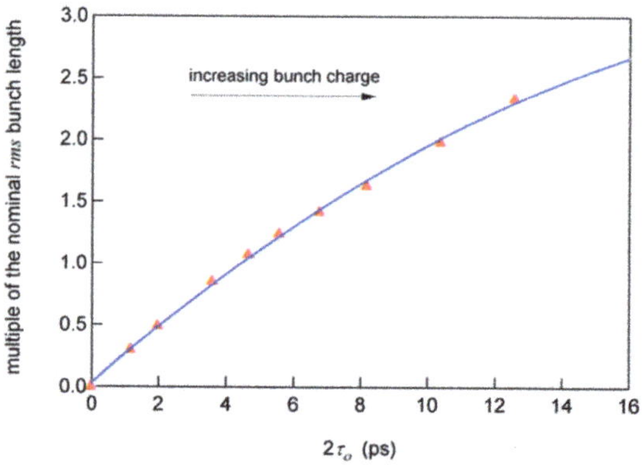

Figure 8. The characteristic length of the oscillating micro bunch for the corresponding bunch charges (red marks) given in *rms* units of the current-dependent nominal bunch length. The blue curve is a fit to guide the eyes.

4. Summary

A simple model is given which provides a tool for studying longitudinal bunch instabilities observed in the coherent emission from short electron bunches. The model is based on the cross-correlation of successive transform-limited Gaussian THz pulse. Such pulses are produced by passing the CSR through a narrow band-pass filter. The model describes the oscillating CSR instability in the THz spectral range by means of oscillating micro bunches and it is justified by the theory described in [21]. These micro bunches are built up by a very small fraction of electrons from the nominal bunch.

The cross-correlation of the THz fields from two successive bunches allows the evaluation of the oscillation phase of the micro bunches in respect to the nominal bunch position. As the result, the measured oscillation phase increases faster than the nominal bunch length with increasing bunch charge in the bunch charge region investigated. These investigations provide not only a picture on the longitudinal phase space but might also be of interest to define the limits of time resolution for pump-probe experiments with THz coherent synchrotron radiation at a storage ring.

Author Contributions: Conceptualization, U.S. and P.K.; Formal analysis, U.S., J.L., B.M. and M.O.; Writing—original draft, U.S.; Writing—review & editing, P.K., J.L., B.M. and M.O. All authors have read and agreed to the published version of the manuscript.

Funding: This research received no external funding.

Acknowledgments: We like to thank G. Wüstefeld, K. Holldack and R. Mitzner for fruitful discussion and acknowledge the support of A. Höhl and T. Neitzke.

Conflicts of Interest: The authors declare no conflict of interest.

References and Note

1. Carr, G.L.; Martin, M.C.; McKinney, W.R.; Jordan, K.; Neil, G.R.; Williams, G.P. High-Power Terahertz Radiation from Relativistic Electrons. *Nature* **2002**, *420*, 153–156. [CrossRef] [PubMed]
2. Williams, G.P.; Hirschmugl, C.J.; Kneedler, E.M.; Takacs, P.Z.; Shleifer, M.; Chabal, Y.J.; Hoffmann, F.M. Coherence Effects in Long-Wavelength Infrared Synchrotron Radiation Emission. *Phys. Rev. Lett.* **1989**, *62*, 261–263. [CrossRef] [PubMed]

3. Nodvick, J.S.; Saxon, D.S. Suppression of Coherent Radiation by Electrons in a Synchrotron. *Phys. Rev.* **1954**, *96*, 180–184. [CrossRef]
4. Abo-Bakr, M.; Feikes, J.; Holldack, K.; Wüstefeld, G.; Hübers, H.-W. Steady-State Far-Infrared Coherent Synchrotron Radiation Detected at BESSY II. *Phys. Rev. Lett.* **2002**, *88*, 254801. [CrossRef] [PubMed]
5. Sannibale, F.; Byrd, J.M.; Loftsdóttir, Á.; Venturini, M.; Abo-Bakr, M.; Feikes, J.; Holldack, K.; Kuske, P.; Wüstefeld, G.; Hübers, H.-W.; et al. A Model Describing Stable Coherent Synchrotron Radiation in Storage Rings. *Phys. Rev. Lett.* **2004**, *93*, 094801. [CrossRef] [PubMed]
6. Heifets, S.; Stupakov, G.; Krinsky, S. Coherent Synchrotron Radiation Instability in a Bunch Compressor. *Phys. Rev. Spec. Top. Accel. Beams* **2002**, *5*, 064401. [CrossRef]
7. Venturini, M.; Warnock, R. Bursts of Coherent Synchrotron Radiation in Electron Storage Rings: A Dynamical Model. *Phys. Rev. Lett.* **2002**, *89*, 224802. [CrossRef]
8. Abo-Bakr, M.; Feikes, J.; Holldack, K.; Kuske, P.; Peatman, W.; Schade, U.; Wüstefeld, G.; Hübers, H.-W. Brilliant, Coherent Far-Infrared (THz) Synchrotron Radiation. *Phys. Rev. Lett.* **2003**, *90*. [CrossRef]
9. Singley, E.; Abo-Bakr, M.; Basov, D.; Feikes, J.; Guptasarma, P.; Holldack, K.; Hübers, H.; Kuske, P.; Martin, M.; Peatman, W.; et al. Measuring the Josephson Plasma Resonance in $Bi_2Sr_2CaCu_2O_8$ Using Intense Coherent THz Synchrotron Radiation. *Phys. Rev. B* **2004**, *69*, 092512. [CrossRef]
10. Müller, A.-S.; Birkel, I.; Gasharova, B.; Huttel, E.; Kubat, R.; Mathis, Y.-L.; Moss, D.A.; Mexner, W.; Rossmanith, R.; Wuensch, M.; et al. Far Infrared Coherent Synchrotron Edge Radiation at ANKA. In Proceedings of the 2005 Particle Accelerator Conference, Knoxville, TN, USA, 16–20 May 2005; pp. 2518–2520.
11. Barros, J.; Manceron, L.; Brubach, J.B.; Creff, G.; Evain, C.; Couprie, M.E.; Loulergue, A.; Nadolski, L.; Tordeux, M.A.; Roy, P. Toward Highly Stable Terahertz Coherent Synchrotron Radiation at the Synchrotron SOLEIL. *J. Phys. Conf. Ser.* **2012**, *359*, 4–8. [CrossRef]
12. Hashimoto, S.; Shoji, Y.; Ando, A.; Takahashi, T. Observation of Coherent Synchrotron Radiation at NewSUBARU. In Proceedings of the 2005 Particle Accelerator Conference, Knoxville, TN, USA, 16–20 May 2005; pp. 4188–4190.
13. Mochihashi, A.; Hosaka, M.; Katoh, M.; Shimada, M.; Kimura, S. UVSOR-II Electron Storage Ring. In Proceedings of the 2006 European Particle Accelerator Conference, Edinburgh, UK, 26–30 Jun 2006; pp. 3380–3382.
14. Martin, I.P.S.; Rehm, G.; Thomas, C.; Bartolini, R. Experience with Low-Alpha Lattices at the Diamond Light Source. *Phys. Rev. Spec. Top. Accel. Beams* **2011**, *040705*, 1–11. [CrossRef]
15. Feikes, J.; von Hartrott, M.; Ries, M.; Schmid, P.; Wüstefeld, G.; Hoehl, A.; Klein, R.; Müller, R.; Ulm, G. Metrology Light Source: The First Electron Storage Ring Optimized for Generating Coherent THz Radiation. *Phys. Rev. Spec. Top. Accel. Beams* **2011**, *14*, 030705. [CrossRef]
16. Schade, U.; Ortolani, M.; Lee, J. THz Experiments with Coherent Synchrotron Radiation from BESSY II. *Synchrotron Radiat. News* **2007**, *20*, 17–24. [CrossRef]
17. Schnegg, A.; Behrends, J.; Lips, K.; Bittl, R.; Holldack, K. Frequency Domain Fourier Transform THz-EPR on Single Molecule Magnets Using Coherent Synchrotron Radiation. *Phys. Chem. Chem. Phys.* **2009**, *11*, 6820. [CrossRef] [PubMed]
18. Tammaro, S.; Pirali, O.; Roy, P.; Lampin, J.F.; Ducournau, G.; Cuisset, A.; Hindle, F.; Mouret, G. High Density Terahertz Frequency Comb Produced by Coherent Synchrotron Radiation. *Nat. Commun.* **2015**, *6*, 1–6. [CrossRef] [PubMed]
19. Wüstefeld, G.; Jankowiak, A.; Knobloch, J.; Ries, M. Simultaneous Long and Short Electron Bunches in the BESSY II Storage Ring. In Proceedings of the 2011 International Particle Accelerator Conference, San Sebastian, Spain, 4–9 September 2011; pp. 2936–2938.
20. Jankowiak, A.; Wüstefeld, G. Low-α Operation of BESSY II and Future Plans for an Alternating Bunch Length Scheme BESSY VSR. *Synchrotron Radiat. News* **2013**, *26*, 22–24. [CrossRef]
21. Evain, C.; Barros, J.; Loulergue, A.; Tordeux, M.A.; Nagaoka, R.; Labat, M.; Cassinari, L.; Creff, G.; Manceron, L.; Brubach, J.B.; et al. Spatio-Temporal Dynamics of Relativistic Electron Bunches during the Micro-Bunching Instability in Storage Rings. *Europhys. Lett.* **2012**, *98*, 40006. [CrossRef]
22. Roussel, E.; Evain, C.; Le Parquier, M.; Szwaj, C.; Bielawski, S.; Manceron, L.; Brubach, J.-B.; Tordeux, M.-A.; Ricaud, J.-P.; Cassinari, L.; et al. Observing Microscopic Structures of a Relativistic Object Using a Time-Stretch Strategy. *Sci. Rep.* **2015**, *5*, 10330. [CrossRef]

23. Evain, C.; Roussel, E.; Le Parquier, M.; Szwaj, C.; Tordeux, M.A.; Brubach, J.B.; Manceron, L.; Roy, P.; Bielawski, S. Direct Observation of Spatiotemporal Dynamics of Short Electron Bunches in Storage Rings. *Phys. Rev. Lett.* **2017**, *118*, 1–6. [CrossRef]
24. Steinmann, J.L.; Boltz, T.; Brosi, M.; Bründermann, E.; Caselle, M.; Kehrer, B.; Rota, L.; Schönfeldt, P.; Schuh, M.; Siegel, M.; et al. Continuous Bunch-by-Bunch Spectroscopic Investigation of the Micro-bunching Instability. *Phys. Rev. Accel. Beams* **2018**, *21*, 110705. [CrossRef]
25. Brosi, M.; Steinmann, J.L.; Blomley, E.; Boltz, T.; Bründermann, E.; Gethmann, J.; Kehrer, B.; Mathis, Y.L.; Papash, A.; Schedler, M.; et al. Systematic Studies of the Microbunching Instability at Very Low Bunch Charges. *Phys. Rev. Accel. Beams* **2019**, *22*, 20701. [CrossRef]
26. Bielawski, S.; Blomley, E.; Brosi, M.; Bründermann, E.; Burkard, E.; Evain, C.; Funkner, S.; Hiller, N.; Nasse, M.J.; Niehues, G.; et al. From Self-Organization in Relativistic Electron Bunches to Coherent Synchrotron Light: Observation Using a Photonic Time-Stretch Digitizer. *Sci. Rep.* **2019**, *9*, 1–9. [CrossRef] [PubMed]
27. Peatman, W.B.; Schade, U. A Brilliant Infrared Light Source at BESSY. *Rev. Sci. Instrum.* **2001**, *72*, 1620–1624. [CrossRef]
28. Feikes, J.; Holldack, K.; Kuske, P.; Wüstefeld, G. Sub-Picosecond Electron Bunches in the BESSY Storage Ring. In Proceedings of the 2004 European Particle Accelerator Conference, Lucerne, Switzerland, 5–9 July 2004; pp. 1954–1956.
29. Martin, D.H.; Puplett, E. Polarised Interferometric Spectrometry for the Millimetre and Submillimetre Spectrum. *Infrared Phys.* **1969**, *10*, 105–109. [CrossRef]
30. Billinghurst, B.E.; Bergstrom, J.C.; Baribeau, C.; Batten, T.; May, T.E.; Vogt, J.M.; Wurtz, W.A. Longitudinal Bunch Dynamics Study with Coherent Synchrotron Radiation. *Phys. Rev. Accel. Beams* **2016**, *19*, 1–11. [CrossRef]
31. Wüstefeld, G. Short Bunches in Electron Storage Rings and Coherent Synchrotron Radiation. In Proceedings of the 2008 European Particle Accelerator Conference, Genoa, Italy, 23–27 June 2008; pp. 26–30.
32. Kuske, P. Investigation of the Temporal Structure of CSR-Bursts at BESSY II. In Proceedings of the 2009 Particle Accelerator Conference, Vancouver, BC, Canada, 4–8 May 2009; pp. 4682–4684.
33. Roussel, E.; Evain, C.; Szwaj, C.; Bielawski, S. Microbunching Instability in Storage Rings: Link between Phase-Space Structure and Terahertz Coherent Synchrotron Radiation Radio-Frequency Spectra. *Phys. Rev. Spec. Top. Accel. Beams* **2014**, *17*, 1–6. [CrossRef]
34. Blochinzew, D.I. *Grundlagen der Quantenmechanik*; Deutscher Verlag der Wissenschaften: Berlin, Germany, 1953; p. 147.
35. Cohen, L. The Generalization of the Wiener-Khinchin Theorem. In Proceedings of the 1998 IEEE International Conference on Acoustics, Speech and Signal Processing, Seattle, WA, USA, 12–15 May 1998; Volume 3, pp. 1577–1580.
36. The time–bandwidth product of a Fourier-transform limited Gaussian-shaped pulse is about 0.08 when the rms criterion is used for the temporal and spectral width
37. Griffiths, P.R.; de Haseth, J.A. *Fourier Transform Infrared Spectroscopy*; John Wiley & Sons, Ltd.: New York, NY, USA, 1986; p. 3.

condensed matter

MDPI

Article

Angular Dependence of Copper Surface Damage Induced by an Intense Coherent THz Radiation Beam

Salvatore Macis [1,2,*], Luca Tomarchio [1], Silvia Tofani [3], S. Javad Rezvani [2], Luigi Faillace [4], Stefano Lupi [1], Akinori Irizawa [5] and Augusto Marcelli [2,6]

1 Department of Physics, Sapienza University, P.le Aldo Moro, 5, 00185 Roma, Italy; luca.tomarchio@uniroma1.it (L.T.); stefano.lupi@roma1.infn.it (S.L.)
2 INFN, Laboratori Nazionali di Frascati, Via Enrico Fermi 40, 00044 Frascati (RM), Italy; javad.rezvani@lnf.infn.it (S.J.R.); augusto.marcelli@lnf.infn.it (A.M.)
3 Department of Information Engineering, Electronics and Telecommunications, Sapienza University, Via Eudossiana, 18, 00184 Roma (RM), Italy; silvia.tofani@uniroma1.it
4 INFN-Milano, Via Celoria 16, 20133 Milano, Italy; Luigi.Faillace@lnf.infn.it
5 The Institute of Scientific and Industrial Research (ISIR), Osaka University, Osaka 565-0871, Japan; irizawa@sanken.osaka-u.ac.jp
6 International Centre for Material Science Superstripes, RICMASS, Via dei Sabelli 119A, 00185 Rome, Italy
* Correspondence: salvatore.macis@roma2.infn.it

Received: 26 January 2020; Accepted: 7 March 2020; Published: 10 March 2020

Abstract: In this work, we show the damage induced by an intense coherent terahertz (THz) beam on copper surfaces. The metallic surface was irradiated by multiple picosecond THz pulses generated by the Free Electron Laser (FEL) at the ISIR facility of the Osaka University, reaching an electric field on the sample surface up to ~4 GV/m. No damage occurs at normal incidence, while images and spectroscopic analysis of the surface point out a clear dependence of the damage on the incidence angle, the electric field intensity, and polarization of the pulsed THz radiation. Ab initio analysis shows that the damage at high incidence angles could be related to the increase of the absorbance, i.e., to the increase of the temperature around or above 1000 °C. The experimental approach we introduced with multiple fast irradiations represents a new powerful technique useful to test, in a reproducible way, the damage induced by an intense electric gradient on copper and other metallic surfaces in view of future THz-based compact particle accelerators.

Keywords: THz; FEL; copper; damage; surface

1. Introduction

The damage induced by a strong EM field on metallic surfaces is a widely studied topic [1–3], where the aim is to understand the response of a material under strong applied electrical forces and the damage generated on the irradiated area. Experiments with intense electromagnetic radiation in the IR-UV range have been performed since the availability of high-intensity lasers [4,5]. Reflection and absorption processes of a high-intensity optical laser pulse by an extremely smooth metal surface have already been studied [5–8]. In particular, laser-induced damaged morphologies and their cumulative effects given by single pulses have already been highlighted [9].

The THz electromagnetic spectral range, typically referred to frequencies between 100 GHz and 10 THz, is fundamental for the spectroscopic analysis of many condensed matter systems [10–16], low dimensional semiconductors and superconducting materials with different electronic properties [17–20]. This frequency range is, in fact, resonant with molecular and phonon excitations and with free electrons in metals [21].

Despite its importance, the latest improvements in THz generation and detection have been hindered by the lack of suitable high-intensity sources. Recent developments have seen the appearance of optical techniques based on non-linear effects, like the optical rectification process in organic and inorganic crystals [22,23]. One of the most efficient ways to generate high-intensity THz pulses is based on relativistic electrons both in a Free Electron Laser (FEL) and a Linac, where the synchrotron radiation emission from the accelerated electron beam is coherent and (as for FELs) nearly monochromatic [13,14,24,25].

High-intensity THz radiation is also useful for producing a high-gradient electric field for particle acceleration [26]. The investigation of the behaviour of metallic cavities [27] under the application of a high-intensity THz field represents an important issue in the development of compact accelerators [22] and to test protective coatings designed to reduce the damaging processes [28,29]. Copper, in particular, represents a strategic material to study under high-gradient electric fields, due to its extensive use in many accelerating components subjected to high electric fields.

Recently Agranat et al. [9] investigated the damage induced on thin metallic films by a high-fluence pulsed THz radiation, while no irradiation studies have been performed on thick metallic samples yet. Here, we highlight the experimental results obtained by illuminating the surface of smooth copper thick samples with intense THz pulses generated at ISIR-FEL, a facility of the Osaka University [25], reaching an electric field on the sample surface up to ~4 GV/m. Using multiple THz pulses, at the fluence below the single-pulse damage threshold, we show evidence of angular dependent damage of copper substrates. We analyze the damage through chemical analysis of the surface by Raman spectroscopy. We used a simple heat model coupled with Fresnel equations to explain the damaging mechanism taking into account the change of the surface reflectance vs. the incidence radiation angle.

2. Material and Methods

The THz-FEL at the quantum beam research facility of the Institute of Scientific and Industrial Research (ISIR) at the Osaka University allows the emission of nearly monochromatic coherent radiation through the acceleration of an electron beam by a LINAC up to 15 MeV. Electrons injected in the undulator emit horizontally polarized synchrotron radiation [25,30]. The optical system described in [25,31] allows the extraction of the coherent THz beam and its transport to the experimental hutch. Finally, the collimated beam is focused on the sample's surface through a parabolic mirror (Figure 1) sets at a distance of 5 cm.

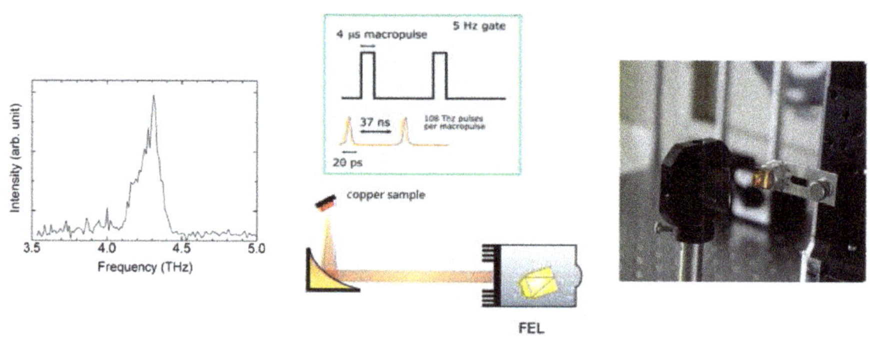

(a) (b) (c)

Figure 1. (**a**) The spectrum of the FEL emission for an undulator gap of 37 mm, (**b**) layout and time structure of the THz pulses, (**c**) photo of the sample holder and of the parabolic mirror used to focus the radiation.

Knife-edge scanning measurements reveal the Gaussian profile of the beam at the focal point. The circular spot has a radius of ~130 μm, close to the diffraction limit [31]. Actually, the time structure of the THz beam is related to the electron beam structure of the LINAC. By fixing the undulator gap at 37 mm, the narrow-band THz emission is centered at ~4.3 THz. With a 5 Hz repetition rate, the time structure composed by a sequence of 4 μs pulses is exploited. At the repetition rate of 27 MHz, each macro-pulse 20 ps long is composed of 108 micro-pulses. Figure 1a shows the typical frequency dispersion of this THz-FEL in the 27 MHz mode, reaching ~3% bandwidth at the best monochromaticity conditions.

As measured by the calibrated energy meter (Multipurpose Energy Max Sensor J-25MB-LE form Coherent®) the FEL provides THz-pulse energy of 4 ± 0.1 mJ on the sample surface. Considering an area with the size of the focus spot (130 μm Gaussian radius), the fluence of the pulse on the sample surface is 12,7 J/cm^2. At the focal point, a peak magnitude of 3.7 GV/m for the THz electric field has been evaluated using the equation:

$$\mathrm{E_{THz}} = \sqrt{\mathrm{Z_0\, I_{THz}}} \tag{1}$$

where the micro-pulse intensity $\mathrm{I_{THz}}$ is

$$\mathrm{I_{THz}} = \frac{W_{THz}}{\tau\, \pi r^2} \tag{2}$$

with W_{THz} the micro pulse energy, τ the micro-pulse time duration, and r the beam Gaussian waist. In our experiment, irradiations have been carried out on 5 mm thick oxygen-free and high thermal conductivity (OFHC) copper samples, polished down to a nominal roughness (rms) <5 nm. Each irradiation, composed of 5000 pulses with the total fluence of 635 J/cm^2, has been delivered in ~15 min. The visible damage on the copper surface has been evaluated with a combined optical and chemical analysis. Images have been collected at different magnifications with a scanning electron microscope "Nanoimages MINI-SEM SNE-3200M", while Raman maps have been collected using the JASCO N5100 Optical and Raman microscope.

3. Results and Discussions

The irradiations have been performed at different angles of incidence, rotating the copper sample from the normal incidence position (0°) up to 60°. As shown in Figure 2a, the rotation axis of the sample is perpendicular to the horizontal plane, i.e., normal to the polarization plane of the electric field. During the irradiations, breakdowns phenomena occur in air near the sample surface (Figure 2b).

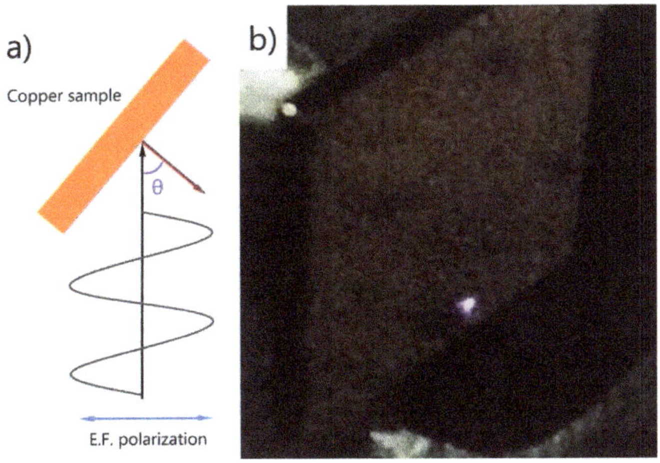

Figure 2. (**a**) Layout of the experimental irradiation at the incidence angle θ. The linear polarized radiation hits the copper surface oriented at the selected angle. (**b**) Image of a spark (bright spot) near the bottom edge occurring on the copper surface irradiated at the incidence angle of 40°.

Figure 3 shows the optical images of the surface damage. In this figure, four optical images taken at different angles after exposition with 5000 pulses are highlighted. At normal incidence, no damage is detectable, while increasing the incidence angle a pattern with evident surface changes appears. This suggests that different phenomena may occur during the irradiation. At this frequency regime, during irradiation, both thermo-mechanical stress and strains are expected to contribute to the surface damage [9]. Using the Raman microscope, we were able to analyze and compare different areas in the damaged regions (see Figure 4). The measurements were performed with a red laser (738 nm) using a spot of ~1 μm (100× magnification).

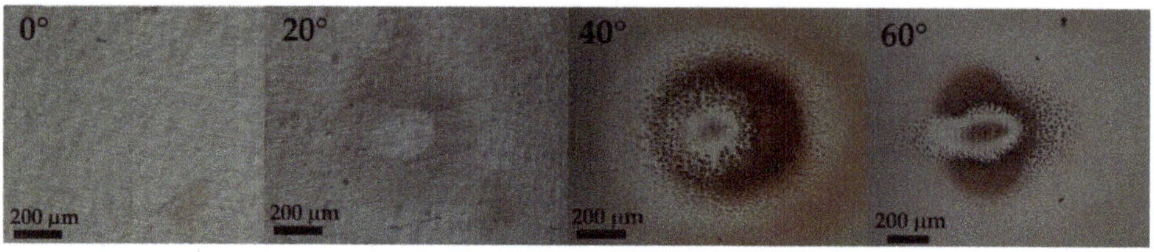

Figure 3. Comparison among optical images of the irradiated copper surface taken at different angles of incidence.

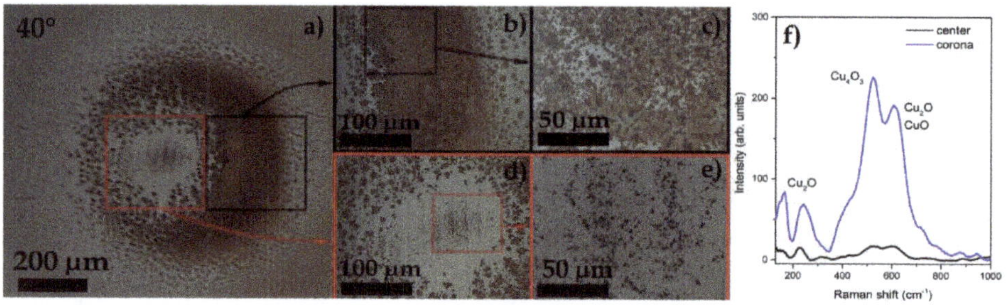

Figure 4. (**a**) 5× magnified image of the damaged region irradiated at 40° after 5000 shots. A central opaque area, an intermediate lighter area, and a dark corona can be recognized. The two squares highlight the 20× magnified region of the corona (black, panel **b**) and the central area (red, panel **d**). The brown corona is due to the oxidation of the copper surface (100× magnification in panel **c**). A 100× magnification of the central area is highlighted in (panel **e**). (**f**) Comparison of the Raman spectra collected in the two magnified areas showing the copper oxides vibrational peaks (mainly CuO and Cu_4O_3) in the corona and a small presence of oxides in the central region.

Looking at Figure 4a, two sub-regions in the damaged area can be distinguished: the center, where the THz electric field reaches its maximum intensity and the corona, with an outer diameter of ~500 μm. The analysis of the optical images and the comparison of the different Raman spectra (Figure 4f) suggest that the surface in the corona region contains copper oxides (mainly Cu_4O_3 and CuO) [32,33], while the center is almost metallic with a weak contribution of copper oxides [32].

In experiments where coherent radiation interacts with a solid system, the fluence shows a threshold value for the generation of plasma on the surface that depends on the wavelength [34]. The measured plasma generation threshold for copper is <2 J/cm^2 at 1 μm, decreasing at longer wavelengths [34]. At this fluence, the irradiation is expected to produce plasma plumes, possibly containing electrons, ions, atoms, molecules, and micro-sized particles, whose dynamics contribute to shaping the damaged pattern [34]. The lack of copper oxide in the central region could be associated with the ablation mechanism induced by the intense THz radiation beam, which continuously removes, at each pulse, part of the oxide layer present on top of the surface.

The spatial distribution of copper oxides is shown in Figure 5. This map describes the copper surface irradiated at 40° and has been obtained collecting a grid of 200 × 200 Raman spectra with a 5 μm spot. It allows to reconstruct the spatial distribution of the copper oxides by integrating, in each of the 40000 Raman spectra, the region from 350 to 800 cm^{-1}, which is associated with CuO and Cu_2O [32]. As already shown in Figure 4f, the corona has the highest concentration of oxides, while in the central region, they are negligible. A further investigation of the pattern has been obtained through SEM microscopy. Three magnified SEM images of the copper surface are shown in Figure 6 at 40° of incidence. The image on the left shows the two regions: The corona well separated by the dark external copper surface.

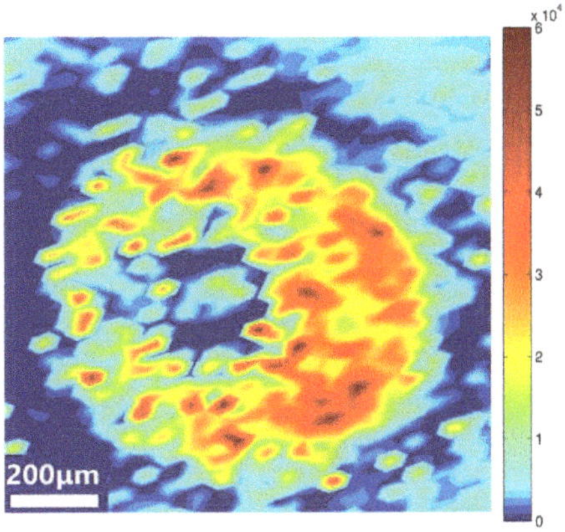

Figure 5. Raman map of the copper oxide contributions related to the copper surface exposed at 40°, and obtained integrating the range 350–800 cm^{-1}. The color bar on the right gives the intensity of the integrated Raman signal.

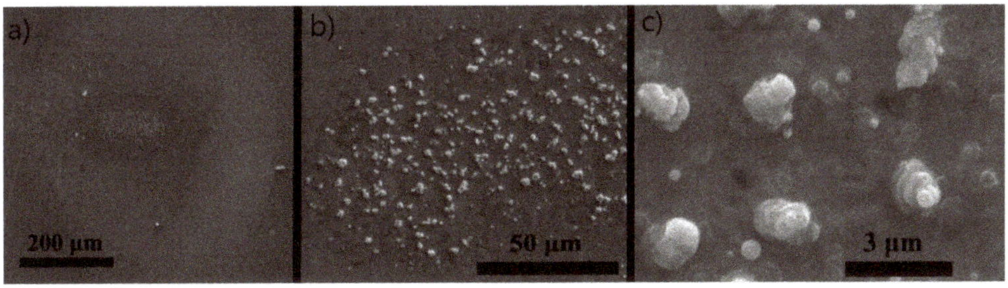

Figure 6. Comparison of SEM images of the irradiated copper surface at the angle of incidence of 40°. By increasing the magnification (from left to right), it is possible to recognize in the central area (**a**), where the highest damage occurs, the presence of hundreds of copper tips (**b**). Their presence (see magnified tips in (**c**)) suggests that in this region, it is possible to reach temperatures comparable to higher than the Cu melting point (~1085 °C).

The most intriguing result of the irradiation we performed with ISIR appears in the center of the pattern, where small structures emerge. Their shape and appearance are highlighted in panel c of Figure 6. In order to generate such microstructures, the central area should be exposed to extreme heat with a possible melting of the upper layers in order to form upward pillars. Similar processes might also emerge from surface instabilities on a rigid substrate, leading to self-organization phenomena and the formation of well-organized nano-micro patterns such as the pillar-like structures shown in Figure 6c [35].

Irradiations have also been performed with a lower number of pulses, resulting in different damaging behaviors. For instance, with an irradiation of 500 pulses no damage could have been resolved on the copper surface at all incidence angles. The result suggests the occurrence of a cumulative thermal effect, most likely a positive feedback process where the growing temperature modifies the absorption efficiency of the surface. Therefore, the rise of the surface temperature is expected for longer exposition times with stronger damage. A similar build-up effect has also been observed in the optical range, as suggested in [36], where a similar behavior of the copper reflectance has been observed when copper was irradiated by multiple intense optical pulses.

Regarding the observed angular dependence, a similar process has been described in the optical range by Miyasaka et al. [37]. Increasing the incidence angle, he revealed for a *p*-polarized high-intensity pulsed laser, a further increase in the absorbance of the copper surface. This latter phenomenon can be modeled through the well-known complex Fresnel equations. The reflected incident radiation with p-polarized light has to take into account the complex form:

$$R_p(\theta) = \left[\frac{-\widetilde{n}_{Cu}{}^2 \cos\theta_i - \sqrt{\widetilde{n}_{Cu}{}^2 - \sin^2\theta_i}}{\widetilde{n}_{Cu}{}^2 \cos\theta_i + \sqrt{\widetilde{n}_{Cu}{}^2 - \sin^2\theta_i}}\right]^2 \quad \widetilde{n}_{Cu} = n_{Cu} + ik_{Cu} \tag{3}$$

where $\widetilde{n}_{Cu}$ is the complex copper refractive index at ~4 THz [38] and θ_i is the incidence angle. From this equation, we may obtain the absorbed fluence F_{abs} by the copper surface as a function of the incidence angle. The incidence fluence can be written as

$$F = \frac{E_{micro}}{A};\ A = \frac{\pi r^2}{\cos\theta_i} \tag{4}$$

where E_{micro} is the energy of a single micro-pulse and A the angle-dependent irradiated area. Indeed, when the beam hits the surface, the irradiated area widens with the angle $\theta_i > 0$. As a consequence, the fluence absorbed by the surface is

$$F_{abs} = \big(1 - R_p(\theta)\big)\cdot F = \big(1 - R_p(\theta)\big)\cdot \frac{E_{micro}\cos\theta_i}{\pi r^2} \tag{5}$$

which gives an increase in the temperature:

$$\Delta T = \frac{F_{abs}}{\rho C \delta_{eff}} \tag{6}$$

where $\rho = 8.96$ g/cm^3 is the copper density at room temperature (RT), $C = 0.385$ J/g °C the specific heat capacity at RT and $\delta_\theta = \delta_{0°}\cdot\cos\theta_i$ is the penetration depth of the radiation in bulk copper. The latter depends on the angle [39], and $\delta_{0°}$~40 nm is the penetration depth at normal incidence in this frequency domain [40]. We obtain the equation:

$$\Delta T_\theta = \big(1 - R_p(\theta)\big)\cdot \frac{E_{micro}}{\pi r^2 \rho C \delta_{0°}} = \big(1 - R_p(\theta)\big)\cdot \Delta T_{0°} \tag{7}$$

The non-homogenous distribution of the radiation inside the spot determines a gradient in the temperature distribution going from the center to the corona region. Figure 7a shows the angular dependence of the absorbed fluence F_{abs} (Equation (5), black), the penetration depth (red), and the temperature increase at the center of the spot (Equation (7), blue). Figure 7b describes the ΔT vs. the radial distance from the center, for different incidence angles (Equation (7)). If we consider the center of the spot, at angles higher than 20°, the temperature increases more than the melting point of copper (~1085 °C), while at 60° the temperature increases above 1800 °C.

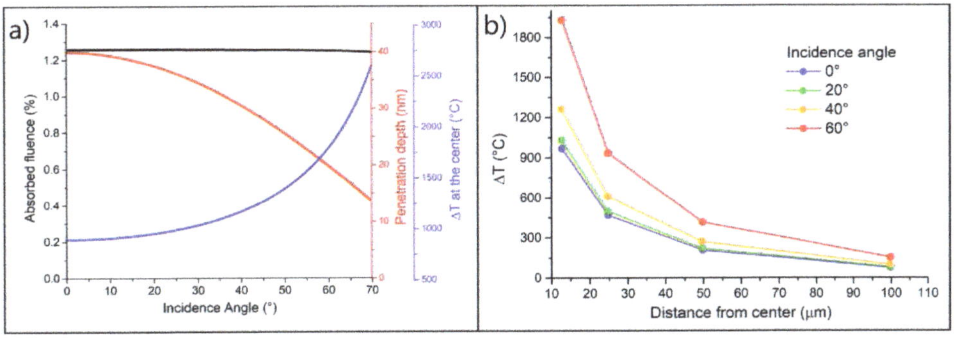

Figure 7. (**a**) Angular dependence of the absorbed fluence F_{abs} (black), penetration depth (red line), and the temperature increase at the center of the spot. (**b**) ΔT as a function of the distance from the center of the spot at different incidence angles.

4. Conclusions

In this work, we showed for the first time the occurrence of an angular dependent reproducible damage induced by THz multiple high-intensity pulses on metallic copper surfaces. We showed that the damage increases for increasing incidence angles with multiple shots (>10^3) and the fluence of 12.7 J/cm^2 per pulse. We estimated the increase of the temperature of the copper surface with a simple model that reasonably matches the experimental results. It suggests that the increase in absorption at high incidence angles may cause damage and oxidation. The model showed that for incidence angles >20°, the final temperature can be higher than the melting temperature of copper. The observed damage is in agreement with previous reports in the optical frequency range, described in terms of higher absorbance, increased angle of incidence [36], and multiple shots irradiation [37].

The phenomenology behind the damaging effects can be explained in terms of a fast local temperature increase. It may induce ablation and copper melting on the surface. However, such a description does not consider electric-related effects, e.g., discharges and/or breakdowns occurring in air due to pulses (Figure 2). The breakdown phenomenon occurring at the copper surface is due to the strong electric field applied (estimated ~4 GV/m) in the central region. Actually, pillars formed in the central region could be related to surface instabilities due to both thermal and electric phenomena. In fact, tips could be generated by the coalescence of the melted copper in the central region, or generated during the strong electric field, during which tips act as point-like structures capable of enhancing local electric fields [41].

This work also points out the possibility to use multiple fast irradiations and high-fluence THz radiation to test in a controlled way the damage induced to small areas of a metallic surface by strong electric fields, i.e., in the range of 4 GV/m. The results also point out the need to characterize better the observed damage patterns and their dynamics following multiple irradiation procedures. This original approach represents an innovative method to test in a reliable and controlled way the damage induced on a smooth surface. The method offers many potential applications in different technologies and studies, e.g., the construction of accelerating cavities and RF devices.

Author Contributions: Conceptualization, methodology, S.M., A.M., A.I. and S.L.; data analysis, S.M., L.T., S.J.R., S.T. and A.M.; Manuscript preparation: S.M., A.M. and L.T.; S.M., L.T., S.J.R., A.M., S.T., A.I., L.F., S.L. have reviewed the manuscript. All authors have read and agreed to the published version of the manuscript.

Funding: We acknowledge the financial support of the Bilateral Cooperation Agreement between Italy and Japan of the Italian Ministry of Foreign Affairs and of the International Cooperation (MAECI) in the framework of the project of major relevance N. PGR0072. This work was also supported by "Dynamic Alliance for Open Innovation Bridging Human, Environment and Materials" from the Ministry of Education, Culture, Sports, Science and Technology of Japan (MEXT).

Acknowledgments: We acknowledge the INFN for the support within DEMETRA and NUCLEAAR, two projects funded by the INFN Vth Committee. We also acknowledge the Osaka University for providing beamtime at the ISIR facility.

Conflicts of Interest: The authors declare no conflict of interest.

References

1. Jee, Y.; Becker, M.F.; Walser, R.M. Laser-induced damage on single-crystal metal surfaces. *J. Opt. Soc. Am. B* **1988**, *5*, 648. [CrossRef]
2. Forno, M.D.; Dolgashev, V.; Bowden, G.; Clarke, C.; Hogan, M.; McCormick, D.; Novokhatski, A.; O'Shea, B.; Spataro, B.; Weathersby, S.; et al. High gradient tests of metallic mm-wave accelerating structures. *Nucl. Instrum. Methods Phys. Res. Sect. A* **2017**, *864*, 12–28. [CrossRef]
3. Forno, M.D.; Dolgashev, V.; Bowden, G.; Clarke, C.; Hogan, M.; McCormick, D.; Novokhatski, A.; O'Shea, B.; Spataro, B.; Weathersby, S.; et al. Measurements of electron beam deflection and RF breakdown rate from a surface wave guided in metallic mm-wave accelerating structures. *Phys. Rev. Accel. Beams* **2018**, *21*, 091301. [CrossRef]
4. Baston, T.J.; Bowden, F.P. Localized Damage of Metal Crystals by Laser Irradiation. *Nature* **1968**, *218*, 150–152. [CrossRef]
5. Fedosejevs, R.; Ottmann, R.; Sigel, R.; Kühnle, G.; Szatmári, S.; Schäfer, F.P. Absorption of subpicosecond ultraviolet laser pulses in high-density plasma. *Appl. Phys. B* **1990**, *50*, 79–99. [CrossRef]
6. Agranat, M.B.; Andreev, N.E.; Ashitkov, S.I.; Veisman, M.E.; Levashov, P.R.; Ovchinnikov, A.V.; Sitnikov, D.S.; Fortov, V.E.; Khishchenko, K.V. Determination of the transport and optical properties of a nonideal solid-density plasma produced by femtosecond laser pulses. *JETP Lett.* **2007**, *85*, 271–276. [CrossRef]
7. Ng, A.; Celliers, P.; Forsman, A.; More, R.M.; Lee, Y.T.; Perrot, F.; Dharma-Wardana, M.W.C.; Rinker, G.A. Reflectivity of intense femtosecond laser pulses from a simple metal. *Phys. Rev. Lett.* **1994**, *72*, 3351. [CrossRef]
8. Cerchez, M.; Jung, R.; Osterholz, J.; Toncian, T.; Willi, O.; Mulser, P.; Ruhl, H. Absorption of ultrashort laser pulses in strongly overdense targets. *Phys. Rev. Lett.* **2008**, *100*, 245001. [CrossRef]
9. Agranat, M.B.; Chefonov, O.V.; Ovchinnikov, A.V.; Ashitkov, S.I.; Fortov, V.E. Damage in a thin metal film by high-power terahertz radiation. *Phys. Rev. Lett.* **2018**, *120*, 085704. [CrossRef]
10. Tan, P.; Huang, J.; Liu, K.; Xiong, Y.; Fan, M. Terahertz radiation sources based on free electron lasers and their applications. *Sci. China Inf. Sci.* **2012**, *55*, 1–15. [CrossRef]
11. Rezvani, J.; Di Gioacchino, D.; Gatti, C.; Poccia, N.; Ligi, C.; Tocci, S.; Guidi, M.C.; Cibella, S.; Lupi, S.; Marcelli, A. Tunable vortex dynamics in proximity junction arrays: A possible accurate and sensitive 2D THz detector. *Acta Phys. Pol. A* **2020**, *137*, 17–20. [CrossRef]
12. Tonouchi, M. Cutting-edge terahertz technology. *Nat. Photon.* **2007**, *1*, 97–105. [CrossRef]
13. Chiadroni, E.; Bacci, A.; Bellaveglia, M.; Boscolo, M.; Castellano, M.; Cultrera, L.; Di Pirro, G.; Ferrario, M.; Ficcadenti, L.; Filippetto, D.; et al. The SPARC linear accelerator based terahertz source. *Appl. Phys. Lett.* **2013**, *102*, 094101. [CrossRef]
14. Chiadroni, E.; Bellaveglia, M.; Calvani, P.; Castellano, M.; Catani, L.; Cianchi, A.; Di Pirro, G.; Ferrario, M.; Gatti, G.; Limaj, O.; et al. Characterization of the THz radiation source at the Frascati linear accelerator. *Rev. Sci. Instrum.* **2013**, *84*, 022703. [CrossRef]
15. Perucchi, A.; Di Mitri, S.; Penco, G.; Allaria, E.; Lupi, S. The TeraFERMI terahertz source at the seeded FERMI free-electron-laser facility. *Rev. Sci. Instrum.* **2013**, *84*, 022702. [CrossRef]
16. Lupi, S.; Nucara, A.; Perucchi, A.; Calvani, P.; Ortolani, M.; Quaroni, L.; Kiskinova, M. Performance of SISSI, the infrared beamline of the ELETTRA storage ring. *J. Opt. Soc. Am. B* **2007**, *24*, 959. [CrossRef]
17. Rezvani, S.J.; Pinto, N.; Enrico, E.; D'Ortenzi, L.; Chiodoni, A.; Boarino, L. Thermally activated tunneling in porous silicon nanowires with embedded Si quantum dots. *J. Phys. D Appl. Phys.* **2016**, *49*, 105104. [CrossRef]
18. Pinto, N.; Rezvani, S.J.; Perali, A.; Flammia, L.; Milošević, M.V.; Fretto, M.; Cassiago, C.; De Leo, N. Dimensional crossover and incipient quantum size effects in superconducting niobium nanofilms. *Sci. Rep.* **2018**, *8*, 4710. [CrossRef]

19. Pinto, N.; Rezvani, S.J.; Favre, L.; Berbezier, I.; Fretto, M.; Boarino, L. Geometrically induced electron-electron interaction in semiconductor nanowires. *Appl. Phys. Lett.* **2016**, *109*, 123101. [CrossRef]
20. Rezvani, S.J.; Perali, A.; Fretto, M.; De Leo, N.; Flammia, L.; Milošević, M.; Nannarone, S.; Pinto, N. Substrate-Induced Proximity Effect in Superconducting Niobium Nanofilms. *Condens. Matter* **2019**, *4*, 4. [CrossRef]
21. Ortolani, M.; Lupi, S.; Baldassarre, L.; Schade, U.; Calvani, P.; Takano, Y.; Nagao, M.; Takenouchi, T.; Kawarada, H. Low energy electrodynamics of superconducting diamonds. *Phys. Rev. Lett.* **2006**, *97*, 097002. [CrossRef] [PubMed]
22. Zhang, X.-C.; Ma, X.F.; Jin, Y.; Lu, T.-M. Terahertz optical rectification from a nonlinear organic crystal. *Appl. Phys. Lett.* **1992**, *61*, 3080–3082. [CrossRef]
23. Zhang, X.; Jin, Y.; Ma, X.F. Coherent measurement of THz optical rectification from electro-optic crystals. *Appl. Phys. Lett.* **1992**, *61*, 2764–2766. [CrossRef]
24. Winnerl, S.; Stehr, D.; Drachenko, O.; Schneider, H.; Helm, M.; Seidel, W.; Michel, P.; Schneider, S.; Seidel, J.; Grafstrom, S.; et al. FELBE Free-Electron Laser: Status and application for time resolved spectroscopy experiments. In Proceedings of the 2006 Joint 31st International Conference on Infrared Millimeter Waves and 14th International Conference on Teraherz Electronics, Shanghai, China, 18–22 September 2006.
25. Kato, R.; Kashiwagi, S.; Morio, Y.; Furuhashi, K.; Terasawa, Y.; Sugimoto, N.; Suemine, S.; Isoyama, G. High power terahertz FEL at ISIR, Osaka University. In Proceedings of the IPAC'10, Kyoto, Japan, 23–28 May 2010.
26. Nanni, E.A.; Huang, W.R.; Hong, K.; Ravi, K.; Fallahi, A.; Moriena, G.; Miller, R.J.D.; Kärtner, F.X. Terahertz-driven linear electron acceleration. *Nat. Commun.* **2015**, *6*, 8486. [CrossRef]
27. Boni, R.; Chimenti, V.; Spataro, B.; Tazzioli, F.; Fernandes, P.; Parodi, R. Design and operation of a multipacting-free 51.4 MHz RF accelerating cavity. *Nucl. Instrum. Methods Phys. Res.* **1989**, *274*, 49–55. [CrossRef]
28. Macis, S.; Rezvani, J.; Davoli, I.; Cibin, G.; Spataro, B.; Scifo, J.; Faillace, L.; Marcelli, A. Structural Evolution of MoO_3 Thin Films Deposited on Copper Substrates upon Annealing: An X-ray Absorption Spectroscopy Study. *Condens. Matter* **2019**, *4*, 41. [CrossRef]
29. Macis, S.; Aramo, C.; Bonavolontà, C.; Cibin, G.; D'Elia, A.; Davoli, I.; De Lucia, M.; Lucci, M.; Lupi, S.; Miliucci, M.; et al. MoO_3 films grown on polycrystalline Cu: Morphological, structural and electronic. *J. Vac. Sci. Technol. A* **2019**, *37*, 021513. [CrossRef]
30. Kwan, T.; Dawson, J.M.; Lin, A.T. Free electron laser. *Phys. Fluids* **1977**, *20*, 581. [CrossRef]
31. Irizawa, A.; Suga, S.; Nagashima, T.; Higashiya, A.; Hashida, M.; Sakabe, S. Laser-induced fine structures on silicon exposed to THz-FEL. *Appl. Phys. Lett.* **2017**, *111*, 251602. [CrossRef]
32. Bommireddy, P.R.; Sivajee-Ganesh, K.; Jayanth-Babu, K.; Hussain, O.M.; Julien, C.M. Microstructure and supercapacitive properties of RF-sputtered copper oxide thin films: Influence of O_2/Ar ratio. *Ionics* **2015**, *21*, 2319–2328.
33. Levitskii, V.S.; Shapovalov, V.I.; Komlev, A.E.; Zav'yalov, A.V.; Vit'ko, V.V.; Komlev, A.A.; Shutova, E.S. Raman spectroscopy of copper oxide films deposited by reactive magnetron sputtering. *Tech. Phys. Lett.* **2015**, *41*, 1094–1096. [CrossRef]
34. Butt, M.Z. Laser ablation characteristics of metallic materials: Role of Debye-Waller thermal parameter. In Proceedings of the IOP Conference Series: Materials Science and Engineering, Metz, France, 30 June–4 July 2014; Volume 60, p. 012068.
35. Antoine, C.Z.; Peauger, F.; Le Pimpec, F. Electromigration occurences and its effects on metallic surfaces submitted to high electromagnetic field: A novel approach to breakdown in accelerators. *Nucl. Instrum. Methods Phys. Res. Sect. A* **2011**, *665*, 54–69. [CrossRef]
36. Vorobyev, A.Y.; Guo, C. Reflection of femtosecond laser light in multipulse ablation of metals. *J. Appl. Phys.* **2011**, *110*, 043102. [CrossRef]
37. Miyasaka, Y.; Hashida, M.; Nishii, T.; Inoue, S.; Sakabe, S. Derivation of effective penetration depth of femtosecond laser pulses in metal from ablation rate dependence on laser fluence, incidence angle, and polarization. *Appl. Phys. Lett.* **2015**, *106*, 013101. [CrossRef]
38. Hagemann, H.J.; Gudat, W.; Kunz, C. Optical constants from the far infrared to the x-ray region: Mg, Al, Cu, Ag, Au, Bi, C, and Al_2O_3. *J. Opt. Soc. Am.* **1975**, *65*, 742. [CrossRef]
39. Liao, Y.-C.; Yu, M.-H. Effects of laser beam energy and incident angle on the pulse laser welding of stainless steel thin sheet. *J. Mater. Process. Technol.* **2007**, *190*, 102–108. [CrossRef]

40. Rakić, A.D.; Djurišić, A.B.; Elazar, J.M.; Majewski, M.L. Optical properties of metallic films for vertical-cavity optoelectronic devices. *Appl. Opt.* **1998**, *37*, 5271–5283. [CrossRef]
41. Grudiev, A.; Calatroni, S.; Wuensch, W. New local field quantity describing the high gradient limit of accelerating structures. *Phys. Rev. Spec. Top. Accel. Beams* **2009**, *12*, 102001. [CrossRef]

condensed matter

MDPI

Article

Sub-THz Waveguide Spectroscopy of Coating Materials for Particle Accelerators

Andrea Passarelli [1,2], Can Koral [2], Maria Rosaria Masullo [2], Wilhelmus Vollenberg [3], Lucia Lain Amador [3] and Antonello Andreone [1,2,*]

[1] Physics Department, University of Naples "Federico II", 80131 Naples, Italy; andrea.passarelli@unina.it
[2] INFN Naples Unit, 80131 Naples, Italy; ckoral@na.infn.it (C.K.); masullo@na.infn.it (M.R.M.)
[3] CERN TE-VSC-SCC, CH-1211 Geneve, Switzerland; Wil.Vollenberg@cern.ch (W.V.); lucia.lain.amador@cern.ch (L.L.A.)
* Correspondence: andreone@unina.it; Tel.: +39-081-2530324

Received: 20 December 2019; Accepted: 15 January 2020; Published: 20 January 2020

Abstract: The electromagnetic characterisation of different materials for the inner wall coating of beam pipes is a long-standing problem in accelerator physics, regardless the purpose they are used for, since their presence may affect in an unpredictable way the beam coupling impedance and therefore the machine performance. Moreover, in particle accelerators and storage rings of new generation very short bunches might be required, extending far in frequency the exploration of the beam spectrum and rendering therefore more and more important to assess the coating material response up to hundreds of GHz. This paper describes a time domain method based on THz waveguide spectroscopy to infer the coating properties at very high frequencies. The technique has been tested on Non Evaporable Getter thick films deposited by DC magnetron sputtering on copper plates.

Keywords: THz; waveguide spectroscopy; coating materials; particle accelerators

1. Introduction

An important step towards the development of a new generation of accelerators and light sources is the special treatment of the vacuum chamber surface, in order to avoid electron cloud (e-cloud) effects that may degrade the machine performance and limit its maximum luminosity.

The e-cloud mechanism starts when the synchrotron radiation, emitted by the beam, creates a large number of photoelectrons at the accelerator wall surface. These primary electrons may cause secondary emission or be elastically reflected [1]. If the value of secondary electron yield (SEY) of the surface material is larger than unity, the number of electrons starts growing exponentially and may lead to beam instabilities and other detrimental side effects [2,3]. It is therefore important to keep the value of SEY as low as possible. Moreover, materials with low photoemission would make not necessary any surface conditioning or in situ heating of the beampipe, which translates in a reduction of machine dead time between experiments.

Reduction of the SEY value in specific sections of the accelerator is therefore mandatory, and an extensive search for the best possible candidates for the pipe internal coating has been extensively carried out in the last years. Amongst other materials, amorphous carbon (a-C) have been thoroughly tested [4] and used [5] at different CERN facilities [6] with very effective results. Another interesting class of materials is non-evaporable getter (NEG) alloys [7], that can be deposited on the inner wall of a vacuum chamber in large accelerators, transforming it from a source of gas into an effective pump. In addition to pumping, NEG films lead to reduced induced gas desorption and secondary electron yields. However, the use of a coating material in an accelerator, being it for the SEY reduction or for the vacuum improvement or both, unavoidably changes the overall surface impedance, possibly producing as adverse effect beam instability because of its electromagnetic interaction with the surroundings. Therefore, before its insertion in the beam pipe, an accurate electromagnetic characterization is required, for building a reliable impedance model and pinpointing possible problems and performance limitations in modern particle accelerators and storage rings [8].

In the last years, the mitigation properties of coatings have been rarely tested under an electromagnetic field and in the microwave region only [9]. Since beam spectrum may extend up to the very high frequency regime, depending on the bunch length, it might be important to perform an in-depth evaluation of the resistive wall impedance up to millimeter waves and beyond.

Recently, the impedance of NEG films has been measured in frequency domain in the sub-THz range, directly depositing 1–2 μm of the material on the lateral walls of a calibrated waveguide [10,11]. This method can be easily extended to the characterisation of other coating in thin film form, however has its own drawbacks, specifically local in-homogeneity with blistering and peel-off, constraints in sample dimensions, and impossibility to re-use the test system (the waveguide) for further measurements.

An alternative approach is time domain waveguide spectroscopy [12], which has been widely used in the past to obtain high resolution absorption spectra of molecular solids [12], or for the characterisation of thin samples [13,14]. This is done by usually resorting to metallic waveguides since, depending on the design, they can provide a long interaction length and a very high confinement of the electromagnetic field [15], resulting in a significant sensitivity enhancement. Besides that, the use of calibrated devices makes possible the development of characterisation techniques that are both precise and reliable.

With this aim, we have developed a tailored waveguide with integrated pyramidal horn antennas and a removable part where the coating is deposited, which can be placed with ease in the optical path of a THz spectrometer. The design allows us to measure in a simple way large area coating deposited on metallic plates as in the case of accelerators, where averaged quantities are needed. This technique has been successfully used to characterize NEG samples deposited on both sides of thin copper slabs inserted in a circular waveguide [16]. In the case of amorphous carbon however, because of the high temperature growth and the demanding sample thickness requirement, two side deposition is not a feasible approach, because the slab will experience mechanical stress followed by sample peel-off.

Moreover, the transition from the horns to the waveguide, because of the different transverse sections, makes the performance of this device very sensitive to mechanical imperfections and deformations. We designed therefore a modified square waveguide where the transition from antenna to the device is ideally removed, since the horn inner aperture and the waveguide transverse section fully overlap. This results in an increased robustness and a better efficiency of the waveguide in the collection of the THz signal. A major drawback is the reduction by more than 50% in cross section when compared with the circular waveguide, and as a consequence a strong decrease of signal transmitted through the device.

Here we present a detailed description of the high resolution waveguide spectroscopy setup and the analytical method developed for the extraction of the sample electromagnetic properties. In order to validate the technique, we measured the sub-THz response of two coating NEG layers about 4 μm thick deposited using DC magnetron sputtering on both sides of copper plates, and evaluate their conductivity.

2. Materials and Methods

2.1. The Device under Test

Ti-Zr-V NEG coatings are grown at the CERN deposition facilities [4] on both sides of copper plates by using a DC magnetron sputtering technique with Krypton as process gas at a working pressure of 7×10^{-4} mbar (see Figure 1a). An alloyed disc cathode with 33.3% atomic nominal relative composition of Titanium, Zirconium and Vanadium is placed at a distance of 200 mm from the substrate. The applied tension and current are 294 V and 750 mA respectively, for a total power of 220 W. The complete deposition, giving on average 4 μm thickness, lasts about two days at a growth rate of 1.4 nm/s. During the process, in order to prevent thermal induced deformations the plate is held in an aluminum frame, that in turn is placed on a rotating axis to ensure a homogeneous deposition on both sides of the slab (see Figure 1b). Local composition and thickness of the coating have been checked using X-ray fluorescence (XRF) measurements along the median line of the slab (the waveguide longitudinal axis), showing that samples keep the target composition within 10% and a uniform profile ±2% with an average surface roughness of 0.2 μm [10,17].

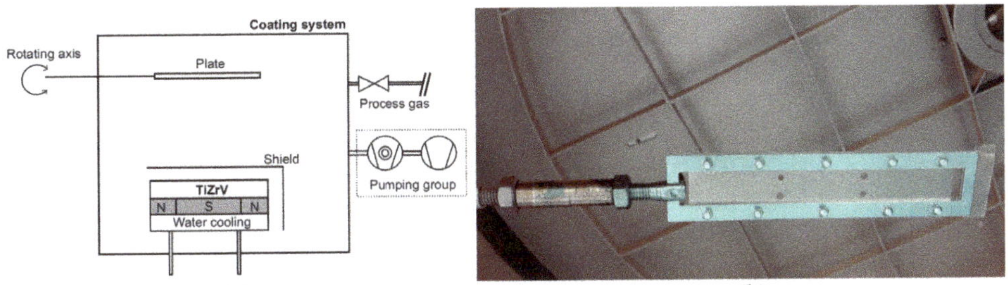

(a) (b)

Figure 1. Non-evaporable getter (NEG) deposition setup. (**a**) Drawing of the system; (**b**) detail of the copper slab placed in the aluminum frame. The four holes shown in the picture are used for the alignment in the waveguide during THz measurements.

For the spectroscopy measurements, we use a gold plated brass device (shown in Figure 2) consisting in a parallelepiped of 16 × 12 × 140 mm^3 machined in two identical pieces. A diagonal waveguide is formed by milling a square cross-section channel, rotated by 45° and 62 mm long, in both halves. Two symmetrical pyramidal horn antennas are embedded in both sides of the structure in order to enhance the electromagnetic signal collection and radiation. Moreover, their inner aperture coincides with the waveguide section, ensuring a smooth transition to the waveguide itself without abrupt changes in the device impedance [18]. Disassembling the device, the thin copper slab with the NEG material deposited on both sides can be easily inserted for the sample characterization. The slab has the same length as the device (140 mm) and thickness 0.050 mm.

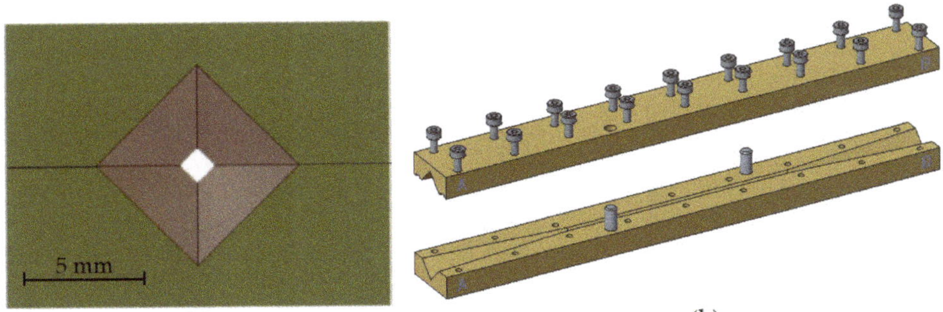

(a) (b)

Figure 2. Sketch of the device used for the spectroscopy measurements, consisting in a diagonal section waveguide ending with two pyramidal horn antennas. (**a**) Front view of the assembled device; (**b**) open view of the waveguide and the embedded antennas. The overall size of the device is 16 × 12 × 140 mm^3.

The dimensions and the material used for the waveguide fabrication are reported in Table 1. For the pyramidal horns, maximum and minimum apertures along their length are indicated.

Table 1. Technical specifications of device under test.

Material	Brass (Au Plated)
Waveguide	Diagonal
Length [mm]	62
Side [mm]	1.1
Transition	Pyramidal
Length [mm]	39
Side [mm]	$6 \rightarrow 1.1$
Total Length [mm]	140

These dimensions have been chosen in order to have a single mode propagation inside the diagonal waveguide and in the two pyramidal transitions. Due to the central slab inserted along the longitudinal direction, the first mode that can propagate through the structure is the sum of the $TE_{1,0}$ and $TE_{0,1}$. The second allowed mode, given the boundary conditions and the waveguide symmetry, is the sum of $TE_{2,1}$ and $TE_{1,2}$ [19]. For an internal side of the diagonal waveguide of 1.1 mm, the usable frequency window for a single mode propagation ranges from 135 GHz to 300 GHz.

2.2. Sub-THz System

Sub-THz measurements are carried out using a time domain spectrometer (TDS) operating in transmission mode. The setup is based on a commercial THz-TDS system (TERA K15 by Menlo Systems) customized for the waveguide characterization. The system is driven by a femtosecond fiber laser @1560 nm with an optical power <100 mW and a pulse duration <90 fs. Fiber-coupled photoconductive antenna modules are utilized for both electric field signal emission and detection. A fast opto-mechanical line with a maximum scanning range of approximately 300 ps is used to control the time delay between the pump and the probe beam.

TPX (polymethylpentene) lenses are used to collimate the short (1–2 ps) linearly polarized pulse on the waveguide, producing a Gaussian-like beam with a waist of approximately 8 mm in diameter and a quasi-plane wave phase front. This is a standard configuration commonly used to perform transmission measurements on a number of different materials [20]. A sketch of the optical setup is shown in Figure 3.

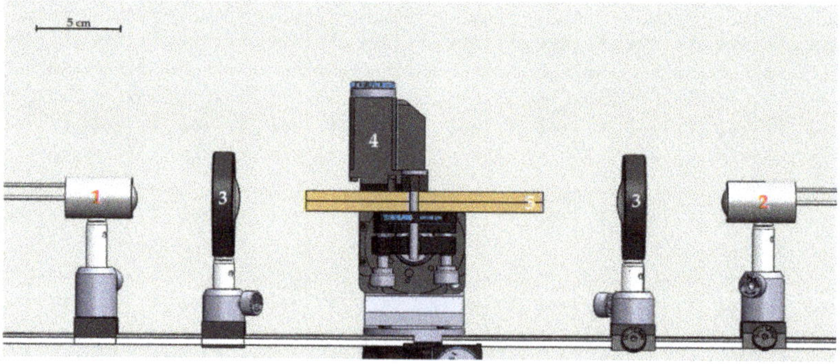

Figure 3. Side view of the opto-mechanical setup utilized for the measurements: (1) emitter, (2) detector, (3) TPX collimating lenses, (4) micrometric alignment system, (5) waveguide with embedded antennas.

For an accurate control of the optical coupling between the free space signal and the input and output horn antennas, the lower part of the waveguide is fixed on a kinematic mount with a micrometric goniometer. Coated slabs are inserted and replaced by removing the upper part only.

The electric field signal as a function of time is recorded for each sample by averaging 1000 pulses for an overall acquisition time of 10 min, in order to minimize the signal-to-noise ratio.

Frequency dependent transmission curves are obtained through the application of a standard FFT algorithm. In the experiment, the frequency resolution is set to about 8 GHz, determined by the scanning range of the delay line.

2.3. The Analytical Method

The conductivity value of the coated material is obtained from the comparison between the signal amplitude transmitted through the waveguide with the coated slab and the one obtained with an uncoated slab used as a reference.

The attenuation both in the diagonal waveguide A_{diag} and in the pyramidal transitions A_{pyr}, considering the propagation of the sum of two modes $\text{TE}_{1,0}$ and $\text{TE}_{0,1}$, is [21,22]:

$$A_{\text{diag}} = A_{\text{pyr}} = \frac{1}{2} Re(Z_S) \frac{\int |\mathbf{n} \times (\mathbf{H}_{1,0} + \mathbf{H}_{0,1})|^2 \, dl}{Re(Z_{1,0}) \, |I_{1,0}|^2 + Re(Z_{0,1}) \, |I_{0,1}|^2} \tag{1}$$

where $Z_{i,j}$ is the i,j mode impedance and $I_{i,j}$ is the relevant excitation current. The field components used in Equation (1) are:

$$H_{x_{1,0}} = I_{1,0} \frac{\sqrt{2}}{a} \cos\left(\frac{\pi}{a} x\right) \tag{2}$$

$$H_{y_{1,0}} = I_{1,0} \frac{\sqrt{2}}{a} \cos\left(\frac{\pi}{a} y\right), \tag{3}$$

where a is the side of the diagonal waveguide. In case of coating material, the expression of Z_S is:

$$Z_S = Z_{\text{coat}} \frac{Z_{\text{cu}} + jZ_{\text{coat}} \tan(k_{\text{coat}} d)}{Z_{\text{coat}} + jZ_{\text{cu}} \tan(k_{\text{coat}} d)}, \tag{4}$$

where d is the coating thickness. When $d = 0$ there is no coating and $Z_S = Z_{\text{cu}}$. The characteristic impedance in the Leontovich approximation for a metallic case ($\epsilon'' \gg \epsilon'$) is [21]:

$$Z = (1+j)\sqrt{\frac{\omega\mu}{2\sigma}} = \frac{1+j}{\sigma\delta} \tag{5}$$

where μ is the total permeability, $\omega = 2\pi f$, and σ the material conductivity.

The propagation constant under the same condition is

$$k = (1-j)\sqrt{\frac{\sigma\omega\mu}{2}} = \frac{1-j}{\delta}, \tag{6}$$

where δ is the skin-depth defined as

$$\delta = \sqrt{\frac{2}{\sigma\omega\mu}} \tag{7}$$

The total attenuation on both sides of the slab in the diagonal waveguide is:

$$A_{diag} = \sqrt{2} \frac{Re(Z_S) k_{z_{\text{diag}}}}{a Z_0 k_0} \left[1 + \frac{2k_{t_{\text{diag}}}^2}{k_{z_{\text{diag}}}^2}\right] l_g \tag{8}$$

where l_g is the length of the waveguide, and

$$k_{t_{\text{diag}}} = \frac{\pi}{a}; \quad k_{z_{\text{diag}}} = \sqrt{k_0^2 - k_{t_{\text{diag}}}^2}$$

We evaluate the relative attenuation in the diagonal waveguide as:

$$RA_{\text{diag}} = A_{\text{diag}}^{\text{coat}} - A_{\text{diag}}^{\text{cu}}. \tag{9}$$

Differently from the contribution in the diagonal section, the attenuation on the slab given by the symmetric input and output transitions is not constant, since the antenna aperture changes along the horn length. The total attenuation in the single pyramidal transition is:

$$\begin{aligned} A_{\text{pyr}} &= \int_0^{l_t} \sqrt{2}\frac{Re(Z_S)k_{z_{\text{pyr}}}(z)}{c(z)Z_0k_0}\left[1+\frac{2k_{t_{\text{pyr}}}^2(z)}{k_{z_{\text{pyr}}}^2(z)}\right]dz \\ &= \frac{1}{\sqrt{2}}\frac{Re(Z_s)}{Z_0}\left\{-\frac{1}{2b_1}\log\left[\frac{\sqrt{1-\left(\frac{\pi}{k_0B}\right)^2}-1}{\sqrt{1-\left(\frac{\pi}{k_0B}\right)^2}+1}\,\frac{\sqrt{1-\left(\frac{\pi}{k_0b}\right)^2}+1}{\sqrt{1-\left(\frac{\pi}{k_0b}\right)^2}-1}\right]\right. \\ &\left.+\frac{2}{b_1}\left[\sqrt{\left(\frac{\pi}{k_0B}\right)^2}-\sqrt{\left(\frac{\pi}{k_0b}\right)^2}\right]\right\} \end{aligned} \tag{10}$$

where

$$k_{t_{\text{pyr}}}(z) = \frac{\pi}{c(z)}; \quad k_{z_{\text{pyr}}}(z) = \sqrt{k_0^2 - k_{t_{\text{pyr}}}^2(z)}$$

and

$$c(z) = b + zb_1 = b + z\frac{B-b}{l_t}$$

expresses how the side of the horn changes along the transition. l_t is the longitudinal length of the transition, B and b are the side dimension at the entrance and exit of the pyramidal horn transition respectively. Since the transitions are connected to the diagonal waveguide, then $b = a$.

The relative attenuation in the pyramidal transition is:

$$RA_{\text{pyr}} = A_{\text{pyr}}^{\text{coat}} - A_{\text{pyr}}^{\text{cu}}. \tag{11}$$

The total relative attenuation is given by the formula:

$$RA_{\text{Total}} = RA_{\text{diag}} + 2\,RA_{\text{pyr}}, \tag{12}$$

resorting to Equations (9) and (11).

3. Results and Discussion

The electromagnetic characterization of the NEG coatings in the sub-THz region is realized performing time domain (TD) measurements of the electromagnetic wave propagating inside the specifically designed waveguide with a thin central copper slab, where the material under test is deposited on both sides. The THz beam is polarized with its electric field parallel to the waveguide slab. The use of a device having a square cross section marks an improvement in terms of beam collection efficiency with respect to previous measurements [16]. However, compared to the free space signal, the transmitted electric field is in the order of 10% only. Figure 4 shows the difference in the signal transmitted in air and in the squared waveguide, without and with slab.

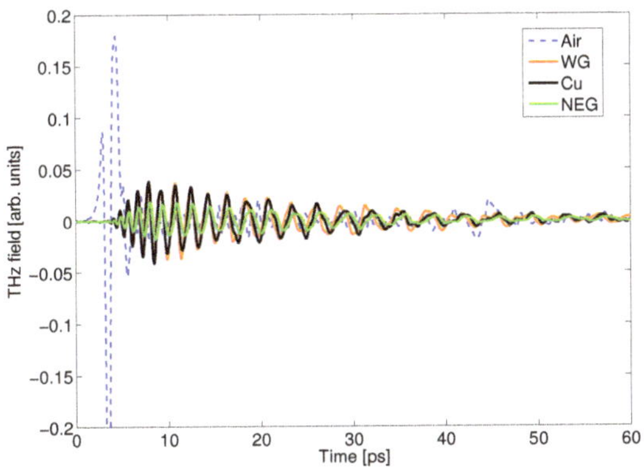

Figure 4. THz time domain averaged signal propagating in air (dashed line), through the waveguide without (orange, continuous line) and with copper (black, continuous line) or NEG coated (green, continuous line) slab.

The TD curves show that the presence of the NEG layer clearly introduces not only a strong attenuation of the signal but also a significant dispersion. When the THz signal passes through the device, the ps-scale input pulse with respect to the case of free space transmission is strongly reshaped by the reflections inside the waveguide and broadened to more than 50 ps. The stretching of the transmitted signal, compared with the free space input pulse, is due to the strongly dispersive character of the waveguide, that acts as a delay line [23]. After the main burst, at around 120 ps interference features (not shown) appear, due to round-trip reflections inside the waveguide, that have been removed in the subsequent frequency analysis.

In frequency domain, the amplitude spectra are obtained using FFT (Fast Fourier Transform) analysis and are presented in Figure 5 for both samples in comparison with the bare copper slab, with the waveguide without the slab, and with air only.

Data are shown up to the frequency where single mode propagation in the waveguide holds. This ensures that there is no interference from higher order modes, which can produce a modification of the field distribution, as discussed in detail in Section 2.3. The cut-off frequency of the first mode propagating inside the waveguide (approximately 150 GHz) can be clearly seen.

In the graph, the NEG coated samples (red and green dots) behave in a similar way, with a marked difference in amplitude (between 4 and 6 dB on average) with respect to copper (black dots) and slightly increasing at higher frequencies. Changing the design of the waveguide from circular to square therefore does imply a better sensitivity to the coating material properties, since the weight of the slab losses relatively to the overall waveguide losses increases, but this is done at the expense of a reduction in the amplitude of the collected signal, raising noise and data fluctuations.

From the comparison with the uncoated slab, the (relative) attenuation given by the losses produced in the device through its overall length (horn antennas and waveguide) by the coating material can be evaluated.

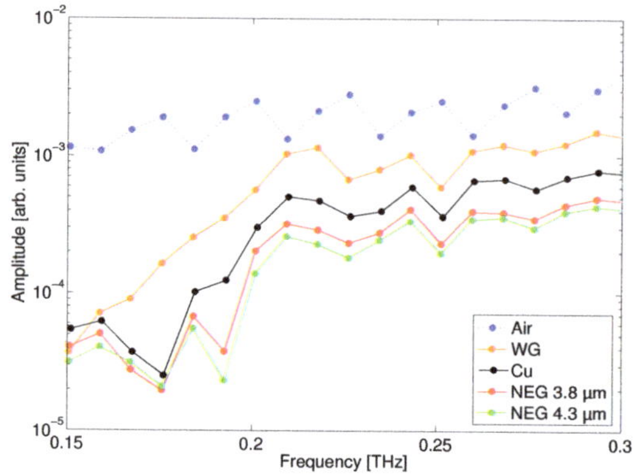

Figure 5. Frequency spectrum showing the averaged amplitude transmission data in air (dashed line-to-point curve), through the waveguide without (orange line-to-point curve) and with copper (black line-to-point curve) or 3.8 μm and 4.3 μm NEG coated (red and green line-to-point curves respectively) slabs.

In Figure 6, the measured relative attenuation due to both the 3.8 μm and 4.3 μm NEG coatings with respect to the copper reference are shown (red and green dots respectively). Data below 200 GHz have been discarded to avoid artifacts in the spectrum due to group and phase velocity dispersion, especially pronounced near the cut off frequency. Therefore, results are presented in the range 200–300 GHz.

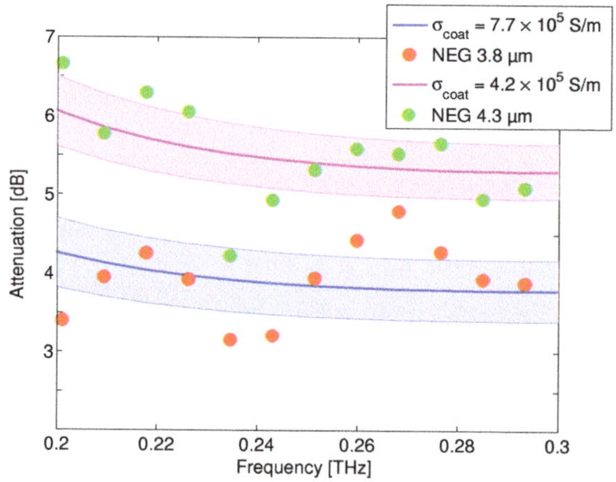

Figure 6. Experimental relative attenuation as a function of frequency on the NEG coated slab of 3.8 μm and 4.3 μm (red and green dots respectively) and best fit curves (blue and magenta lines respectively).

The conductivity value σ_{coat} of the coating material is obtained resorting on the analytical tool detailed in Section 2.3. From the best fit of the analytical formula, we yield $\sigma_{coat} = (7.7 \pm 1.1) \times 10^5$ S/m for the 3.8 μm sample, and $\sigma_{coat} = (4.2 \pm 0.5) \times 10^5$ S/m for the 4.3 μm sample. Magenta and blue continuous lines in Figure 6 show the analytically evaluated attenuation for the two estimated conductivities, taken from Equation (12) in Section 2. For both curves the 95% confidence interval on the attenuation due to the conductivity evaluation uncertainties is also displayed as shaded area. The difference observed between the two samples might be an indication of an increased disorder, that translates in poorer transport properties and therefore lower conductivity, caused by the larger thickness.

Resorting on [24], we can also evaluate the effect of sample roughness on the NEG conductivity. Using the average roughness value for our samples (0.2 μm), we estimate a maximum conductivity reduction of the order of 8%, that lies within the measurement error band in our frequency range.

Results agree fairly well both with previous data obtained on different NEG samples using the circular waveguide [16] and DC (direct current) conductivity values extracted using the frequency domain approach [10]. Nevertheless, differently to these latter measurements, a TD method allows to evaluate the electromagnetic properties of coatings in a reliable and simple way exploiting tailored (and reusable) waveguides. The knowledge of σ_{coat} under operating conditions (coating deposited on a metallic slab) is extremely useful for the evaluation of the real part of the surface impedance as a function of frequency, that is currently used for modeling the resistive wall component of the beam impedance in modern accelerators.

Author Contributions: A.P. and A.A. conceived the experiment. A.P., M.R.M. and A.A. developed the theory for the electromagnetic characterization of coating material in a waveguide. A.P. and C.K. performed the THz measurements. W.V. and L.L.A. realized the NEG samples. A.P. ad A.A. wrote the manuscript, with contributions from all authors. A.A. supervised the activity. All authors have read and agreed to the published version of the manuscript.

Funding: This research has been funded by the CLIC project in the framework of the CERN-INFN Naples collaboration (KN4542/BE Addendum no.13 to Agreement KN3083). Partial support from INFN Projects "TERA" and "MICA" is gratefully acknowledged.

Acknowledgments: The authors thank S. Calatroni, P. Costa Pinto, R. Corsini, Y. Papaphilippou, and M. Taborelli from CERN for their support.

Conflicts of Interest: The authors declare no conflict of interest.

References

1. Furman, M.A. Electron Cloud Effects in Accelerators. *arXiv* **2013**, arXiv:1310.1706.
2. Rumolo, G.; Ruggiero, F.; Zimmermann, F. Simulation of the electron-cloud build up and its consequences on heat load, beam stability, and diagnostics. *Phys. Rev. Spec. Top.-Accel. Beams* **2001**, *4*, 012801.10.1103/PhysRevSTAB.4.012801. [CrossRef]
3. Cimino, R.; Gonzalez, L.A.; Larciprete, R.; Di Gaspare, A.; Iadarola, G.; Rumolo, G. Detailed investigation of the low energy secondary electron yield of technical Cu and its relevance for the LHC. *Phys. Rev. Spec. Top.-Accel. Beams* **2015**, *18*, 051002.10.1103/PhysRevSTAB.18.051002. [CrossRef]
4. Yin Vallgren, C.; Arduini, G.; Bauche, J.; Calatroni, S.; Chiggiato, P.; Cornelis, K.; Pinto, P.C.; Henrist, B.; Métral, E.; Neupert, H.; et al. Amorphous carbon coatings for the mitigation of electron cloud in the CERN Super Proton Synchrotron. *Phys. Rev. Spec. Top.-Accel. Beams* **2011**, *14*, 071001.10.1103/PhysRevSTAB.14.071001. [CrossRef]
5. Taborelli, M.; Chiggiato, P.; Costa Pinto, P.; Cruikshank, P. *Nine Years of Carbon Coating Development for the SPS Upgrade: Achievements and Heritage*; Technical Report CERN-ACC-2016-0010; CERN: Geneva, Switzerland, 2015.
6. Palmer, M.; Livezey, J.; Wolski, A.; Zwaska, R.; Sagan, D.; Savino, J.; Williams, H.; Antoniou, F.; Tan, C.Y.; Sikora, J. Electron Cloud at Low Emittance in CesrTA. In Proceedings of the International Particle Accelerator Conference, Kyoto, Japan, 23–28 May 2010.
7. Benvenuti, C.; Chiggiato, P.; Pinto, P.C.; Santana, A.E.; Hedley, T.; Mongelluzzo, A.; Ruzinov, V.; Wevers, I. Vacuum properties of TiZrV non-evaporable getter films. *Vacuum* **2001**, *60*, 57–65. [CrossRef]
8. Migliorati, M.; Belli, E.; Zobov, M. Impact of the resistive wall impedance on beam dynamics in the Future Circular e^+e^- Collider. *Phys. Rev. Accel. Beams* **2018**, *21*, 041001.10.1103/PhysRevAccelBeams.21.041001. [CrossRef]
9. Federmann, S.; Caspers, F.; Mahner, E. Measurements of electron cloud density in the CERN Super Proton Synchrotron with the microwave transmission method. *Phys. Rev. Spec. Top.-Accel. Beams* **2011**, *14*, 012802.10.1103/PhysRevSTAB.14.012802. [CrossRef]

10. Koukovini-Platia, E.; Rumolo, G.; Zannini, C. Electromagnetic characterization of nonevaporable getter properties between 220–330 and 500–750 GHz for the Compact Linear Collider damping rings. *Phys. Rev. Accel. Beams* **2017**, *20*, 011002.10.1103/PhysRevAccelBeams.20.011002. [CrossRef]
11. Koukovini Platia, E. High Frequency Effects of Impedances and Coatings in the CLIC Damping Rings. Ph.D. Thesis, Ecole Polytechnique, Lausanne, Switzerland, 2015.
12. Theuer, M.; Melinger, J.S. High Resolution Waveguide Terahertz Time-Domain Spectroscopy. *J. Infrared Millim. Terahertz Waves* **2011**, *32*, 1267–1284.10.1007/s10762-011-9816-3. [CrossRef]
13. Razanoelina, M.; Kinjo, R.; Takayama, K.; Kawayama, I.; Murakami, H.; Mittleman, D.M.; Tonouchi, M. Parallel-Plate Waveguide Terahertz Time Domain Spectroscopy for Ultrathin Conductive Films. *J. Infrared, Millim. Terahertz Waves* **2015**, *36*, 1182–1194.10.1007/s10762-015-0194-0. [CrossRef]
14. Khachatrian, A.; Melinger, J.S.; Qadri, S.B. Waveguide terahertz time-domain spectroscopy of ammonium nitrate polycrystalline films. *J. Appl. Phys.* **2012**, *111*, 093103. [CrossRef]
15. Zhang, J.; Grischkowsky, D. Adiabatic compression of parallel-plate metal waveguides for sensitivity enhancement of waveguide THz time-domain spectroscopy. *Appl. Phys. Lett.* **2005**, *86*, 061109.10.1063/1.1863439. [CrossRef]
16. Passarelli, A.; Bartosik, H.; Rumolo, G.; Vaccaro, V.G.; Masullo, M.R.; Koral, C.; Papari, G.P.; Andreone, A.; Boine-Frankenheim, O. Novel measurement technique for the electromagnetic characterization of coating materials in the sub-THz frequency range. *Phys. Rev. Accel. Beams* **2018**, *21*, 103101.10.1103/PhysRevAccelBeams.21.103101. [CrossRef]
17. Prodromides, A.E.; Levy, F. Non-Evaporable Getter Thin Film Coatings for Vacuum Applications. Ph.D. Thesis, Ecole Polytechnique, Lausanne, Switzerland, 1 September 2002; pp. 95–96.
18. Johansson, J.F.; Whyborn, N.D. The diagonal horn as a sub-millimeter wave antenna. *IEEE Trans. Microw. Theory Tech.* **1992**, *40*, 795–800.10.1109/22.137380. [CrossRef]
19. Lee, C.S.; Lee, S.W.; Chuang, S.L. Plot of Modal Field Distribution in Rectangular and Circular Waveguides. *IEEE Trans. Microw. Theory Tech.* **1985**, *33*, 271–274.10.1109/TMTT.1985.1132998. [CrossRef]
20. Papari, G.P.; Silvestri, B.; Vitiello, G.; De Stefano, L.; Rea, I.; Luciani, G.; Aronne, A.; Andreone, A. Morphological, Structural, and Charge Transfer Properties of F-Doped ZnO: A Spectroscopic Investigation. *J. Phys. Chem. C* **2017**, *121*, 16012–16020.10.1021/acs.jpcc.7b04821. [CrossRef]
21. Marcuvitz, N. *Waveguide Handbook*; IEE: London, UK, 1986.
22. Franceschetti, G. *Electromagnetics: Theory, Techniques, and Engineering Paradigms*; Springer: New York, NY, USA, 1997.
23. Mendis, R.; Mittleman, D.M. Comparison of the lowest-order transverse-electric (TE1) and transverse-magnetic (TEM) modes of the parallel-plate waveguide for terahertz pulse applications. *Opt. Express* **2009**, *17*, 14839–14850.10.1364/OE.17.014839. [CrossRef] [PubMed]
24. Hammerstad, E.; Jensen, O. Accurate Models for Microstrip Computer-Aided Design. In Proceedings of the 1980 IEEE MTT-S International Microwave symposium Digest, Washington, DC, USA, 28–30 May 1980; pp. 407–409.10.1109/MWSYM.1980.1124303. [CrossRef]

condensed matter

MDPI

Article

The TeraFERMI Electro-Optic Sampling Set-Up for Fluence-Dependent Spectroscopic Measurements

Nidhi Adhlakha [1], Paola Di Pietro [1], Federica Piccirilli [2], Paolo Cinquegrana [1], Simone Di Mitri [1], Paolo Sigalotti [1], Simone Spampinati [1], Marco Veronese [1], Stefano Lupi [2,3] and Andrea Perucchi [1,*]

1 Elettra-Sincrotrone Trieste S.C.p.A, S.S.14 km 163.5, Basovizza, 34149 Trieste, Italy; nidhi.adlakha@elettra.eu (N.A.); paola.dipietro@elettra.eu (P.D.P.); paolo.cinquegrana@elettra.eu (P.C.); simone.dimitri@elettra.eu (S.D.M.); paolo.sigalotti@elettra.eu (P.S.); simone.spampinati@elettra.eu (S.S.); marco.veronese@elettra.eu (M.V.)

2 CNR-IOM, Area Science Park Basovizza, 34146 Trieste, Italy; piccirilli@iom.cnr.it (F.P.); stefano.lupi@roma1.infn.it (S.L.)

3 Dipartimento di Fisica, Università "La Sapienza", P.le A. Moro 2, 00185 Roma, Italy

* Correspondence: andrea.perucchi@elettra.eu

Received: 10 December 2019; Accepted: 19 January 2020; Published: 20 January 2020

Abstract: TeraFERMI is the THz beamline at the FERMI free-electron-laser facility in Trieste (Italy). It uses superradiant Coherent Transition Radiation emission to produce THz pulses of 10 to 100 μJ intensity over a spectral range which can extend up to 12 THz. TeraFERMI can be used to perform non-linear, fluence-dependent THz spectroscopy and THz-pump/IR-probe measurements. We describe in this paper the optical set-up based on electro-optic-sampling, which is presently in use in our facility and discuss the properties of a representative THz electric field profile measured from our source. The measured electric field profile can be understood as the superimposed emission from two electron bunches of different length, as predicted by electron beam dynamics simulations.

Keywords: THz spectroscopy; free-electron-lasers; nonlinear optics

1. Introduction

The interest in high power pulsed THz sources has seen a steady increase during the last decade. This is due to the great potential which is offered by THz control of matter [1,2]. THz light can indeed be used to tune and modify material's properties in several ways, from the strong acceleration of free charge carriers, to anharmonic lattice distortion and even to ultrafast magnetic switching since magnetic fields in the Tesla range can be accompanying strong THz pulses.

While synchrotrons have always represented a significant source of high brightness THz light for spectroscopy, their use in non-linear THz studies is limited by their relatively low electric fields (<kV/cm). On the other hand, single pass accelerators can store high charge in sub-ps bunches. For this reason, their use for THz control of matter is becoming widespread [3–9].

TeraFERMI is the THz facility based on the FERMI free-electron-laser (FEL) located in Trieste (Italy). The electron bunches that are used by the FERMI free-electron-laser to produce UV and soft X-ray radiation [10] are further exploited by TeraFERMI [11] in order to produce THz waves. The THz pulses emitted by TeraFERMI are coherent due to the so-called superradiance phenomenon. This occurs when the separation between the radiation-emitting electrons is shorter than the wavelength of the light.

In the first part of this paper we will review the basic characteristics and figures of merit of the TeraFERMI source, while on the second part we will mainly describe the electro-optic-sampling (EOS)

based experimental set-up that we are routinely exploiting for nonlinear THz spectroscopies at TeraFERMI. In the third section we discuss a typical electro-optic sampling profile measured at TeraFERMI. Interestingly, the measured electric field profile is compatible with the emission from a 100 fs-long current spike in the bunch head of the ps-long entire electron bunch, as previously predicted by electron beam dynamics simulations [12]. The presence of this current spike is due to Coherent Synchrotron Radiation (CSR) wakefields in the dump line and is the origin of the multi-THz emission at TeraFERMI even in the absence of a sub-ps compression of the total electron bunch.

2. The TeraFERMI Beamline

The FERMI FEL operates alternatively with two different FEL lines called FEL1 and FEL2, operating between 20–100 nm and 4–20 nm respectively. After the FEL process, both electron lines are merging into one unique beamline before the electron dump. The TeraFERMI source intercepts the electrons at this position, thus allowing operation simultaneously with both FERMI FELs. The THz photons emitted by a Coherent Transition Radiation (CTR) source (see Section 2.1) are propagated along the THz beamline into the FERMI experimental hall, where the THz light is delivered to a dedicated laboratory.

The THz pulses from TeraFERMI are then used to induce non-linear changes in matter. To probe the effects of the THz pulse we use either the same THz photons emitted by TeraFERMI or another probe synchronized to the FERMI master-clock. At present the laboratory is equipped with a femtosecond infrared fiber laser (MENLO C-Fiber 780), which can be operated either on the fundamental at 1560 nm or on its second harmonic at 780 nm. As we will see in the following, the laser can also be used to perform electro-optic sampling measurements thereby probing the time-resolved evolution of the THz electric fields, as well as their spectral content.

2.1. Source

Coherent Transition Radiation is the phenomenon occurring when a relativistic electron crosses the boundary between two media of different refractive index. The emitted THz light is particularly bright and displays cylindrical symmetry and radial polarization. Its energy distribution as a function of frequency (ω) and emission angle (θ) can be described by the Ginzburg-Frank formula [13]:

$$\frac{d^2U}{d\omega d\Omega} = \frac{e^2}{4\pi^3\epsilon_0 c}\frac{\beta^2 sin^2\theta}{(1-\beta^2 cos^2\theta)^2}, \tag{1}$$

where β is the relativistic factor v/c. To get an estimate of the emitted intensity one should multiply the Ginzburg-Frank equation (or its generalized version for the near-field case [14]) by the so-called coherence enhancement factor:

$$N[1+N|\int_{-\infty}^{+\infty}\rho(t)exp(-i\omega t)dt|^2]. \tag{2}$$

The TeraFERMI source consists in a 1µm-thick Al membrane of 38 mm diameter, oriented at 45° with respect to the electron beam. A wedged diamond window with 20 mm clear aperture is located 80 mm far from the source. The Ginzburg Frank equation allows estimating typical TeraFERMI emission energies ranging from 30 to 100 µJ per pulse, in good agreement with experimental findings [12]. A pictorial view of the TeraFERMI CTR extraction scheme is illustrated in Figure 1.

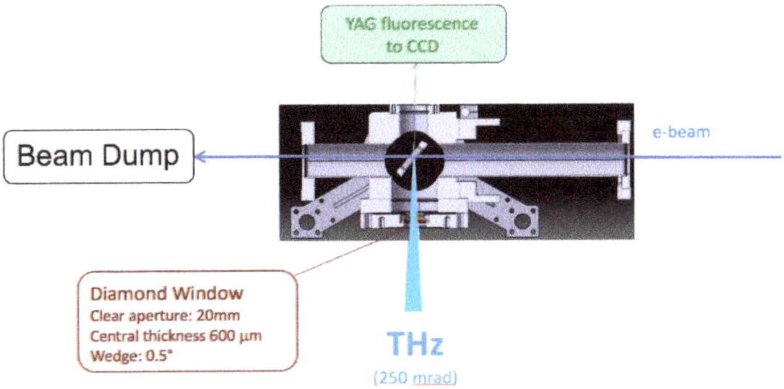

Figure 1. Schematics of the TeraFERMI Coherent Transition Radiation (CTR) extraction. The Al membrane intercepts the electron beam at a 45° angle, shortly before the electron beam dump. A diamond window separates the ultra-high vacuum of the electron chamber with respect to the low-vacuum of the TeraFERMI beamline transport system (see Section 2.2). A fluorescent Yttrium Aluminium Garnet (YAG) screen can also be inserted instead of the CTR source (forming a 90° angle with respect to the Al CTR screen). This allows to image the electron beam profile through an additional viewport located opposite with respect to the diamond window.

Under the present (∼1 GeV) energy conditions, the intensity emitted by CTR is comparable to the one that would be obtained through CSR (see Figure 2 in Reference [11]). The main difference between the two types of emission resides in the polarization properties (linear for CSR, radial for CTR) and in the spatial profile. CSR originates from an extended source in the horizontal direction, which may give rise to optical aberrations, unless the vertical and horizontal source emissions are independently focused, by an optimized set of conical and cylindrical mirrors [15]. On the other hand the cylindrical symmetry of CTR radiation can be more easily transported with a simple set of toroidal mirrors (see next Section).

2.2. Transport System

The THz beam emitted at the TeraFERMI source needs to be transported along a distance of about 30 m, to reach the dedicated THz laboratory located in the FERMI experimental hall. In order to transport a THz beam over such a considerable length, the beam can not be propagated as a collimated beam but needs to be continuously refocused by a set of 6 toroidal mirrors [16]. The entire beamline, which is separated from the source by an initial diamond window (see Section 2.1), is kept under low vacuum conditions in order to avoid water vapour absorptions. The final window of the beamline is interchangeable so that one can choose the more suitable material (z-cut quartz, sapphire, TPX, etc.) according to the requirements of the experiment. Above 0.3 THz the losses along transport can be totally ascribed to the transmission of the two optical windows. When the TPX window is mounted, the overall beamline transmission is of about 55–60%, which roughly corresponds to the product of the 70% transmission of the initial diamond window and that of the final TPX window (90–95%).

Once reaching the optical table the beam is further refocused through a series of parabolic mirrors finally allowing to steer the THz beam at sample position with the best possible focusing properties. The beam size at focus is 800 μm diameter FWHM, as measured by a pyroelectric camera (Pyrocam IIIHR). This focus was characterised in ambient humidity conditions, when the frequency content at the camera is almost completely restricted below 1 THz.

By inserting a microbolometric camera (i2S TZcam) slightly out of focus it is possible to characterize the typical doughnut shape expected from a CTR source. The left/right asymmetry (see Figure 2), which is also expected from the THzTransport [14] simulation [16], is due to the combination of the 45° orientation of the source radiator and of the 90° off-axis arrangement of the various optical components.

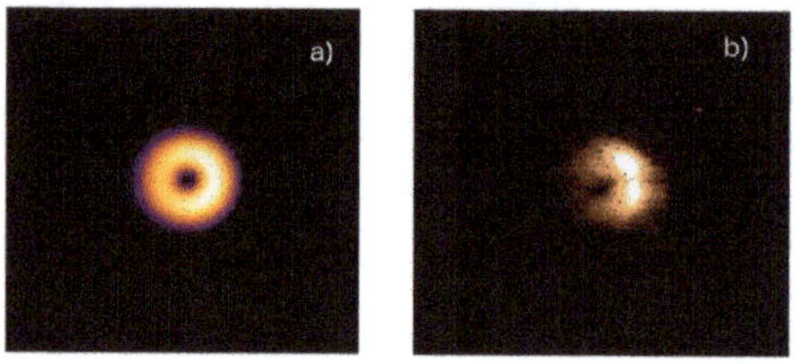

Figure 2. (**a**) Simulated beam profile at 1 THz, at focus position. (**b**) Measured beam profile with a microbolometric camera. The typical CTR doughnut shape and the left/right asymmetry is in good agreement with the simulation. In order to achieve the proper spatial resolution needed to visualize the doughnut shape the measurement was performed out of focus, with a larger beam, so that the present comparison is only qualitative. The measurement in (**b**) can not be used to estimate the real beam dimension which was previously measured with the Pyrocam IIIHR THz camera (see text).

2.3. Beamline Performance

The overall TeraFERMI performances, were are already summarized in Reference [17]. During standard user operation conditions the beamline produces THz pulses whose energy ranges from 15 to 60 μJ per pulse, with a bunch charge of ~700 pC and a bunch length of about 1 ps, as measured at the end of the LINAC. This results in energies at sample up to 35 μJ per pulse. The optical spectrum, as measured by a Michelson interferometer, extends up to 4 THz. However, upon optimization of the electron beam compression specific for TeraFERMI, energies at source up to 100 μJ and a spectral extent up to 12 THz were measured [17]. Two characteristic spectra illustrating the performances under standard and THz beamline-dedicated machine conditions are shown in Figure 3.

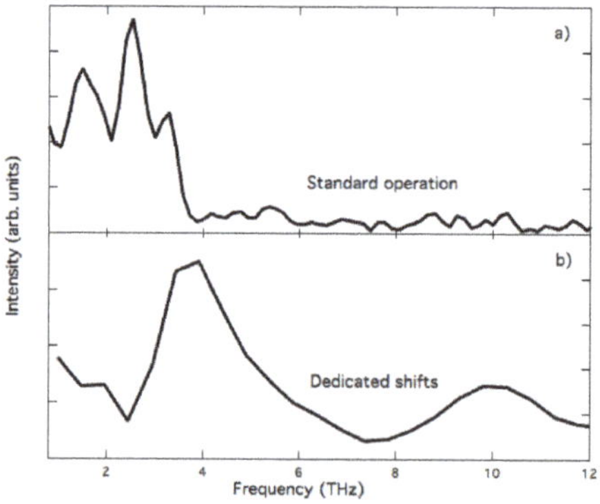

Figure 3. Characteristic TeraFERMI spectra, as measured with a Michelson interferometer in (**a**) standard operation conditions and (**b**) dedicated shifts. Figure adapted from Reference [17].

2.4. Optical Scheme

The full optical scheme for transmission spectroscopy experiments is depicted in Figure 4. The THz pulse is first polarized and attenuated by a series of three polyethylene grating polarizers. The first and third polarizers select the polarization required for the experiment under study, while the role of the second polarizer is to act as an attenuator, to allow studying the material's fluence dependence.

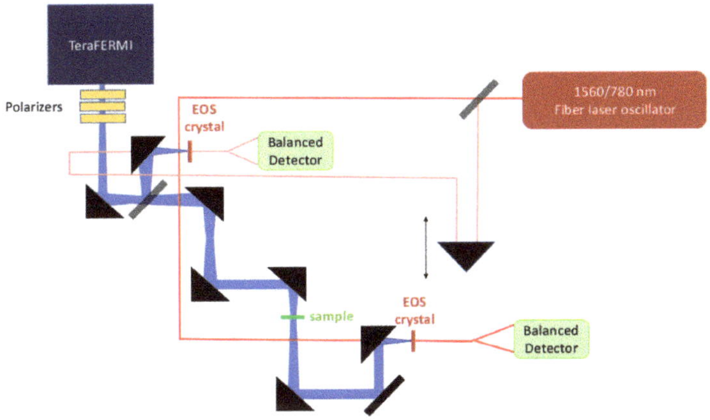

Figure 4. Schematics of the THz and optical beam paths. The three initial polarizers are used to set the polarization and attenuate the beam according to experimental requirements. The beam undergoes a series of expansions and recollimations in order to decrease its final size at sample position. Both THz and laser paths are splitted in two to allow for the simultaneous measurement of a reference spectrum.

After being transmitted by the sample the beam is first re-collimated and then focused by a slotted parabolic mirror on the electro-optic sampling crystal. At this point the THz pulse interacts with an optical pulse coming from an infrared fs laser synchronized to the free-electron-laser, as discussed in the next Section.

The main concept underlying EOS is that the THz electric field of the pulse under investigation induces a birefringence in the EOS crystal through the Pockels effect [18]. Thus, if the incoming laser pulse displays circular polarization, which can easily be achieved through a $\lambda/4$ waveplate, after the interaction with a crystal (in our case, a 1 mm thick ZnTe crystal) in presence of THz fields, the polarization will be turned into elliptical. This polarization change can be probed with a Wollaston prism (spatially separating the two orthogonal polarization components) and a balanced detector. Under appropriate approximations, the measured difference signal will then be proportional to the incoming THz field.

In order to allow measuring an online reference of the THz emission spectrum for the source, it is possible to split the THz beam as shown in Figure 4, with the help of a pellicle beam splitter. The infrared laser beam is also being splitted in two different components, one of which is delayed by a suitable amount in order to perform EOS on the reference channel.

3. The TeraFERMI Electro-Optic Sampling Set-Up

3.1. Laser Synchronization

In order to perform electro-optic sampling measurements (but also to provide time-resolved THz pump-IR probe information), TeraFERMI has been equipped with a commercial mode-locked fiber laser with ultrashort pulses (<100 fs FWHM) and two output wavelength, 780 nm and 1560 nm. A tight synchronization is required to perform time resolved experiments with FERMI THz light. To this purpose the highest performance time reference (LINK) is provided to the beamline optical table, for example, a stabilized optical pulsed signal from the Optical Master Oscillator (OMO) of FERMI.

A Phase Locked Loop (PLL) is used to stabilize the laser phase with respect to the reference. Two error signals feed the loop, one coming from an RF unit (TMU-RF) and one coming from a Balanced Optical Cross-Correlator (BOCC). The RF phase error signal is used for a first, rough synchronization thus making it possible to find the optical phase error signal, with a duration in the order of 1 ps (<0.1% of the pulse period). The optical signal is then used to achieve the final synchronization with few fs RMS jitter in the 10 Hz–10 MHz range.

Inside the BOCC setup we have installed a delay line to provide 6 ns delay (the period of the LINK pulses) with steps of 5 fs. This device is inside the synchronization loop because both the wavelengths of the pump probe laser are derived from the same oscillator, so that when the translation stage is moved the loop corrects the movement by acting on the piezoelectric motor inside the cavity. As a result, the phase of the 1560 nm and 780 nm signals are shifted exactly by the same delay imposed by the delay line. This technique is used to avoid perturbations due to moving elements on the optical path of the beam towards the sample.

The optimization of the LINK dispersion allows to obtain a single clean pulse to be used in the Sum Frequency Generation (SFG) process together with the 1550 nm pump-probe laser pulses inside the BOCC. To this purpose we have developed an optical scheme to measure trough cross-correlation the shape of the LINK pulses. This scheme will be used as BOCC for the synchronization of the laser.

3.2. Fast Pulse Detection System

Because of the mismatch between the 78.895 MHz repetition rate of the laser and the 50 Hz repetition rate of the FERMI facility, the detection system employed at TeraFERMI should be able to detect separately all the pulses from the infrared laser. To this aim we are employing a balanced photodetector with fast monitor output up to 350 MHz. The reading of the detector is then performed with the help of a 12 bit, 1 GHz digital oscilloscope. The sampling rate of 10 GS/s allows probing the signal from the photodetector with a 100 ps resolution.

At the maximum sampling rate the oscilloscope acquires the detector signal over a time range of 50 ns. This means that, due to the 78.895 MHz rep rate, we can continuously monitor the signal coming from 4 consecutive laser pulses. Obviously, only one out of the 4 pulses which are present on the oscilloscope's display can be affected by THz light. This means that the remaining three can be used for laser diagnostics. Their signal can be averaged and subtracted to the one which is affected by THz light. In this way we can cancel the residual signal due to laser beam fluctuations and non-perfect balancing between the two detector's channels.

4. Electro-Optic Sampling Results

Measured Electric Field and Spectrum

An example of EOS measurement during standard beamtime operations is shown in Figure 5. This profile was obtained by scanning the infrared laser with respect to the THz beam at 100 fs steps. For each position of the scanning delay line we have averaged over 10 shots. However, even with a lower (down to 1) number of averages the electric field profile can be clearly reconstructed, thus showing that jitter between THz and laser is not exceeding our step size. We will come back to this point in the following paragraph.

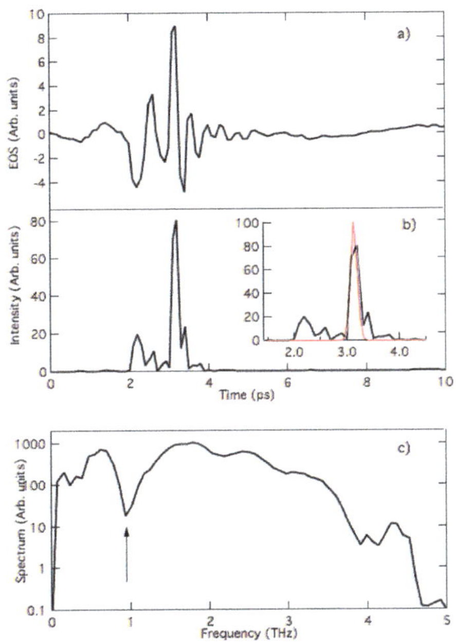

Figure 5. (**a**) Representative electro-optic-sampling (EOS) measurement, acquired by averaging 10 shots for each position of the delay line. (**b**) Intensity time profile obtained by evaluating the square of the EOS profile shown in (**a**). The red curve is a Gaussian fit to the main peak. (**c**) Intensity spectrum obtained by calculating the magnitude squared of the Fourier Transform of the spectrum in (**a**).The arrow indicates the dip in the spectrum at about 1 THz which may result from the interference of a long wavelength (~0.9 THz) and shorter wavelength (~2 THz) components.

Interestingly, the full energy profile (see Figure 5b) has a duration of about 2 ps, with a main sharp peak containing 60% of the energy of the full pulse. This THz peak is the proof of the presence of a complex electron bunch profile, compatible with the formation of a strong current spike due to the presence of Coherent Synchrotron Radiation (CSR) wakefields, as predicted in Reference [12]. The measured THz peak can be fitted with a gaussian profile, corresponding to a $\Delta t_{FWHM} = 112$ fs FWHM.

The power spectrum related to this electric field shape is reported in Figure 5c, in a logarithmic scale. The spectrum extends up to about 4.5 THz, which roughly corresponds to the maximum detectable signal in a ZnTe EOS crystal, due to the presence of an optical phonon centered at 5.3 THz [18]. The spectrum shows a pronounced dip at about 1 THz.

The electric field intensity can be determined by employing the standard relationship [19]:

$$E[MV/cm] = 0.39\sqrt{\frac{P[W]}{\pi(r_0[\mu m])^2}}. \tag{3}$$

The total intensity measured at focus with a calibrated pyroelectric detector in that configuration was $I = 4.8$ µJ. We can then calculate the peak pulse power as $P = 0.58 * I/\Delta t_{FWHM} = 25$ MW. By inserting this value in Equation (3), together with the proper r_0 for the measured spatial pulse profile (400 µm FWHM), we end up with an estimate of the THz electric field at 1.65 MV/cm. However it is important to keep in mind that the focus, as measured with the Pyrocam IIHR was characterised in atmospheric humidity conditions, when the presence of spectral components above 1 THz is almost totally absent due to water vapor absorptions. If we assume that the size of the focus scales linearly with the wavelength of the radiation, according to diffraction laws, the spatial profile of the spectral components between 1 and 4 THz may present a radius down to about 100–150 µm FWHM. According to Equation (3), this would imply reaching electric field values as high as 4.5–6 THz. This estimate clearly requires further investigations, for instance by characterising the beam profile under appropriate N2-purging conditions.

The knowledge of the field shape and value is of the highest importance for the interpretation of non-linear phenomena such as saturable absorption or harmonic generation, which depend directly on the electric field, through appropriate scaling laws. Also in pump-probe experiments, the knowledge of the exact shape of the field is important to precisely single out the rise-time and relaxation phenomena occurring as a consequence of the THz excitation.

To achieve a better understanding of the peculiar spectrum and electric field profile of TeraFERMI, we consider two THz single-cycle pulses centered at 0.9 and 2 THz, as depicted in red and blue respectively in Figure 6a. These two pulses mimic the expected composite shape of the bunch profile, being made up of a ∼100 fs high current spike superimposed on a ps-long electron bunch. With a suitable, phenomenological, choice of the relative time delay and phase, the sum of the two pulses yields the electric field profile of Figure 6b, which roughly reminds the EOS result already shown in Figure 5a and reproduced for clarity in Figure 6d. Interestingly the Fourier Transform of this composite pulse displays a remarkable dip at about 1 THz, in agreement with the experimental data, which is due to the interference of the electric fields from the two pulses at distinct frequencies. It is worth emphasizing that the presence of the short current spike is crucial, since it allows extending the TeraFERMI emission in the multi-THz range even when the compression of the electron bunch in the LINAC is not optimized for THz emission, as it is normally the case in parasitic mode operation.

Different machine conditions and in particular different LINAC settings can change the overall bunch compression and the electron energy spread at the entrance of the FEL. This will affect both the duration of the overall electron bunch and the duration and intensity of the current peak. While this can create a different balance in the spectrum between lower and higher frequency components, the presence of a dip at about 1 THz in the TeraFERMI spectra is very common, during normal user operation, thus underlining the validity of the two-component scenario. An EOS characterization of the TeraFERMI emission under optimized conditions would also be of high interest, to provide a better understanding of the spectral structures observed in Figure 3b) with the Michelson interferometer but is unfortunately still missing.

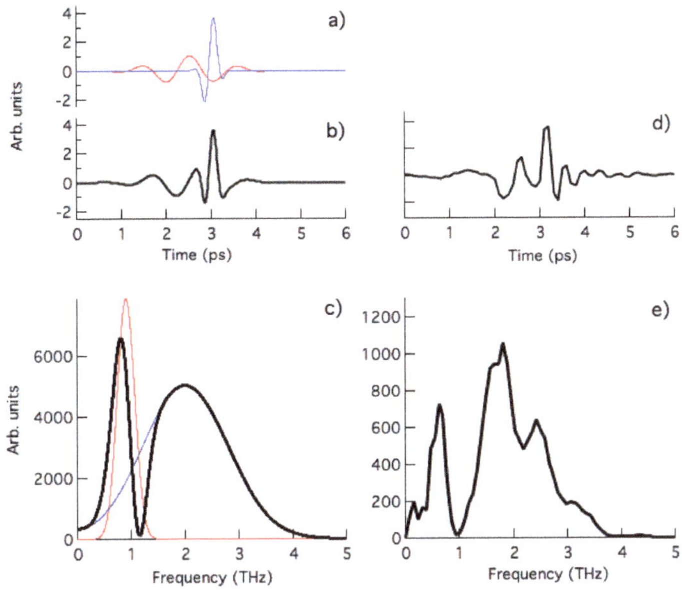

Figure 6. (**a**) Simulated THz waveforms centered at 0.9 (**red**) and 2 THz (**blue**), with a Gaussian envelope of width 1 ps and 200 fs respectively. (**b**) Simulated THz waveform resulting from the sum of the two previous contributions (**c**) Intensity spectrum of the 0.9 THz and 2 THz pulses separately and of the sum pulse black. An interference is clearly seen in the black spectra slightly above 1 THz. (**d**) Experimentally measured EOS profile (same as in Figure 5a). (**e**) Experimental spectrum (same as in Figure 5c).

The measured electric field profile as shown above is affected by many broadening effects which are hampering the detection of high frequency components. In particular the laser pulse duration (90 fs RMS) and the jitter in the electron bunch arrival time (<65 fs) are both affecting the measured electric field line-shape. This effect can be mimicked in terms of Gaussian convolutions of the measured electric field profiles. To better quantify the phenomenon we have modelled the electric field shape in terms of discontinuous flat lines, such as their Gaussian convolution provides an electric field profile similar to the one we measured (see Figure 7). The spectrum associated with the modelled electric field provides a much more enhanced signal from 1 to 3 THz. Even though this is obviously an oversimplified model, it may give an idea of the amount of light produced from our source which could still go undetected by our set-up.

Besides the Gaussian broadening effects discussed above other limiting factors could be the velocity mismatch between the THz and optical pulses in the 1 mm thick ZnTe crystal [18] and the already mentioned absorption from ZnTe itself due to the presence of its 5.3 THz optical phonon.

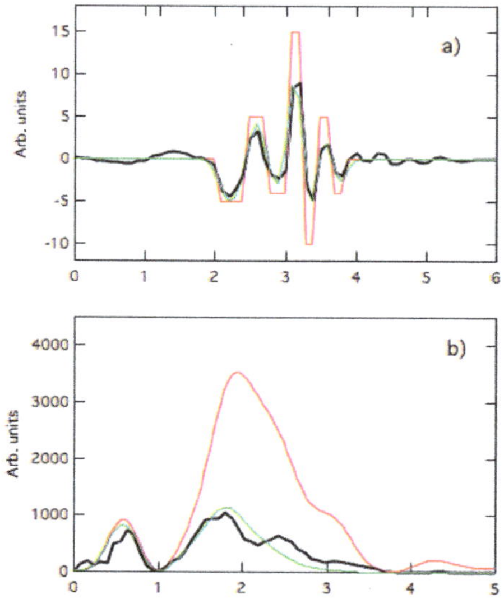

Figure 7. (**a**) Experimental EOS profile (black). Model electric field in the form of broken flat lines (red). Convolution of the model field with a 90 fs Gaussian function (green) (**b**) Intensity spectra calculated from the experimental EOS profile (black), from the broken line (red) and from its Gaussian convolution (green). The components at higher frequencies (i.e., from 1 to 4 THz) are clearly suppressed due to the Gaussian convolution which is mimicking the effects of the laser broadening.

5. Conclusions

We have reviewed in this paper the properties of the TeraFERMI beamline for THz non-linear and pump-probe studies. In particular we have presented the set-up presently in use for electro-optic-sampling measurements. This allowed characterizing in detail the THz electric field emission from TeraFERMI, which is a crucial information for the interpretation of non-linear THz spectroscopy and pump-probe dynamics. We have found that the usual THz emission from TeraFERMI is the combination of two main components originating by the THz emission from the ps-long electron bunch profile summed to the one from a ~100 fs long high current spike. This finding is in agreement with previous electron beam dynamics studies [12], pointing to the essential role played by CSR induced wakefields in the high frequency emission properties of TeraFERMI.

Author Contributions: Conceptualization, S.L. and A.P.; Investigation, N.A., P.D.P., S.D.M. and A.P.; Methodology, P.C., P.S., S.S. and M.V.; Software, F.P.; Writing—original draft, A.P.; Writing—review and editing, S.L. All authors have read and agreed to the published version of the manuscript.

Funding: This research received no external funding.

Acknowledgments: We acknowledge F. Vitucci from Crisel Instruments and i2S for providing the TZcam. We are also indebted with F. Novelli for useful discussions and suggestions in the preparation of the optical set-up.

Conflicts of Interest: The authors declare no conflict of interest.

Abbreviations

The following abbreviations are used in this manuscript:

FEL	Free Electron Laser
EOS	Electro Optic Sampling
CSR	Coherent Synchrotron Radiation
CTR	Coherent Transition Radiation
YAG	Yttrium Aluminium Garnet
OMO	Optical Master Oscillator
PLL	Phase Locked Loop
BOCC	Balanced Optical Cross-Correlator
SFG	Sum Frequency Generation

References

1. Salén, P.; Basini, M.; Bonetti, S.; Hebling, J.; Krasilnikov, M.; Nikiting, A.Y.; Shamuilov, G.; Tibaid, Z.; Zhaunerchyki, V.; Goryashko, V. Matter Manipulation with Extreme Terahertz Light: Progress in the enabling THz technology. *Phys. Rep.* **2019**, *836–837*, 1–74. [CrossRef]
2. Kampfrath, T.; Tanaka, K.; Nelson, K.A. Resonant and nonresonant control over matter and light by intense terahertz transients. *Nat. Photonics* **2013**, *7*, 680. [CrossRef]
3. Carr, G.L.; Martin, M.C.; McKinney, W.R.; Jordan, K.; Neil, G.R.; Williams, G.P. High-power terahertz radiation from relativistic electrons. *Nature* **2002**, *420*, 153. [CrossRef] [PubMed]
4. Shen, Y.; Watanabe, T.; Arena, D.A.; Kao, C.-C.; Murphy, J.B.; Tsang, T.Y.; Wang, X.J.; Carr, G.L. High-power terahertz radiation from relativistic electrons, Nonlinear Cross-Phase Modulation with Intense Single-Cycle Terahertz Pulses. *Phys. Rev. Lett.* **2007**, *99*, 043901. [CrossRef] [PubMed]
5. Gensch, M.; Bittner, L.; Chesnov, A.; Delsim-Hashemi, H.; Drescher, M.; Faatz, B.; Feldhaus, J.; Fruehling, U.; Geloni, G.A.; Gerth, C.; et al. New infrared undulator beamline at FLASH. *Infrared Phys. Technol.* **2008**, *51*, 423. [CrossRef]
6. Chiadroni, E.; Bacci, A.; Bellaveglia, M.; Boscolo, M.; Castellano, M.; Cultrera, L.; Di Pirro, G.; Ferrario, M.; Ficcadenti, L.; Filippetto, D.; et al. The SPARC linear accelerator based terahertz source. *Appl. Phys. Lett.* **2013**, *102*, 094101. [CrossRef]
7. Green, B.; Kovalev, S.; Asgekar, V.; Geloni, G.; Lehnert, U.; Golz, T.; Kuntzsch, M.; Bauer, C.; Hauser, J.; Voigtlaender, J.; et al. High-Field High-Repetition-Rate Sources for the Coherent THz Control of Matter. *Sci. Rep.* **2016**, *6*, 22256. [CrossRef] [PubMed]
8. Gensch, M. Super-radiant THz facilities: New opportunities for studying selective THz control of matter. *Proc. SPIE* **2018**, *10756*, 1075606.
9. Zapolnova, E.; Golz, T.; Pan, R.; Klose, K.; Schreiber, S.; Stojanovic, N. THz pulse doubler at FLASH: Double pulses for pump-probe experiments at X-ray FELs. *J. Synch. Rad.* **2017**, *25*, 39. [CrossRef] [PubMed]
10. Allaria, E.; Appio, R.; Badano, L.; Barletta, W.A.; Bassanese, S.; Biedron, S.G.; Borga, A.; Busetto, E.; Castronovo, D.; Cinquegrana, P.; et al. Highly coherent and stable pulses from the FERMI seeded free-electron laser in the extreme ultraviolet. *Nat. Photonics* **2012**, *6*, 699. [CrossRef]
11. Perucchi, A.; Di Mitri, S.; Penco, G.; Allaria, E.; Lupi, S. The TeraFERMI terahertz source at the seeded FERMI free-electron-laser facility. *Rev. Sci. Instr.* **2013**, *84*, 2. [CrossRef] [PubMed]
12. Di Mitri, S.; Perucchi, A.; Adhlakha, N.; Di Pietro, P.; Nicastro, S.; Roussel, E.; Spampinati, S.; Veronese, M.; Allaria, E.M.; Badano, L.; et al. Coherent THz Emission Enhanced by Coherent Synchrotron Radiation Wakefield. *Sci. Rep.* **2018**, *8*, 11661. [CrossRef] [PubMed]
13. Landau, L.D.; Lifshitz, E.M. *Electrodynamics of Continuous Media*; Pergamon: New York, NY, USA, 1960.
14. Casalbuoni, S.; Schmidt, B.; Schmüser, P.; Steffen, B. Far-Infrared Transition and Diffraction Radiation Part I: Production, Diffraction Effects and Optical Propagation. *TESLA Rep.* **2005**, *15*, 2012.

15. Moreno, T.; Westfahl, H.; de Oliveira Freitas, R.; Petroff, Y.; Dumas, P. Optical layouts for large infrared beamline opening angles. *J. Phys. Conf. Ser.* **2013**, *425*, 142003. [CrossRef]
16. Svetina, C.; Mahne, N.; Raimondi, L.; Perucchi, A.; Di Pietro, P.; Lupi, S.; Schmidt, B.; Zangrando, M. Photon transport of the superradiant TeraFERMI THz beamline at the FERMI free-electron laser. *J. Synch. Rad.* **2016**, *23*, 106. [CrossRef] [PubMed]
17. Di Pietro, P.; Adhlakha, N.; Piccirilli, F.; Capasso, L.; Svetina, C.; Di Mitri, S.; Veronese, M.; Giorgianni, F.; Lupi, S.; Perucchi, A. TeraFERMI: A superradiant beamline for THz nonlinear studies at the FERMI free-electron-laser facility. *Sync. Rad. News* **2017**, *30*, 36–39. [CrossRef]
18. Casalbuoni, S.; Schlarb, H.; Schmidt, B.; Schmüser, P.; Steffen, B.; Winter, A. Numerical Studies on the Electro-Optic Sampling of Relativistic Electron Bunches. In Proceedings of the 2005 Particle Accelerator Conference, Knoxville, TN, USA, 16–20 May 2005.
19. Van Tilborg, J. Coherent terahertz radiation from laser-wakefield-accelerated electron beams. *Eindhoven Tech. Univ. Eindhoven* **2006**. doi:10.6100/IR609970

condensed matter

MDPI

Article

Novel Schemes for Compact FELs in the THz Region

Andrea Doria *, Gian Piero Gallerano and Emilio Giovenale

ENEA, Fusion Physics Division, C.R. Frascati, via E. Fermi 45, I-00044 Frascati (Roma), Italy; gianpiero.gallerano@enea.it (G.P.G.); emilio.giovenale@enea.it (E.G.)
* Correspondence: andrea.doria@enea.it

Received: 23 October 2019; Accepted: 15 November 2019; Published: 19 November 2019

Abstract: The rapid advance of terahertz technologies in terms of radiation generators, systems, and scientific or industrial applications has put a particular focus on compact sources with challenging performances in terms of generated power (peak and/or average), radiation time structure, and frequency band tunability. Free electron laser (FEL)-based sources are probably the best candidates to express such a versatility; there are a number of schemes that have been investigated over the years to generate coherent radiation from free electrons in the mm-wave and terahertz regions of the spectrum, covering a wide frequency range from approximately 100 GHz to 10 THz. This paper proposes novel schemes for exploring the limits in the performance of radio frequency-driven free-electron devices in terms of ultrashort pulse duration, wide bandwidth operation, and energy recovery for near continuous wave (CW) operation. The aim of the present work is to demonstrate the feasibility of an FEL achieving performance comparable to a conventional photoconductive THz source, which is commonly used for time-domain spectroscopy (TDS), in terms of bandwidth and pulse duration. We will also demonstrate that a THz FEL could be very powerful and flexible in terms of tailoring its spectral features.

Keywords: terahertz; free electron laser; energy recovery

1. Introduction

Time domain-based terahertz (THz) sources have gained more and more attention during the past 15 or 20 years [1], and these systems are now commercially available, reliable, and easy to use. Such THz sources are mainly divided in two categories that are both characterized by broadband emission. The first is the Auston switch, which is named after the American researcher that initiated this research field [2]; the device is made up by a coplanar strip antenna deposited on a nonabsorbing photo-conducting substrate. The circuit, biased by an appropriate voltage, can be closed by a focalized powerful short infrared laser pulse on a gap of few microns. The transient current gives rise to a time-dependent electric field, and the consequent polarization vector follows the envelope of the laser pulse. The most important drawback of the Auston switch is its limited bandwidth and the frequency dispersion that results in a rapid lengthening of this pulse before the launching in the vacuum. The limit bandwidth for these devices is of the order of 1 THz. To overcome this limitation, a different approach has been developed. The idea is to use an electro-optic device realized with a nonabsorbing material [3]. The interaction mechanism is now a nonlinear second-order process: the rapidly oscillating electric field of the laser pulse excites in the medium a polarization vector which is proportional to the modulus of the electric field itself. This polarization vector is the source for a Hertzian vector potential that generates a radiation field. This is a rectification process because the rapid oscillation of the electric field of the laser pulse is averaged in the frequency domain and only the envelope of this pulse remains. Being the medium weakly absorbing, the polarization vector can be very fast in following the laser profile; this means that there are a few limits in the bandwidth of the THz pulse that can be generated with such devices. The main drawback is that, since this is a second-order process, the intensities of the THz radiation are reduced with respect to the Auston switch, but high intensities can be achieved by the more powerful short-pulse infrared laser source now available also commercially. A complementary and alternative approach, based on free electron devices, is discussed in the present paper.

2. The Free Electron Laser

In order to understand how a free electron laser can perform in a way similar to the time-domain THz sources, let us recall the main elements constituting a free electron laser (FEL). An FEL uses an electron beam generated by a particle accelerator that, due to the interaction with an external element, can emit radiation. To be more explicit, the electrons—which have several MeV of energy, properly aligned, and focused by steering coils and quadrupoles—are transported into the interaction region that in "conventional" FELs it constituted by a permanent magnet device called a magnetic undulator. Here, the electron kinetic energy is transformed in a radiation field by means of the synchrotron emission mechanisms, which can be considered a spontaneous emission process. With the addition of an optical cavity, the radiation is stored into a volume; therefore, the emission mechanism becomes stimulated by the simultaneous presence of the electrons, the magnetic field, and the stored radiation.

$$\begin{cases} \lambda = \frac{\lambda_u}{2\gamma^2}(1+K^2) & \text{Resonance Condition} \\ \eta \propto \frac{1}{2N} & \text{Efficiency} \\ G \propto \frac{N^3 I}{\gamma^5} g(\theta) & \text{Gain} \end{cases} , \quad (1)$$

Relevant FEL relations are reported in Equation (1) indicating that the resonant wavelength is inversely proportional to the second power of the electron energy (expressed by the relativistic factor γ) and directly proportional to the undulator period λ_u and the magnetic field strength contained in the undulator parameter K. The efficiency is inversely proportional to the number of the undulator period N, and lastly, the gain is proportional to the electron beam current I and to the undulator length at the third power, but also inversely proportional to the electron energy at the fifth power. Moreover, the gain is proportional the function $g(\theta)$ of the detuning parameter $\theta = 2\pi N(\omega - \omega_0)/\omega_0$ (where ω_0 is the spontaneously emitted frequency). It is important to point out that the synchrotron emission is not the exclusive emission mechanism for free electron devices, but is the one for which the interaction between the particle and field is more effective.

Therefore, it is possible to analyze the characteristics that an FEL should satisfy if designed to operate in the THz region. First of all, looking at the first part of Equation (1), the electron beam energy can be quite low, with respect to the visible and UV operations, and this allows the design of a compact FEL layout with small particle accelerators. A broadband emission that is typical of the conventional solid-state THz sources can be obtained by means of short interaction regions, which clearly add compactness to the overall device. This can be acceptable because the high gain is ensured by the low energy beam (see the third part of Equation (1)). To maintain the bandwidth as broad as possible, we must avoid the use of an optical cavity, which certainly reduces the frequency components, and obtain emission in a single electron passage. The lack of the optical cavity allows a straight-line e-beam propagation. In order to extract as much power as possible in a single-pass device, we have to work in the coherent spontaneous emission regime and super-radiant regime.

3. Coherence and Order in FELs

The previous section discussed the necessity of exploiting all the coherent mechanisms in order to avoid the use of an optical cavity for a THz FEL. The first degree of coherence is obtained when the electron bunch length is comparable to the wavelength of the radiation to be generated by the device. This mechanism was firstly proposed by Schwinger for the analysis of synchrotron radiation [4] and by V.L.Ginzburg to explain strong radiation emissions from astronomic objects [5]. From the classical electromagnetic theory, it can be demonstrated [6] that the emission into a solid angle $d\Omega$ in the frequency interval $d\omega$ is

$$\frac{dI}{d\omega d\Omega} = [N + N(N-1)f(\omega)]\frac{q^2\omega^2}{4\pi^2 c}\left|\int_{-\infty}^{\infty} \hat{n} \times \left(\hat{n} \times \vec{\beta}\right) e^{i\omega[t - \frac{\hat{n}\cdot\vec{r}(t)}{c}]} dt\right|^2. \quad (2)$$

The function $f(\omega)$ in Equation (2) describes the coherence of the emitted radiation; its values range from 0 to 1. The limit $f(\omega) = 0$ indicates an incoherent emission from the bunch that means that the power emitted by N particles is just the sum of N individual contributions. The upper limit $f(\omega) = 1$ represents the total coherence, when the power emitted by N charged particles is given by N^2 times the emission of the single particle. This result is fully equivalent to the emission of a single particle having a charge $Q = Nq$, because looking at Equation (2), power is proportional to $Q^2 = N^2q^2$. In order to establish a suitable value for $f(\omega)$, we can observe from Equation (2) that coherent effects dominate when $N << N^2 f(\omega)$, which implies $f(\omega)N >> 1$; for the typical case of an S-band radio-frequency (RF) accelerator, the number of electrons in a single bunch is about $N \sim 10^8$; therefore, a suitable value range for the $f(\omega)$ function that guarantees high coherence in the emission starts from $f(w) >> \sim 10^{-4}$. This implies that the wavelength of the radiation generated should be longer than a fraction of the bunch length in order to have a coherent enhancement of the emission. Considering again an S-band accelerator, for which the bunch length is of the order of few millimeters, the result is that coherent emission dominates in the THz or at longer wavelengths.

A second degree of coherence can be exploited taking into account the relationship among all the electron bunches. In fact, an RF accelerator generates a train of pulses, and if the correlation among all the consecutive pulses is good, the radiation will be emitted at discrete frequencies that are harmonics of the RF [7]. This can be easily seen by making a Fourier expansion of the current, as shown in Equation (3)

$$\vec{J}_T = \sum_{l}^{\infty} \vec{J}_{Tl} \exp(-i\omega_l t) \ ; \ \omega_l = 2\pi \frac{l}{T_{RF}}, \quad (3)$$

where T_{RF} is the RF period and l is the harmonic number. The spectral width of the single harmonic is related to the overall length of the bunch train, the so-called macropulse, as well as to the amplitude and phase stability of the bunches over the macropulse. Each harmonic component of the current acts as an isolated source that couples with the electric field of the radiation, which is expanded over the free space, or guiding structure modes E_λ [7]

$$A_{l,\lambda} = -\frac{Z_0}{2\beta_g}\int_V \vec{J}_{Tl} \cdot \vec{E}_\lambda \, d^3x, \quad (4)$$

where β_g is the group velocity of the radiation pulse and Z_0 is the vacuum impedance. The radiated power is calculated by means of the flux of the Poynting vector: $P_{l,0,n} = \beta_g/(2Z_0)\left|A_{l,0,n}\right|^2$.

A further degree of coherence can be added if we treat the bunch as a collection of N particles distributed in the longitudinal phase space [8]. Each electron carries its own energy γ_j and a proper position or phase ψ_j equal to the RF times the time t_j along the bunch. Using the previous expression (4), the total radiated power can be calculated summing over the particle distribution

$$A_{l,0,n} = -\frac{Z_0}{\beta_g}\frac{q}{T_{RF}}\frac{KL}{\sqrt{\Sigma}}F\sum_{j=1}^{N_e}\frac{1}{\beta_{zj}\gamma_j}\frac{\sin(\theta_{lj}/2)}{\theta_{lj}/2} i e^{i(\frac{\theta_{lj}}{2} + l\psi_j)} \ ; \ \theta_{lj} = \left(\frac{\omega_l}{c\beta_{zj}} - \frac{2\pi}{\lambda_u} - k_{0,n}\right), \quad (5)$$

where Σ is the radiation mode size and $k_{0,n}$ is the transverse mode momentum. From Equation (5), we see quite explicitly that the presence of a phase factor gives rise to an interference mechanism; the contributions to the total power, given by the sum over all electrons, indicated by the index "j", may result in constructive or destructive interference. To maximize the power extraction, by means of the

Poynting theorem, it is possible to minimize the negative interference. This goal can be achieved when the phase term in Equation (5) is overall constant with respect to the single electron emission and will only depend by the harmonic number l influence. In conclusion, the ideal phase-matching distribution, which maximizes the power extraction by means of the minimization of the destructive interference, is obtained by putting the phase term equal to a constant φ_l that depends only by the harmonic number. The ideal distribution for the electrons, in the longitudinal phase space, is therefore as follows:

$$\left(\frac{\theta_{lj}}{2} + l\psi\right) = \phi_l. \tag{6}$$

When the electrons in a bunch are distributed in the longitudinal phase space as in Equation (6), the radiation emission due to each particle is added in phase, as indicated by Equation (5), maximizing the power extraction during the interaction process. A different way to understand the meaning of the ideal energy-phase correlation curve is that, during the passage of the electrons inside the magnetic undulator, the combined ballistic and oscillating motion is such that the maximum bunching (and consequently, the maximum energy spread and the minimum temporal width) is obtained when the bunch reaches the center of the interaction region. So, the ideal phase-matching curve is the one for which the distribution realizes the best bunching at the undulator center [8] (see Figure 1).

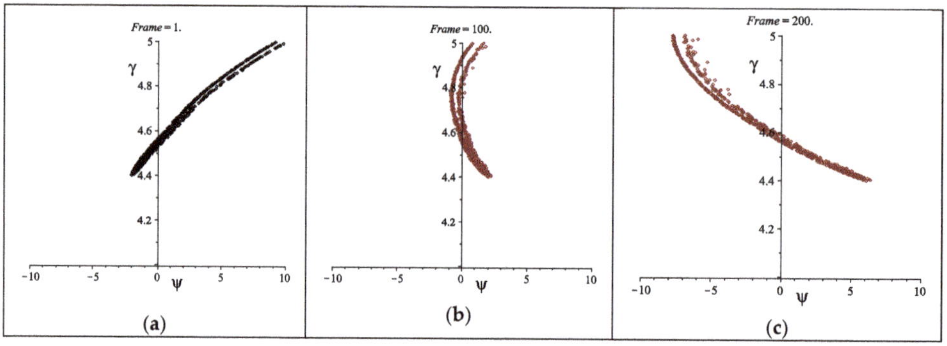

Figure 1. Electron distribution in the longitudinal phase space while traveling along the magnetic undulator. Frame = 1 (**a**) represents the undulator entrance; Frame = 100 (**b**) represents the undulator center; and Frame = 200 (**c**) represents the undulator end.

In the ENEA Research Centre in Frascati, two FEL compact devices have been realized [9,10], in order to test the three described mechanisms for the generation of coherent radiation from free electrons. In particular, the FEL-CATS (Compact Advanced THz Source) [10] has been realized to test the energy-phase correlation mechanism for the emission of coherent radiation, up to saturation levels, without the use of any optical cavities. This FEL present an element called a phase-matching device (PMD) that is capable of manipulating the electron beam in the phase space. The PMD is an RF device located downstream of the Linac accelerator (2.3 MeV) in which the RF field amplitude and phase, relative to the Linac, can be controlled. This control allows the reference electron passing through the center of the PMD with a phase close to zero. The result is that the higher energy electrons, arriving first at the PMD, are decelerated, while the lower energy electrons are accelerated and emerge first from the PMD. This is exactly the required rotation of the electron distribution in the phase space. The FEL-CATS operates from 450 to 800 μm (corresponding to frequencies from 0.4 to 0.7 THz).

4. The Two-Frequency Cavity

In order to realize FEL sources generating radiation in the THz range with better performance than FEL-CATS, it is necessary to increase the electron energy (see Equation (1)) and to shorten the bunch duration (see Equation (2)). The PMD realized in ENEA for the FEL-CATS experiment is no longer adequate for such a task. A possible solution to this problem is the use of a two-frequency RF device [11,12]. In fact, such cavities offer appealing possibilities to control the bunch length of an electron beam generated by an RF accelerator. The use of a double-frequency cavity requires the second frequency to be a harmonic of the fundamental one, in order to be resonant. The electrons passing through the cavity will see a field that is the result of the sum of the two field components: the fundamental and its harmonic.

$$V(\phi) = V_0[\sin(\phi + \phi_s) + k\sin(n\phi + \phi_n)] \tag{7}$$

There are some design parameters that can be set, such as the harmonic number n and the relative amplitude ratio k that contributes to the bunch length, but the most relevant is the relative phase (φ_s-φ_n) between the fundamental and the harmonic. These parameters establish the slope of the total field in the cavity right where the electron bunch passes. Two different regimes can be in principle realized: one in which the particles gather with respect to the reference electron, and a second in which one can realize a pulse lengthening, whose spread can help dampen coherent instabilities that often cut down the bunch lifetime in recirculated accelerators. In Figure 2, the behavior of the total field in the two-frequency cavity is reported, as indicated by Equation (7), together with the fundamental frequency field and the harmonic frequency field.

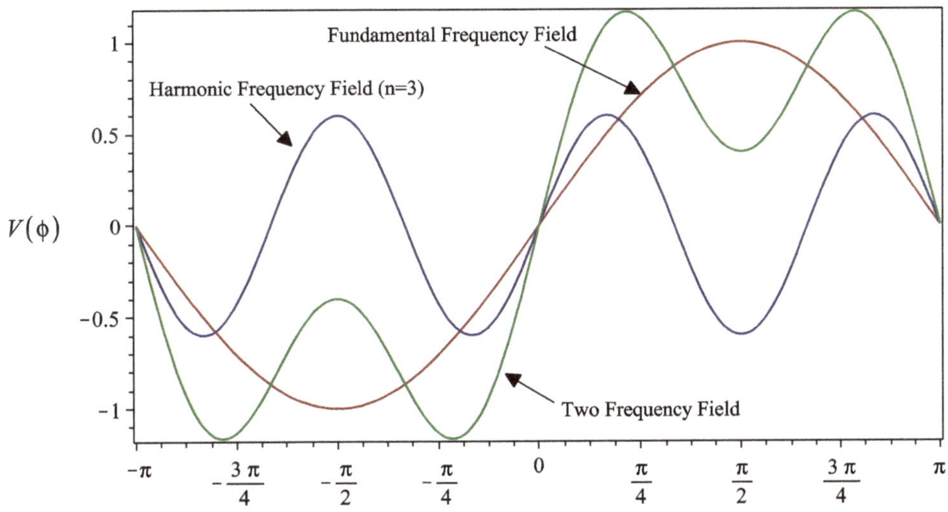

Figure 2. Electric potential in a two-frequency cavity as a function of the phase φ for the fundamental frequency, the third harmonic, and the resultant field given by Equation (7).

The figure reports the ideal conditions to obtain the best bunching close to the reference electron, for a third harmonic field (n = 3) and a relative amplitude between the fundamental and harmonic of $k = 0.6$; the phases for the two fields are $\varphi_s = 0$ and $\varphi_n = \pi$. In this case, the total field results in a slope around the reference particle, whose steepness can be controlled by the amplitude of the third harmonic. This is the required solution for the bunching of the electrons up to the ideal phase-matching distribution reported in Equation (6).

From the calculation of the separatrix of the electron distribution in the longitudinal phase space, it can be demonstrated [11] that the bunch profile is given by the following expression.

$$I(\phi,n) = A\exp\left\{\frac{1}{\sigma_{\phi 0}\cos\phi_S}\left[\cos(\phi+\phi_S) - \cos\phi_S + \phi\sin\phi_S - \frac{n}{k}(\cos(n\phi)-1)\right]\right\} \approx \approx A\exp\left[-\left(1-\frac{kn}{\cos\phi_S}\right)\frac{\phi^2}{\sigma_{\phi 0}}\right] \quad (8)$$

The final expression is obtained by a Taylor expansion of the trigonometric functions around the bunch center; the profile, therefore, results in a more simple Gaussian form. With the parameters used for the evaluation of Figure 2 in terms of phase difference and amplitude ratio, we display in Figure 3 the bunch profile for the fundamental harmonic only and for the simultaneous presence of the third or fifth harmonics. It is clear that the higher the harmonics content, the shorter the bunch will be at the cavity exit, due to the increased steepness of the resultant field around the reference electron.

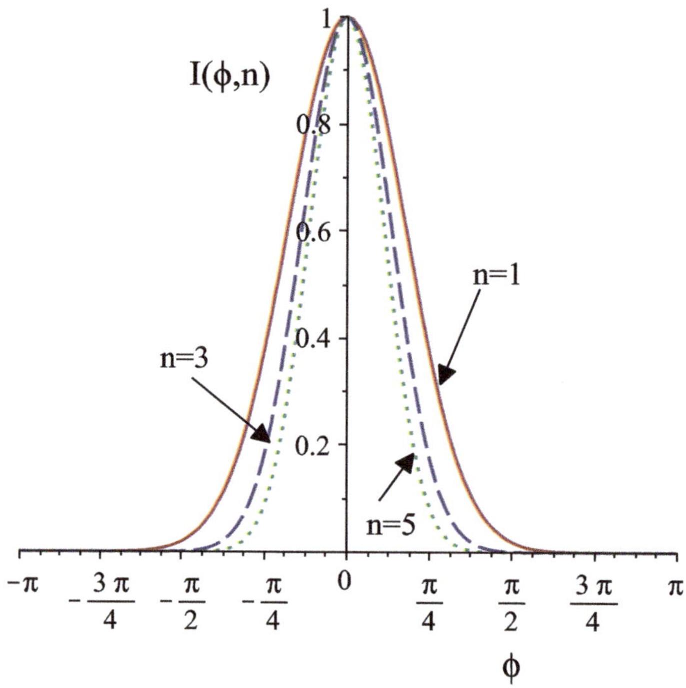

Figure 3. Longitudinal electron bunch profile expressed in terms of the phase φ, for the fundamental frequency and for the third and fifth added harmonics.

5. FEL as a THz Radiator

In order to design an FEL operating at THz frequencies, simulation software that is capable of evaluating several of the characteristics of FEL sources has been developed. This code makes use of the electric field generated inside the double-frequency cavity, which is used as a phase-matching device, as described in the previous paragraph. The simulation considers an FEL device based on the main parameters reported in Table 1. The electrons generated and accelerated by a Linac are injected into the two-frequency cavity that is fed with the fundamental (3 GHz) and fifth harmonic and that is set, according to Table 1, in order to distribute the electrons according to Equation (6).

The simulation reported in Figure 4 shows the electron distribution in the longitudinal phase space at the cavity exit. During the transport toward the magnetic undulator, the particles continue to bunch due to ballistic effects; therefore, the peak current continues to increase. We have previously seen that the electron should enter the wiggler with the ideal distribution in the phase space to reduce the negative interference in order to enhance the extraction efficiency during its passage in the interaction region. To do so, a variable free propagation space may be considered a further optimization tool.

Table 1. Main parameters of the free electron laser (FEL) simulation.

Parameter Definition	Parameter Value
Electron Energy	$E = 7.5$ [MeV] ($\gamma = 15$)
Average Current (over the Macropulse)	$I_{AV} = 560$ [mA]
Peak Current (over the Microbunch after the Two-Frequency Cavity)	$I_p = 77$ [A]
Microbunch Duration	$\tau_{el} = 700$ [fs]
Radio Frequency (RF)	$\nu_{RF} = 3$ [GHz]
Harmonic Number	$n = 5$
Harmonic Relative Amplitude	$k = -0.6$
Phase of the Fundamental (RF)	$\emptyset_s = 0$
Phase of the Harmonic	$\emptyset_n = \pi$
Undulator Period	$\lambda_u = 2.5$ [cm]
Undulator Parameter	$K = 1.45$
Number of Undulator Periods	$N = 5$
Waveguide Vertical Gap	$b = 0.154$ [cm]

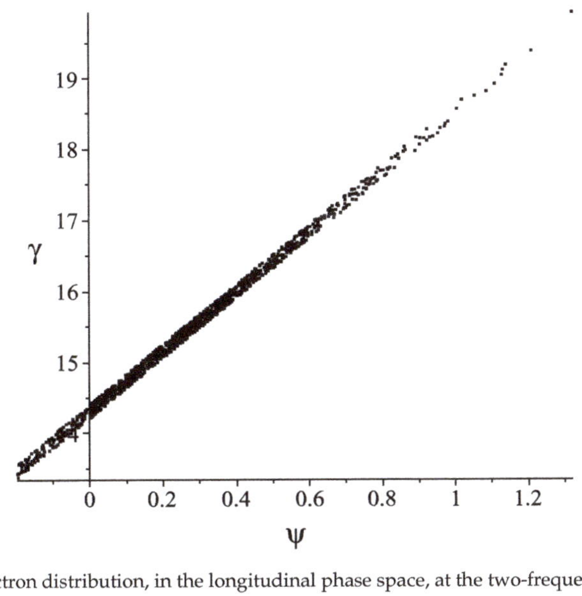

Figure 4. Electron distribution, in the longitudinal phase space, at the two-frequency cavity exit.

To clearly understand the effects of the drift space, we have analyzed the modification of the output power spectrum, averaged over the macropulse, during the propagation in a drift space, between the end of the two-frequency cavity and the undulator entrance, over a range of 36 cm. Analyzing Figure 5 (column (a)), we can observe that degradation occurs starting from about 7–8 cm from the cavity exit, and that at the end of the drift, it is quite significant. This phenomenon can also be observed by analyzing the single radiation peak pulse and the associated electric field burst; a reduction in the power of about 50%, in the worst situation, is expected (see Figure 5 column (b)). An optimization can be performed acting on the parameters of the two-frequency cavity, such as the voltage on the fundamental and its harmonic, or their relative phase, in order to reduce the rotation in the phase space and let the ballistic propagation complete such rotation and realize the ideal phase-matching distribution. After the proper optimization, we obtain a spectrum that ranges from 0.5 to 1.5 THz with an integrated power over the macropulse duration and over the total spectral bandwidth, of about 90 kW (see Figure 6). It is very important to stress that with this device, due to the RF properties of the accelerator, which are expressed also by Equations (3)–(5), it is possible to isolate the single harmonic with an interferometer, for instance, still having an average power for the single frequency of the order of hundreds of watts. This is not possible with conventional THz sources. Moreover, another interesting result is that the single frequency, being a harmonic of the RF, has a temporal structure equal to that of the RF macropulse. In addition, we have to refer to the RF macropulse for its temporal coherence, which for conventional magnetrons and klystrons is usually quite good. On the other hand, if we look at the whole bandwidth, the temporal structure is the well-known train of microbunches separated by the RF period.

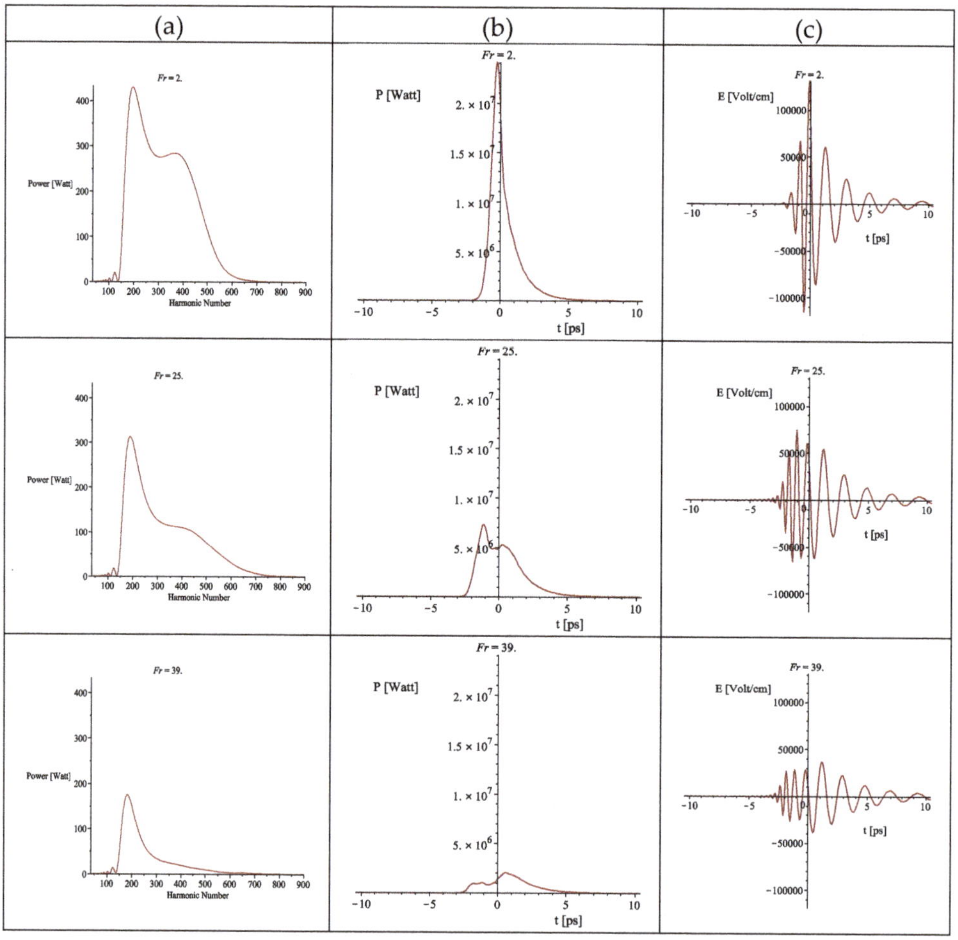

Figure 5. Simulation of the effect of a drift space in the terahertz (THz)-FEL emission. The three rows refer to three cases: Fr = 2 denotes 0 cm of drift space; Fr = 25 indicates 18 cm of drift space, and Fr = 39 represents 36 cm of drift space. Column (**a**) reports the radiation power spectrum; column (**b**) indicates the radiation peak power time profile, and column (**c**) shows the peak electric field as a function of time.

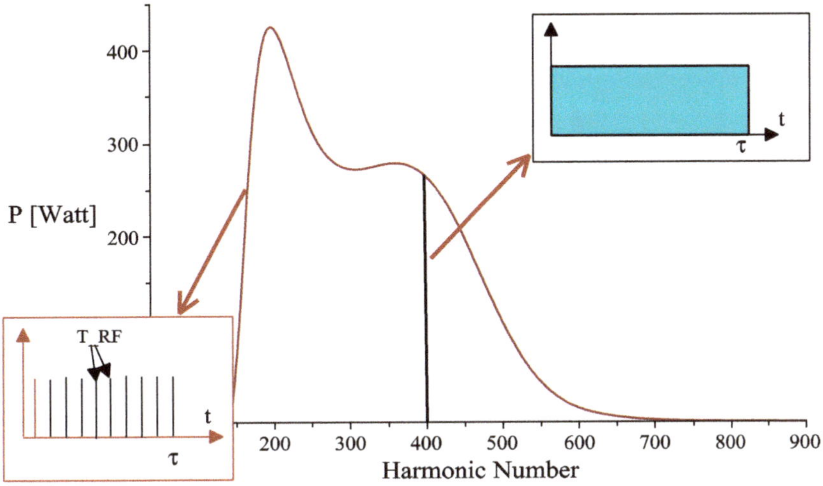

Figure 6. Simulation of the optimized radiation power spectrum emitted as a comb of frequencies related to the harmonics of the fundamental RF (ν_{RF} = 3 GHz).

In conclusion, an FEL THz radiator can operate as a natural "frequency-comb" emitter or a single-frequency emitter; therefore, it can be considered as a convenient, flexible, and powerful source for the generation of coherent radiation in the THz spectral region.

6. Energy Recovery

Another way to improve the performances of broadband THz FELs is related to their efficiency. One of the problems suffered by RF accelerator-based FELs, both normal conducting and superconducting, is their relatively low overall efficiency. Considering the basic generation mechanism, the power extraction from the electron beam in favor of the radiation never exceeds 1/2*N*, where *N* is the number of periods of the undulator for the low-gain regime, and ρ is the Pierce parameter for the high-gain regime (see [6,13]). In both cases, the best expected efficiency is of the order of few percent. A possibility to increase the efficiency of an FEL is to design a scheme for recovering the electron beam kinetic energy remaining after the undulator interaction. The energy recovered can be used to accelerate a new bunch of electrons, thus lowering the accelerator radio-frequency power requirements. The possibility of managing a lower RF power, and at the same time an exhausted electron beam of low energy, is very useful when designing both the thermal load of the device and the beam dump for the electron exhausting. Generally speaking, there are two possibilities for energy recovery (ER). One is to recycle the electron beam after the undulator interaction. The second possibility is to recover only the energy of the electrons before the beam dumps.

This second technique is particularly useful when the accelerator is designed to be a superconductor due to the reduced losses related to the shunt impedance. This is a relevant point because it allows presenting some interesting solutions. Any RF cavity exhibits power loss related to the ohmic dissipation through the shunt impedance. Considering standard cavity geometry, it can be demonstrated [14] that the shunt impedance per unit length times the cell dimension over the Q factor of the cavity is of the order of 150 Ohm ($(r_S/Q) \cdot d \approx 150\,ohm$). This leads us to a simplified but reasonable expression for the power loss as a function of frequency and surface resistance, which is again related to the frequency and conductivity of the material:

$$P_{Loss} = \frac{\pi c}{\omega} \frac{1}{L \cdot 150\,ohm} \frac{V^2 R_{Surf}}{G} \;\; ; \;\; R_{Surf} = \sqrt{\frac{\pi \mu \nu}{\sigma}}. \tag{9}$$

Here, V is the accelerating voltage, L is the accelerator length, R_{Surf} is the surface resistance, G is a factor, expressed in ohm, that depends on the RF cavity geometry, ν and ω are the frequency and the pulsation, respectively, and σ is the conductivity of the metal with which the Linac is realized.

Let us first analyze the case of a copper Linac operating in a normal conducting regime ($\sigma = 5.8 \times 10^7$ ohm^{-1}m^{-1}) at room temperature, and the cavities' geometry for which $G = 280$ ohm; the results are reported in Table 2 for two different operating RF frequencies.

Table 2. Linac power losses, at room temperature, for two radio frequency (RF) values.

Case 1:	$\nu = 3$ GHz	$L = 1$ m	$V = 7.5$ MV	$R_{Surf} = 1.43 \times 10^{-2}$ ohm	P_{Loss} **= 0.96 MW**
Case 2:	$\nu = 8$ GHz	$L = 1$ m	$V = 7.5$ MV	$R_{Surf} = 2.33 \times 10^{-2}$ ohm	P_{Loss} **= 0.58 MW**

For Case 1, at 3 GHz, we have about 1 MW of power loss, due to ohmic losses across the structure, before any power could be delivered for the acceleration of the beam. With the same parameters, but at 8 GHz of RF (Case 2), we get about one-half of the previous value. This means that if we want to accelerate a beam to a certain energy value, we first need to spend such an amount of power to charge the structure; the rest of the power of the RF source is delivered to the e-beam. The power loss scales with the inverse of the square root of the frequency.

If a Linac based on superconducting material, as niobium, is considered, we know that the two models, the Bardeen–Cooper–Schrieffer (BCS) theory and the two-fluid model [15], are well established and report the same results up to frequencies of about 10 GHz. The surface resistance is now expressed by the following relation:

$$R_{Sup} = A\frac{\omega^2}{T}e^{-\alpha\frac{T_c}{T}} + R_{res}. \tag{10}$$

From the experimental data and simulations reported in [15], the A and α parameters can be deduced, and therefore the surface resistance for the two RF regimes can be deduced as well. The power loss formal expression is the same with the only difference that the surface resistance increases more rapidly with the frequency; consequently, the power losses increase with the frequency as well. Table 3 reports the power loss data for the superconducting Linac, with the same gradient, at 4.2 K.

Table 3. Linac power losses, at liquid helium temperature, for two RF values.

Case 1:	$\nu = 3$ GHz	$L = 1$ m	$V = 7.5$ MV	$R_{Surf} = 2.50 \times 10^{-6}$ ohm	P_{Loss} **= 166 W**
Case 2:	$\nu = 8$ GHz	$L = 1$ m	$V = 7.5$ MV	$R_{Surf} = 1.75 \times 10^{-5}$ ohm	P_{Loss} **= 445 W**

The values reported are greatly reduced with respect to the normal conducting operation. By the RF source point of view, this amount of loss is negligible; all the RF power would be delivered to the beam in this case. The only problem is that this power must be thermodynamically managed at 4.2 K.

A third alternative is the use of normal-conducting metals, such as copper, at cryogenic temperatures. In fact, at low temperatures, the mean free path of electrons in metals increases; consequently, the electrons will not experience a constant electric field in a coherence length, but rather a variable one. These and other complex effects associated with the frequency, going under the name of an anomalous skin effect [16]. The result is that the classic ohm law, as we know it, is no longer valid, and the current must be calculated over the actual electron path. A simulation for a Linac with the previously reported values in Table 2 is displayed in Figure 7 for three frequencies.

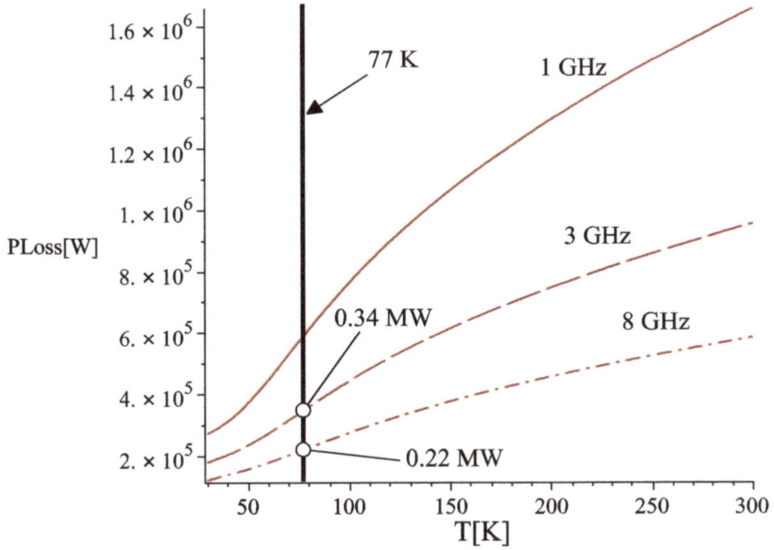

Figure 7. Linac power loss behavior as a function of temperature and frequencies as follows from the anomalous skin effect theory.

The power losses decrease with the temperature, as expected, but again, higher frequencies exhibit lower losses. At liquid nitrogen temperatures, the power losses amount to about one-third with respect to the room temperature case. This is a quite good improvement for any energy recovery regime.

Once the amount of ohmic losses on the RF cavities that we have to deal with (both for the Linac and for the two-frequency cavity used as PMD) is clear, it is possible to face the problem about which kind of ER schemes might be appropriate for an FEL operating in the THz range. The most used is the so-called "same-cell" scheme in which a single RF structure acts as an accelerator in a first passage and a decelerator in a second passage. This scheme is usually the most favored choice due to economical reasons; in fact, the costs of the RF structure, especially if realized with superconductors, are quite high, and any saving in the RF power should not relapse on the RF structures (see Figure 8a).

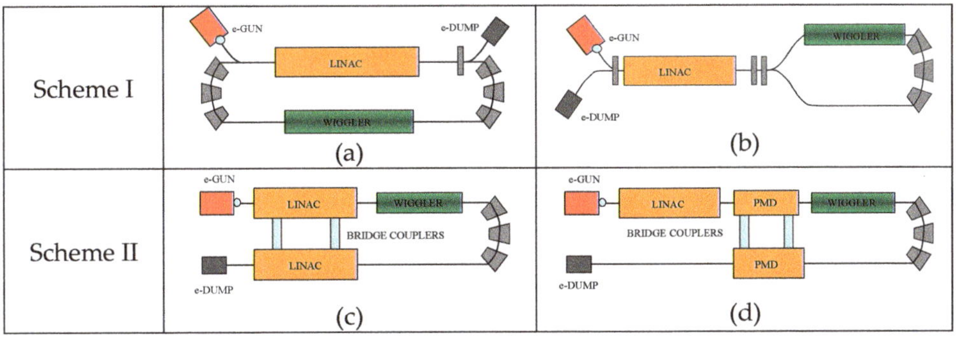

Figure 8. Energy recovery schemes suitable for THz-FELs. (**a**) "Same-cell" scheme. (**b**) Alternative "same-cell" scheme. (**c**) Two-cavity scheme with Bridge Couplers. (**d**) Two-cavity scheme with PMDs.

An alternative to the same-cell scheme is the one reported in Figure 8b; in this variant, the electrons, after the Linac, follow a symmetric path where the magnetic undulator for the FEL radiation generation is placed in one of the two branches.

A second ER geometry is obtained when the beam, after the FEL interaction, is directed through a different decelerating structure. Generally, this second RF structure is similar to the accelerating Linac to which it must be strongly coupled in order to efficiently transfer the RF energy (see Figure 8 (c)). The "different-structure" scheme appears to be expensive because of the use of a second Linac structure. However, due to what has been discussed in the previous sections, THz generation, by means of free electron devices, requires low energy electron beams; therefore, the costs of the RF structures can still be affordable. For this reason, this scheme is under investigation at Frascati for the FEL-CATS. Moreover, this second scheme offers us the possibility of coupling the power of the exhausted beam directly into the two-frequency RF cavity: the PMD (see Figure 8d). This is a very interesting opportunity because the electrons, after the passage into the cavity and into the undulator, carry a memory of such an interaction by means of the shape of the current. Furthermore, by the use of the presented model of harmonic expansion (see Equations (3)–(5)), the shape of the field generated inside the PMD will be better tailored for the enhancement of the efficiency in the radiation generation mechanism. A semianalytical and numerical model is under study.

7. Conclusions

Innovative design schemes for the realization of a compact FEL source for operation in the THz spectral region have been discussed in this work. The proposed device, unlike conventional FELs, is not tuned by varying the electron energy or the undulator gap; instead, it is designed to emit several frequencies, which are the harmonics of the accelerator RF, within a broad spectral band. These characteristics are obtained through working on the coherent properties of the electron beam generated in a RF accelerator together with a proper manipulation of the electron distribution. The goal is to get the ideal energy-phase correlation that allows saturation-level emission without the use of an optical cavity. The features obtained are unique: the presence of the RF harmonics, which are spectrally very narrow, and that can be easily extracted make this device suitable for any spectroscopic application that requires a high spectral power density. A significant example is the diagnostic of magnetically confined plasma. Moreover, the presence of many narrow spectral components (related to the fundamental RF that can be as stable as 1 part over 10^6) allows creating a combination of them in a frequency "comb" that finds many applications in metrology (optical clock) in spectral phase interferometry and spectroscopy. A wide variety of features in a single device represent uniqueness in the laser source scenario.

Author Contributions: A.D. and G.P.G. have developed the theory about the coherent emission also in the case of energy-phase correlated electron beam. A.D. realised all the simulation codes for the electron dynamics, radiation emission and power losses in the RF structures. E.G. has controlled and verified the experiment feasibility of the device proposed.

Funding: This research received no external founding.

Conflicts of Interest: The authors declare no conflict of interest.

References

1. Coutaz, J.L.; Garet, F.; Wallace, V.P. *Principles of Terahertz Time-Domain Spectroscopy*, 1st ed.; Jenny Stanford Publishing Pte. Ltd.: Singapore, 2018.
2. Auston, D.H.; Glass, A.M. Optical generation of intense picosecond electrical pulses. *Appl. Phys. Lett.* **1972**, *20*, 398–399. [CrossRef]
3. Wu, Q.; Zhang, X.C. Free space electro optic sampling of terahertz beams. *Appl. Phys. Lett.* **1995**, *67*, 3523–3525. [CrossRef]
4. Schwinger, J. On the Classical Radiation of Accelerated Electrons. *Phys. Rev.* **1949**, *75*, 1912–1925. [CrossRef]
5. Ginzburg, V.L.; Zheleznyakov, V.V. On the possible mechanisms of sporadic solar radio emission (radiation in an isotropic plasma). *Sov. Astron.* **1958**, *2*, 653–678.
6. Dattoli, G.; Doria, A.; Sabia, E.; Artioli, M. *Charged Beam Dynamics, Particle Accelerators and Free Electron Lasers, 1st ed*; IOP Publishing Ltd.: Bristol, UK, 2017.

7. Doria, A.; Bartolini, R.; Feinstein, J.; Gallerano, G.P.; Pantell, R.H. Coherent Emission and Gain from a Bunched Electron Beam. *IEEE JQE* **1993**, *29*, 1428–1436. [CrossRef]
8. Doria, A.; Gallerano, G.P.; Giovenale, E.; Letardi, S.; Messina, G.; Ronsivalle, C. Enhancement of Coherent Emission by Energy-Phase Correlation in a Bunched Electron Beam. *Phys. Rev. Lett.* **1998**, *80*, 2841–2844. [CrossRef]
9. Ciocci, F.; Bartolini, R.; Doria, A.; Gallerano, G.P.; Giovenale, E.; Kimmitt, M.F.; Renieri, G.M.A. Operation of a Compact Free Electron Laser in the Millimeter Wave Region with a Bunched Electron Beam. *Phys. Rev. Lett.* **1993**, *70*, 928–931. [CrossRef] [PubMed]
10. Doria, A.; Gallerano, G.P.; Giovenale, E.; Messina, G.; Spassovsky, I. Enhanced Coherent Emission of Terahertz Radiation by Energy-Phase Correlation in a Bunched Electron Beam. *Phys. Rev. Lett.* **2004**, *93*, 264801. [CrossRef] [PubMed]
11. Chin, Y.H. *Double RF System for Bunch Shortening*; ESG 108 LBL 29622; Lawrence Berkeley National Laboratory: Berkeley, CA, USA, 1990; pp. 1–17.
12. Hofmann, A.; Myers, S. Beam Dynamics in a Double RF System. In Proceedings of the 11th Internationl Conference on High Energy Accelerators, CERN, Geneva, Switzerland, 7–11 July 1980; pp. 610–614.
13. Dattoli, G.; Renieri, A. Experimental and theoretical aspects of free electron laser. In *Laser Handbook Vol 4*; Stitch, M.L., Bass, M., Eds.; North Holland Physics Publishing: Amsterdam, The Netherlands, 1985; pp. 1–141.
14. Piel, H. *Superconducting Cavities*; CERN 89-04; CERN Accelerator School: Geneva, Switzerland, 1989; pp. 149–196.
15. Matula, R.A. Electrical Resistivity of Copper, Gold, Palladium and Silver. *J. Phys. Chem. Ref. Data* **1979**, *8*, 1147–1298. [CrossRef]
16. Reuter, G.E.H.; Sondheimer, E.H. The Theory of the Anomalous Skin Effect in Metals. *Proc. R. Soc. A* **1948**, *195*, 336–364. [CrossRef]

condensed matter

Review

A Versatile THz Source from High-Brightness Electron Beams: Generation and Characterization

Enrica Chiadroni [1,*], Alessandro Cianchi [2], Massimo Ferrario [1], Andrea Mostacci [3], Riccardo Pompili [1] and Vladimir Shpakov [1]

1 Laboratori Nazionali di Frascati (INFN-LNF), Via Enrico Fermi 54, 00044 Frascati, Italy; massimo.ferrario@lnf.infn.it (M.F.); riccardo.pompili@lnf.infn.it (R.P.); vladimir.shpakov@lnf.infn.it (V.S.)
2 University of Rome Tor Vergata and INFN Tor Vergata, Via Ricerca Scientifica 1, 00133 Rome, Italy; alessandro.cianchi@roma2.infn.it
3 Sapienza University of Rome, Department of Basic and Applied Sciences for Engineering, via A. Scarpa 16, 00161 Rome, Italy; andrea.mostacci@uniroma1.it
* Correspondence: enrica.chiadroni@lnf.infn.it

Received: 24 April 2020; Accepted: 4 June 2020; Published: 8 June 2020

Abstract: Ultra-short electron bunches, such as those delivered by a high-brightness photo-injector, are suitable to produce high peak power THz radiation, both broad and narrow band, with sub-picosecond down to femtosecond pulse shaping. The features of this kind of source in the THz range of the electromagnetic spectrum are extremely appealing for frequency- and time-domain experiments in a wide variety of fields. The present manuscript will overview the method of generation and characterization of THz radiation produced by high-brightness electron beams, as those available at the SPARC_LAB test facility.

Keywords: THz source; high-brightness electron beams; laser comb technique

1. Introduction

Terahertz (THz) science [1,2] is recently arousing interest worldwide for its impact in scientific research and technological progress because of the unique characteristics of this radiation. Indeed, THz radiation is non-ionizing, it penetrates in most insulating materials, e.g., plastic, ceramics, paper, and it is reflected by metals; moreover, it is absorbed by polar liquids, showing absorption spectra peculiar of the physical process. As a matter of fact, powerful THz sources are highly demanded, since they can provide efficient tools to investigate matter in the non-linear regime. Moreover, both biomedical and security applications can profit from the development of innovative THz sources, detection devices and characterization techniques.

THz radiation lies between the photonics and electronics bands of the electromagnetic spectrum. Effectively, this frequency region extends from 300 GHz up to 30 THz, corresponding to energies, in the meV range, of great interest for many physical, chemical and biological processes. Furthermore, molecular spectroscopy and imaging [3] can take advantage of a high peak power, tunable and narrow-band THz radiation.

Unlike conventional THz sources, such as those based on diodes, a new generation of sources, based on particle accelerators, allows a raising of the power, both average and peak, by many orders of magnitude. In particular, linac-based coherent THz radiation from relativistic high-brightness electron beams is a cutting-edge source [4–6] providing an efficient tool to investigate matter in the non-linear regime. Indeed, if the electron bunch longitudinal structure does not change during the emission process, the THz radiation is characterized by pulses whose temporal shape resembles the electron bunch one and with associated peak electric fields greater than 1–10 MV/cm [7].

The great advantage of a linac-based THz source lies in its versatility; indeed, the electron beam longitudinal distribution can be properly shaped to generate tunable, few-cycle and multi-cycle, coherent THz radiation with both broad and narrow spectral bandwidth. In particular, to enable the generation of narrow-band THz radiation, different approaches that foresee the longitudinal manipulation of the electron beam are currently used: temporal shaping of the photocathode drive laser pulse [8–11], mask insertion in a high dispersion, low beta function region of a beamline dogleg [12], transverse-to-longitudinal phase space exchange technique [13] to turn transverse modulation from a multi-slit into a longitudinal one, and magnetic undulator radiation [14–19]. Here, we report on the first approach, being the one developed and fully studied at the SPARC_LAB test facility [20].

2. Materials and Methods

The key for high efficiency in a beam-based radiation source is to exploit the coherence enhancement effect boosted by tailoring the electron beam profile. For instance, coherent radiation in the THz range can be generated through ultra-short, i.e., sub-ps, relativistic electron bunches hitting a metal target as Transition Radiation (TR) [21]. Indeed, when a charged particle crosses the interface between two different refraction index media, TR is emitted both in forward and backward direction. The radiation results from the prompt change of the boundary conditions for the electromagnetic (EM) field carried by the particle in the first and second media. When a charged particle goes through a slit, opened in a metal target, only part of the particle field interacts with the screen, and the emitted radiation is called Diffraction Radiation (DR). Since the particle goes through the aperture without interacting with the screen, the emission process is not intercepting and non-invasive for the particle itself. Coherent Diffraction Radiation (CDR) is particularly appealing as non-disruptive longitudinal diagnostics [22].

When a bunch of N charged particles is considered, TR (DR) emission comes from each particle with the same intensity, but the position of each particle in the bunch has an impact on the emission, contributing with a phase term. Therefore, two regimes can be identified depending on the bunch length, σ_z, with respect to the emission wavelength, λ: at $\lambda \ll \sigma_z$, each particle behaves independently from each other, resulting in incoherent emission, proportional to the number of electrons in the bunch, N; at $\lambda \gg \sigma_z$, all electrons emit in phase and the amplitudes add constructively, resulting in coherent emission, proportional to N^2. Therefore, the total emitted energy can be written as

$$\frac{dU}{d\lambda} = \frac{dU_{sp}}{d\lambda}[N + N(N-1)|F(\lambda)|^2]\,. \tag{1}$$

$F(\lambda)$ is the longitudinal bunch form factor, defined as $F(\lambda) = \int_{-\infty}^{\infty} S(z)e^{-2\pi i z/\lambda}dz$, with $S(z) = \int S_{3D}(\vec{r})\; dxdy$ the longitudinal charge distribution, i.e., the projection of the three-dimensional normalized particle density distribution onto the z axis. The bunch longitudinal form factor sets the emission regime: at wavelengths longer than the bunch length, $F(\lambda) \to 1$ and the spectrum is dominated by coherent emission, which extends to THz frequencies for sub-ps electron beams, with a huge intensity gain with respect to most existing THz sources thanks to the N^2 dependency.

2.1. Broad Band Radiation

Short, i.e., sub-ps, electron bunches drive the generation of high peak power radiation in the THz range. In linac-based facilities short electron bunches are typically produced by either magnetic or radio-frequency (RF) compression techniques. Magnetic compression, performed at high energy to overcome space charge effects, foresees a bunch with a time-energy correlation (or chirp) moving along an energy-dependent path length, represented by a dispersive, non-isochronous beam transport section. As alternative at lower energies, in the few MeV range, the RF compression scheme, known as velocity bunching [23] (VB), exploits the interaction with the electromagnetic fields of an accelerating cavity. A time-velocity correlation (or chirp) is provided to the electron bunch, inducing a longitudinal phase space rotation in the traveling RF wave potential due to the fact that electrons in the bunch tail are faster than those in the bunch head. Since the beam is slightly relativistic, being slower than the phase velocity of the RF wave, when injected at the zero crossing field phase, it slips back to phases where the field is accelerating and, simultaneously, it is chirped and compressed. Therefore, compression and acceleration take place at the same time within the same accelerating section. Beyond the longitudinal dimension, the electron beam longitudinal shape also plays a significant role in the spectral content of the emission. In particular, asymmetric current profiles, caused by both longitudinal and transverse distortions of the phase space, and typical of compression techniques, promote emission at higher THz frequencies. Indeed, Figure 1 shows the longitudinal form factor as function of the frequency for four different beam longitudinal distributions, with the same RMS pulse duration (inset): Gaussian (black line), uniform (red line), triangular (blue line), and real measured one (green asterisks). The triangular-like longitudinal distribution results as the most efficient one to generate THz radiation, thanks to the steep rising edge, which is also shown in the longitudinally compressed beams (see green asterisks curve in the inset of Figure 1).

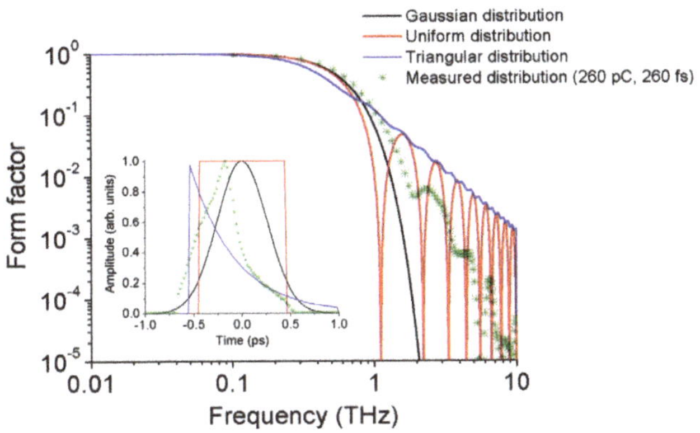

Figure 1. Longitudinal bunch profile (inset) and corresponding form factors for three different particle distribution with the same RMS pulse duration, i.e., 260 fs.

2.2. Narrow-Band and Tunable Radiation

One of the methods used to produce narrow-band and tunable THz radiation is based on the generation of Coherent Transition Radiation (CTR) from a train of sub-ps electron bunches hitting a metal target [10,24].

The generation of a train of sub-picosecond electron bunches with picosecond spacing has been conceived and demonstrated at the SPARC_LAB test facility [25], applying the VB compression technique to a longitudinally modulated electron bunch as generated by a train of UV laser pulses hitting the photocathode, the so-called comb laser beam [26].

One way to generate comb laser pulses relies on birefringent crystals: an input laser pulse is decomposed into two orthogonally polarized pulses with a time separation proportional to the crystal length. A multi-peak laser profile is produced by stacking more birefringent crystals in the laser beam path. The so-generated comb laser pulse, illuminating a metallic photocathode in a RF gun, produces a train of short electron bunches thanks to the prompt temporal response of the metal, which allows for transverse and longitudinal shaping of the electron beam through laser manipulation. Downstream the photo-injector, the beam acquires an energy modulation because of the space charge forces and, if injected in a RF-compressor operating in the over-compression regime, the energy modulation turns back into density modulation. The train parameters, i.e., bunch charge, length, and time separation, can be fully controlled by properly adjust both laser and linac settings [27]. A train of sub-ps relativistic electron bunches intercepting an aluminum screen, emits CTR. Since the CTR emission is a surface phenomenon, it happens with femtosecond timescale and the electron bunch structure does not change during the emission process; in addition, no interaction occurs between particles and radiation, resulting in a radiation pulse structure that resembles the multi-bunch longitudinal one with a spike-like spectrum at the frequency of the bunch distance and higher harmonics, as depicted in Figure 2.

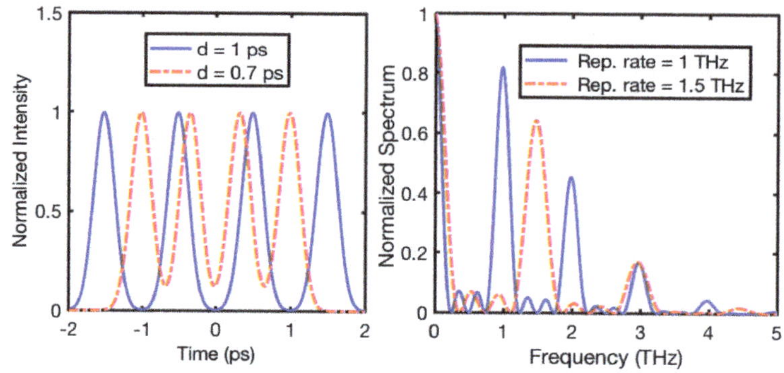

Figure 2. Left: 4-bunches comb beam with different time separation (blue solid line, 1 ps; red dashed line, 0.7 ps). Right: Corresponding form factors highlighting the tunability of the narrow-band THz source.

Decreasing the sub-pulse length, the single pulse spectrum becomes larger and more harmonics of the sub-pulse repetition frequency appears in the comb spectrum, with the advantage that the energy at single frequency is the same as all the charge would be condensed in a single micro-pulse. In addition, the central THz frequency, and its harmonics, can be tuned by changing the micro-bunches time separation.

2.3. Experimental Layout

CTR is generated, at the SPARC_LAB test facility, by highly relativistic electron beams hitting a 40 × 40 mm, aluminum coated silicon screen (380 μm thick with 40 nm coating). A proper optical system is needed to extract the CTR and transport it down to the detection apparatus and, eventually, to the sample. A sketch of the experimental layout, with light optics showing the radiation path, is depicted in Figure 3. Radiation is extracted from the vacuum pipe through a crystalline diamond window (700 μm thick, 0.3 deg wedge, 40 mm clear aperture), and collected by aluminum coated parabolic mirrors.

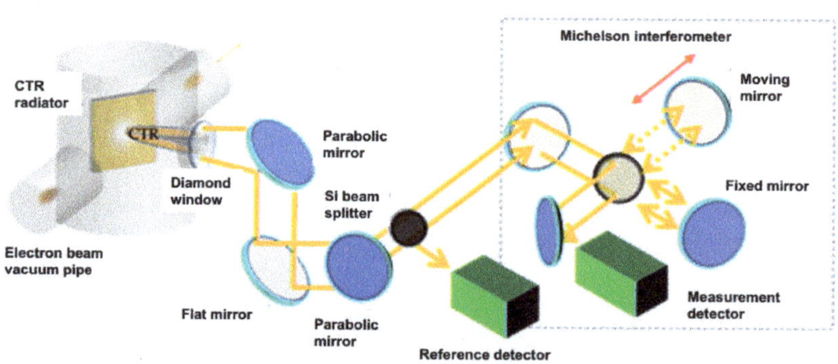

Figure 3. Experimental layout.

The coherent THz source is currently characterized by measuring the radiation spectrum in air (or dry air environment, i.e., 12% of relative humidity at 297 K) both through a set of discrete band-pass THz filters and an interferometer. The dry air system, a Kaeser Membrane Module dryer (FE-6D D-pack) working with compressed air, contributes to reduce water absorption in the Perspex box embedding the experimental apparatus (overall volume of about 200 litres), allowing measurement of radiation from very low charge bunches, of the order of 10 pC, needed for the production of ultra-short, fs-scale, electron bunches.

A two-beam interferometer is usually used for frequency domain measurements. For ps and sub-ps electron bunches a Martin–Puplett interferometer (MPI) [28] is preferred because of the beam splitter properties. Indeed, despite a Michelson interferometer, the beam splitter is a wire grid whose reflectivity coefficients depend on wires diameter and spacing. Furthermore, it splits polarizations, thus measuring the intensity of both polarizations: the difference interferogram can be normalized to the sum. Correlated fluctuations due to electron beam current instabilities are then canceled. The main limitation is however due to the high-frequency suppression introduced by the spacing between wires, which prevents the measurement of ultra-short bunches. Therefore, for bunches whose duration is from a few hundreds of fs down to a few tens of fs, the most suitable choice is a Michelson interferometer, which allows the covering of a wide spectral range, from 0.5 to 30 THz, by properly changing the thickness of the beam splitter from 24 μm down to 3.5 μm.

The autocorrelation measurements reported here have been performed using a Michelson interferometer with an additional detector (reference detector in Figure 3), which collects the radiation after a silicon beam splitter, placed before the Michelson interferometer box. The intensity measured by the measurement detector is normalized to the one measured by the reference detector, contributing to an increase of the signal to noise ratio; as a matter of fact this upgrade to the conventional Michelson interferometer allows the cancelation of correlated fluctuations due to electron beam current instabilities as in the two-beam MPI.

3. Results and Discussion

The versatility of linac-based THz sources stands in the possibility to tune the THz spectral content by properly setting the photo-injector regime and the UV photocathode laser longitudinal distribution for either narrow or broad band THz emission. In case of a longitudinally modulated electron beam, whose current profile and longitudinal phase space are shown in Figure 4 (left and right, respectively), the measured interferogram (Figure 5, left) presents the peculiar multi-peaks pattern, in particular $(2n-1)$ peaks, with $n = 5$ the number of bunches in the train. At zero optical path difference each sub-pulse within the two comb-like beamlet completely overlaps, resulting in the central maximum in the interferogram. As the path difference increases, the interferogram intensity decreases until four of the five pulses overlap, corresponding to the first side maxima. Further increasing the distance, the interferogram intensity keeps on decreasing resulting in the farther side maxima.

The spectrum corresponding to the autocorrelation function of the radiation produced by the 5 bunches beam is shown in Figure 5 (right). The main peak is at 0.65 THz corresponding to a 1.54 ps time separation between two of the five bunches, with a bandwidth (BW) of around 30%, whereas commercial THz filters have around 15–20% BW. Since the single bunch length in the train is of the order of 100 fs and less, a second and third harmonics arouse at around 1.25 THz and 1.95 THz. The peak at around 1 THz corresponds to the 0.8–0.9 ps distance of two of the four bunches, as shown in the electron beam current profile and longitudinal phase space.

Therefore, paying careful attention to a regular stacking of similar electron bunches, produces the suppression of all the frequencies apart the one corresponding to the succession itself, reaching in intensity at most the value of the single sub-pulse case. It is worth clarifying that to obtain the same radiation intensity at this frequency, in the case of a single bunch all the electrons N must be concentrated in the single bunch itself, with the drawback that space charge usually prevents to collect high charge in <100 fs. The width of this line depends on the number of pulses that constitutes the comb. Therefore the radiation produced by a longitudinally modulated beam (comb beam) is emitted at the frequency of comb and higher harmonics depending on the sub-bunch length, and the intensity at those frequencies is as if all electrons N would be concentrated in the single bunch whose duration is of the order of 100 fs or less. The transform limit is not violated, since the overall bandwidth is wide due to the fact that not only the fundamental frequency is populated, but also the higher harmonics. In addition, the shorter the pulse, the wider is the THz spectrum, because more harmonics show up. As a matter of fact, the photons of the total bandwidth are distributed differently, based on the temporal structure of the bunch, as its Fourier transform shows (Figure 5, right). Therefore the total peak power must be evaluated assuming that all the charge is collected in the micro-bunch duration, since it is the duration of the single micro-bunch in the train that sets the extension of the spectrum, thus the number of harmonics. Therefore, the term *narrow* for the frequency bandwidth of radiation from a train of micro-bunches refers to the bandwidth of each line of the frequency spectrum, which is naturally filtered from both the number and duration of bunches in the train.

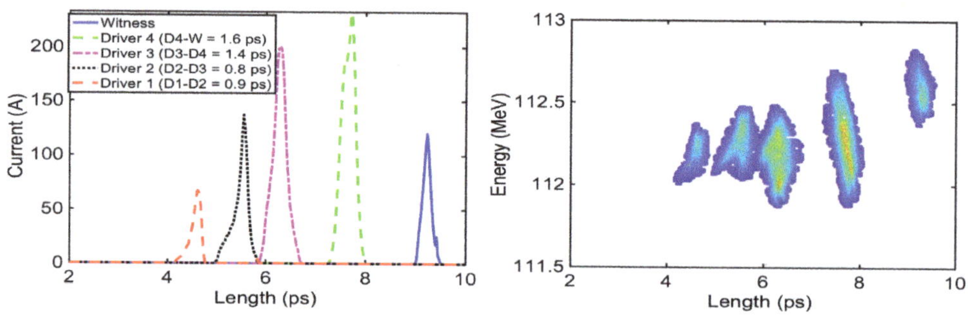

Figure 4. Electron beam longitudinal profile (left) and longitudinal phase space of a 5-bunches ramp train (right).

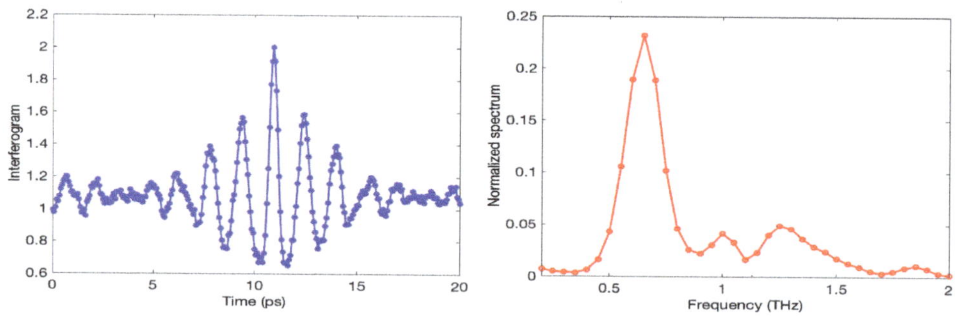

Figure 5. Autocorrelation function (left) and relative spectrum (right) of a 5-pulses generated radiation.

The spectral properties of the THz radiation, as generated by a single sub-ps electron bunch can be also investigated sampling the signal by means of commercial band-pass THz filters [29]. Highly intense and broadband CTR has been produced, for instance, by a 110 MeV, 500 pC, 180 fs RMS beam impinging on an aluminum coated silicon screen, and its spectrum measured up to 5 THz, as shown in Figure 6.

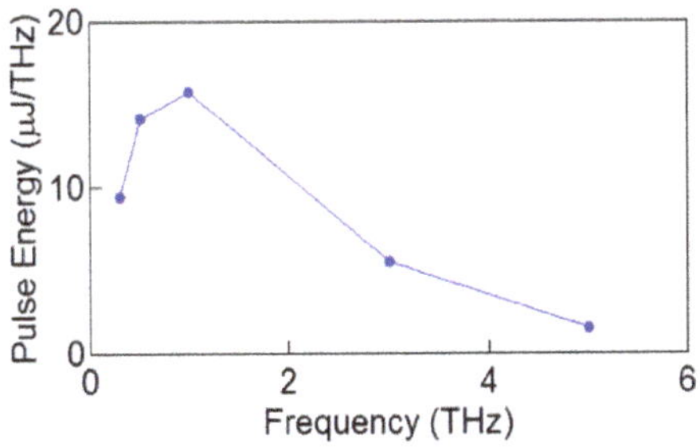

Figure 6. Measured CTR spectrum sampled with THz band-pass filters center at 0.3–0.5–1–3–5 THz [11].

The drop below 0.3 THz in Figure 6 is due to the effect of the screen finite size with respect to the source size [30], being of the order of $\gamma\lambda/2\pi$ (γ the Lorentz factor and λ the emitted wavelength), combined with the fall of responsivity of the pyroelectric detector. On the other hand, the longitudinal extension of the bunch limits the generation above 5 THz. The integrated energy per pulse reaches up to 35 uJ as measured by a pyroelectric detector, whose responsivity and absorption curves have been provided by manufacturers. The pyrodetector is a 5 mm diameter THz-I-BNC GENTEC-EO, operating in the spectral range 0.1–30 THz, with responsivity ≈140 kV/W and a Noise Equivalent Power at 20 Hz $\approx 4 \times 10^{-10}$ W/Hz$^{0.5}$.

The main figures of merit of the SPARC_LAB THz source for a single bunch to generate broadband THz radiation and five-ramp comb train to generate a narrow-band THz radiation are reported in Table 1 [11].

Table 1. SPARC_LAB THz sources: radiation and electron beam parameters for a single bunch and a ramped five-bunch train ($\sigma_t(D1)$ = 88 fs, $\sigma_t(D2)$ = 127 fs, $\sigma_t(D3)$ = 151 fs, $\sigma_t(D4)$ = 165 fs, $\sigma_t(W)$ = 100 fs).

THz Radiation Parameters			Electron Beam Parameters		
	Single Bunch	**Ramped Comb**		**Single Bunch**	**Ramped Comb**
Energy per pulse (μJ)	35 †	~1	Charge (pC)	500	220
Peak power (MW)	~80 †	~3	Energy (MeV)	121	110
Electric field (MV/cm)	>1 †	-	RMS Bunch duration (fs)	180	††
Bandwidth * (THz) $\Delta\nu$	~2	0.25	Rep. Rate (Hz)	10	10
RMS Pulse duration t_p (ps)	~0.18	~1.23 **	Comb distance (ps)	-	1.3

† Systematic uncertainty due to missing detector calibration below 0.61 THz; †† the RMS duration, σ_t, for each bunch in the train is reported in the table caption; * Defined as the FWHM; ** From measured results the time-bandwidth product is $\Delta\nu\ t_p$ = 0.72.

Linac-based THz sources are characterized by pulse energies reaching the multi-hundreds μJ regime [31] by increasing the charge, thanks to the N^2 dependence, and/or properly shaping the bunch longitudinal profile. In addition, taking profit of the beam longitudinal manipulation, the spectral bandwidth can be either tunable narrow or broad, making it extremely versatile. Finally, the repetition rate can span from a few Hz up to GHz, depending on the accelerator repetition rate, paving the way to an average power of the order of 10 W [16]. On the other hand, laser-driven THz sources, in particular those based on Optical Rectification [32] with organic crystals, have demonstrated to achieve hundred μJ pulses in tight focus resulting in electric and magnetic field larger than 10 MV/cm and 3 Tesla, over a broad 0.1–8 THz frequency range [33].

Compared to other linac-based sources [16], and based on the measured parameters, listed in Table 1, the SPARC_LAB THz radiation is strongly competitive and extremely suitable for investigating non-linear THz spectroscopy and for testing novel detectors and materials. Indeed, thanks to MV/cm electric fields, which can excite fundamental quantum statuses of a given system, it allows for non-linear optic experiments [34].

4. Conclusions

In photo-injector-based facilities, THz radiation can be produced under different regimes, by a proper control of the bunch shaping and charge distribution. Indeed, taking profit of the prompt photoemission of electrons at the metal cathode, different electron longitudinal distribution can be tailored and manipulated by properly setting the UV photocathode laser and the photo-injector parameters.

In particular, the present manuscript reports the most valuable results obtained at SPARC_LAB with both ultra-short single bunch and multi-bunches electron comb beams to provide high energy per pulse and broad and narrow spectral bandwidth THz radiation, respectively for non-linear and pump-probe experiments in solid-state physics and material science.

Author Contributions: E.C. planned and managed the THz activities at SPARC_LAB, and wrote the manuscript; V.S. was involved in the project and setup of the experimental apparatus; M.F. coordinated the SPARC_LAB test facility; A.C. took part to the preparation of the experiments; A.M. and E.C. carried out the data analysis; R.P. provided numerical simulations for photo-injector settings. All authors were involved in the operation of the experiments, extensively discussed the results and reviewed the manuscript. All authors have read and agreed to the published version of the manuscript.

Funding: This research was partially funded by CSN5 of INFN.

Acknowledgments: Authors would like to acknowledge all the colleagues at the SPARC_LAB test facility of the INFN Frascati and colleagues within the collaboration, in particular, S. Lupi (Sapienza University of Rome) and F. Giorgianni (PSI).

Conflicts of Interest: The authors declare no conflict of interest.

References and Note

1. Dhillon, S.S.; Vitiello, M.S.; Linfield, E.H.; Davies, A.G.; Hoffmann, M.C.; Booske, J.; Paoloni, C.; Gensch, M.; Weightman, P.; Williams, G.P.; et al. The 2017 terahertz science and technology roadmap. *J. Phys. D: Appl. Phys.* **2017**, *50*, 043001. [CrossRef]
2. Lee, Y.-S. *Principles of THz Science and Technology*; Springer Science+Business Media, LLC: New York, NY, USA, 2009; ISBN 978-0-387-09539-4.
3. Sherwin, M.S.; Schmuttenmaer, C.A.; Bucksbaum, P.H. Opportunities in THz Science. In Proceedings of the Report of a DOE-NSF-NIH Workshop, Arlington, VA, USA, 12–14 February 2004.
4. Hoffmann, M.C.; Schulz, S.; Wesch, S.; Wunderlich, S.; Cavalleri, A.; Schimdt, B. Coherent single-cycle pulses with MV/cm field strengths from a relativistic transition radiation light source. *Opt. Lett.* **2011**, *36*, 4473–4475. [CrossRef] [PubMed]
5. Chiadroni, E.; Bacci, A.; Bellaveglia, M.; Boscolo, M.; Castellano, M.; Cultrera, L.; Di Pirro, G.; Ferrario, M.; Ficcadenti, L.; Filippetto, D.; et al. The SPARC linear accelerator based terahertz source. *Appl. Phys. Lett.* **2013**, *102*, 094101. [CrossRef]
6. Wu, Z.; Adli, E.; Fisher, A.; Loos, H.; Hogan, M. First Measurements of the FACET Coherent Terahertz Radiation Source. In Proceedings of the 3rd International Particle Accelerator Conference (IPAC2012), New Orleans, LA, USA, 20–25 May 2012.
7. van Tilborg, J.; Schroeder, C.B.; Tóth, C.; Geddes, C.G.R.; Esarey, E.; Leemans, W.P. Single-shot spatiotemporal measurements of high-field terahertz pulses. *Opt. Lett.* **2007**, *32*, 313–315. [CrossRef] [PubMed]
8. Neumann, J.G.; Fiorito, R.B.; O'Shea, P.G.; Loos, H.; Sheehy, B.; Shen, Y.; Zu, W. Terahertz laser modulation of electron beams. *J. Appl. Phys.* **2009**, *105*, 053304. [CrossRef]
9. Shen, Y.; Yang, X.; Carr, G.L.; Hidaka, Y.; Murphy, J.B.; Wang, X. Tunable Few-Cycle and Multicycle Coherent Terahertz Radiation from Relativistic Electrons. *Phys. Rev. Lett.* **2011**, *107*, 204801. [CrossRef]
10. Chiadroni, E.; Bellaveglia, M.; Calvani, P.; Castellano, M.; Catani, L.; Cianchi, A.; Di Pirro, G.; Ferrario, M.; Gatti, G.; Limaj, O.; et al. Characterization of the THz radiation source at the Frascati linear accelerator. *Rev. Sci. Instrum.* **2013**, *84*, 022703. [CrossRef]
11. Giorgianni, F.; Anania, M.P.; Bellaveglia, M.; Biagioni, A.; Chiadroni, E.; Cianchi, A.; Daniele, M.; Del Franco, M.; Di Giovenale, D.; Di Pirro, G.; et al. Tailoring of Highly Intense THz Radiation Through High Brightness Electron Beams Longitudinal Manipulation. *Appl. Sci.* **2016**, *6*, 56. [CrossRef]

12. Anderson, S.G.; Musumeci, P.; Rosenzweig, J.B.; Brown, W.J.; England, R.J.; Ferrario, M.; Jacob, J.S.; Thompson, M.C.; Travish, G.; Tremaine, A.M.; et al. Velocity bunching of high-brightness electron beams. *Phys. Rev. ST Accel. Beams* **2005**, *8*, 014401. [CrossRef]
13. Piot, P.; Sun, Y.-E.; Maxwell, T.J.; Ruan, J.; Lumpkin, A.H.; Rihaoui, M.M.; Thurman-Keup, R. Observation of Coherently-Enhanced Tunable Narrow-Band Terahertz Transition Radiation from a Relativistic Sub-Picosecond Electron Bunch Train. *Appl. Phys. Lett.* **2011**, *98*, 261501. [CrossRef]
14. Bocek, D.; Hernandez, M.; Kung, P.; Lihn, H.C.; Settakorn, C.; Wiedemann, H. Observation of coherent undulator radiation from sub?picosecond electron pulses. *AIP Conf.* **1996**, *367*, 381–390. [CrossRef]
15. Zhang, J.; Deng, H.; Lin, X.; Dai, D.; Sun, Q.; Lu, S.; Yu, T.; Zhao, H.; Yang, H.; Dai, Z. Tunable few-cycle coherent terahertz radiation with watt-level power from relativistic femtosecond electron beam. *Nucl. Instrum. Methods Phys. Res. A* **2012**, *693*, 23–25. [CrossRef]
16. Green, B.; Kovalev, S.; Asgekar, V.; Geloni, G.; Lehnert, U.; Golz, T.; Kuntzsch, M.; Bauer, C.; Hauser, J.; Voigtlaender, J.; et al. High-Field High-Repetition-Rate Sources for the Coherent THz Control of Matter. *Sci. Rep.* **2016**, *6*, 22256. [CrossRef]
17. Wen, X.; Huang, S.; Lin, L.; Wang, F.; Zhu, F.; Feng, L.; Yang, L.; Wang, Z.; Fan, P.; Hao, J.; et al. Superradiant THz undulator radiation source based on a superconducting photo-injector. *Nucl. Instrum. Meth. Phys. Res. A* **2016**, *820*, 75–79. [CrossRef]
18. Kashiwagi, S.; Abe, T.; Saito, H.; Hinode, F.; Kanomata, K.; Miura, S.; Muto, T.; Nagasawa, I.; Nanbu, K.; Ninomiya, S.; et al. Observation of coherent undulator radiation in THz region. *Infrared Phys. Technol.* **2018**, *93*, 335–339. [CrossRef]
19. Krainara, S.; Zen, H.; Chatani, S.; Kii, T.; Ohgaki, H. Properties of THz coherent undulator radiation generated from a compact accelerator source at Kyoto University. *Rev. Sci. Instrum.* **2019**, *90*, 103307. [CrossRef]
20. Ferrario, M.; Alesini, D.; Anania, M.; Bacci, A.; Bellaveglia, M.; Bogdanov, O.; Boni, R.; Castellano, M.; Chiadroni, E.; Cianchi, A.; et al. SPARC_LAB present and future. *Nucl. Instrum. Meth. Phys. Res. B* **2013**, *309*, 183–188. [CrossRef]
21. Ginzburg, V.L.; Frank, I.M. Radiation of a uniformly moving electron due to its transition from one medium into another. *J. Phys. USSR* **1945**, *9*, 353–362.
22. Chiadroni, E. Bunch Length Characterization at the TTF VUV-FEL. Ph.D. Thesis, Tor Vergata University of Rome, Rome, Italy, 2006; TESLA-FEL Report No. 2006-09.
23. Serafini, L.; Ferrario, M. Velocity bunching in photo-injectors. *AIP Conf. Proc.* **2001**, *581*, 87–106.
24. Marchetti, B.; Bacci, A.; Chiadroni, E.; Cianchi, A.; Ferrario, M.; Mostacci, A.; Pompili, R.; Ronsivalle, C.; Spataro, B.; Zagorodnov, I.; et al. Novel schemes for the optimization of the SPARC narrow band THz source. *Rev. Sci. Instrum.* **2015**, *86*, 073301. [CrossRef]
25. Ferrario, M.; Alesini, D.; Bacci, A.; Bellaveglia, M.; Boni, R.; Boscolo, M.; Calvani, P.; Castellano, M.; Chiadroni, E.; Cianchi, A.; et al. Laser comb with velocity bunching: Preliminary results at SPARC. *Nucl. Instr. Methods Phys. Res. Sect. A Accel. Spectrom. Detect. Assoc. Equip.* **2011**, *637*, S43–S46. [CrossRef]
26. Villa, F.; Anania, M.P.; Bellaveglia, M.; Bisesto, F.; Chiadroni, E.; Cianchi, A.; Curcio, A.; Galletti, M.; Di Giovenale, D.; Di Pirro, G.; et al. Laser pulse shaping for high gradient accelerators. *Nucl. Instrum. Methods Phys. Res. Sect. A* **2016**, *829*, 446–451. [CrossRef]
27. Mostacci, A.; Alesini, D.; Antici, P.; Bacci, A.; Bellaveglia, M.; Boni, R.; Castellano, M.; Chiadroni, E.; Cianchi, A.; Di Pirro, G.; et al. Advanced beam manipulation techniques at SPARC. In Proceedings of the IPAC2011, THYB01, San Sebastian, Spain, 4–9 September 2011.
28. Martin, D.H.; Puplett, E. Polarized interferometric spectrometry for the millimeter and sub-millimeter spectrum. *Infrared Phys.* **1970**, *10*, 105–109. [CrossRef]
29. All THz band-pass filters but the 5 THz one are produced by Tydex, whose part number designation is: BPF0.3-47, BPF0.5-47, BPF1.0-47, BPF3.0-47, with 47 the clear aperture in mm; the 5 THz band-pass filter is manufactured by Lake Shore Cryotronics, Inc. (https://www.lakeshore.com/about-us/press-releases/2007/11/05/new-far-infrared-thz-band-pass-filters), and has been also characterized by measuring with a Michelson interferometer the globar spectrum at 5 THz.
30. Castellano, M.; Cianchi, A.; Orlandi, G.; Verzilov, V.A. Effects of diffraction and target finite size on coherent transition radiation spectra in bunch length measurements. *Nucl. Instrum. Methods Phys. Res. A* **1999**, *435*, 297–307. [CrossRef]

31. Stojanovic, N.; Drescher, M. Accelerator- and laser-based sources of high-field terahertz pulses. *J. Phys. B At. Mol. Opt. Phys.* **2013**, *46*, 192001. [CrossRef]
32. Bass, M.; Franken, P.A.; Ward, J.F.; Weinreich, G. Optical Rectification. *Phys. Rev. Lett.* **1962**, *9*, 446. [CrossRef]
33. Vicario, C.; Jazbinsek, M.; Ovchinnikov, A.V.; Chefonov, O.V.; Ashitkov, S.I.; Agranat, M.B.; Hauri, C.P. High Efficiency THz Generation in DSTMS, DAST and OH1 pumped by Cr:forsterite laser. *Opt. Express* **2015**, *23*, 4573–4580. [CrossRef] [PubMed]
34. Giorgianni, F.; Chiadroni, E.; Rovere, A.; Cestelli-Guidi, M.; Perucchi, A.; Bellaveglia, M.; Castellano, M.; Di Giovenale, D.; Di Pirro, G.; Ferrario, M.; et al. Strong nonlinear terahertz response induced by Dirac surface states in Bi2Se3 topological insulator. *Nat. Commun.* **2016**, *7*, 11421. [CrossRef] [PubMed]

condensed matter

MDPI

Review

THz Pulsed Imaging in Biomedical Applications

Annalisa D'Arco [1,*], Marta Di Fabrizio [2], Valerio Dolci [1,2,3], Massimo Petrarca [1,3] and Stefano Lupi [2,4]

1 INFN-Section of Rome 'La Sapienza', P.le Aldo Moro 2, 00185 Rome, Italy; valerio.dolci@roma1.infn.it (V.D.); massimo.petrarca@uniroma1.it (M.P.)
2 Physics Department, University of Rome 'La Sapienza', Piazzale Aldo Moro 5, 00185 Rome, Italy; martadifabrizio@gmail.com (M.D.F.); stefano.lupi@roma1.infn.it (S.L.)
3 SBAI-Department of Basic and Applied Sciences for Engineering, Physics, University of Rome 'La Sapienza', Via Scarpa 16, 00161 Rome, Italy
4 INFN-LNF Via E. Fermi 40, 00044 Frascati, Italy
* Correspondence: annalisa.darco@roma1.infn.it

Received: 13 December 2019; Accepted: 24 March 2020; Published: 1 April 2020

Abstract: Recent advances in technology have allowed the production and the coherent detection of sub-ps pulses of terahertz (THz) radiation. Therefore, the potentialities of this technique have been readily recognized for THz spectroscopy and imaging in biomedicine. In particular, THz pulsed imaging (TPI) has rapidly increased its applications in the last decade. In this paper, we present a short review of TPI, discussing its basic principles and performances, and its state-of-the-art applications on biomedical systems.

Keywords: terahertz spectroscopy; terahertz time-domain spectroscopy; terahertz pulsed imaging; near-field imaging; biomedical imaging; cancer diagnosis; endoscopy

1. Introduction

The increasing request for diagnosis techniques in the biomedical area, able to provide morphological, chemical and/or functional information for early, noninvasive and label-free detection, has led to hard cooperation in different scientific fields. Driven by technological advances, a plethora of new modern diagnostics systems is investigated and validated, as complementary biomedical techniques to the conventional ones, covering various scales, from macromolecules to tissues/organs. Among them, worthy of special mention are: infrared (IR) imaging [1,2], scanning near-field microscopy [3,4], photoacoustic microscopy [5], ultrasonic imaging [6], optical coherence tomography [7,8], digital holography microscopy [9–11], Raman scattering microscopy [12], Coherent Raman Scattering (CRS) spectroscopic imaging [13–23], two-photon fluorescence (TPF) [19,24] and second-harmonic generation (SHG) imaging [25–27], and super-resolved imaging techniques [28–31].

Many efforts were done in THz technology, improving THz sources and detectors' responses [32–38], and ensuring devices' flexibility and portability, in recent decades. This has stimulated a wide diffusion of THz systems for spectroscopic and imaging purposes, applicable in various science fields like biology and medicine [39–44], gas sensing [45], chemical analysis [46,47], new materials characterization in low-frequency range and non-destructive evaluation of composite materials and constructions [48,49], astronomy [50], microelectronics and security [51–53], agri-food industry [54], art conservation [55], etc. Thus, alongside these techniques, TPI has been developed and applied in the field of biomedicine.

THz radiation has a variety of properties that make this spectral region a viable spectroscopic imaging technique. First of all, (i) THz radiation has low photon energy (4 meV at 1 THz), coinciding with the energy levels related to rotational and vibrational molecule modes and intermolecular vibrations, such as hydrogen bonds. These low-frequency motions allow the identification of biomolecules, characterizing their spectral features in THz region [56,57]. (ii) In addition, the low photon energy is insufficient to cause atoms ionization. (iii) For this reason, it is suitable and attractive for noninvasive biomedical imaging, has a great potentiality and could be applied in vivo real-time studies without causing biological ionizing damage, unlike X- and γ-rays [42,43]. Whereas extremely low frequency electromagnetic fields have been long studied in relation to their possible human health effects, the biological effects of radiofrequency and THz signals have come under scrutiny. In particular, the effects of THz radiation are of interest because of the expanding THz technologies in biomedical applications. The prevailing view is that THz radiation, if thermal effects are not considered, does not damage directly the DNA, but it could act as a co-inductor for non-thermal effects [58]. Recent studies show evidence of potential thermal and non-thermal effects induced by exposure to THz radiation [59,60]. The biological response to this stimulus is due to different parameters: the physical settings of the THz radiation (frequency, mean and peak power, radiation intensity, continuous or pulsed radiation, the degree of coherence, etc.,); the physical and biological properties of the irradiated biological object (refractive index, absorption features, type of cells, tissues, organs) and the design of the exposure [60]. (iv) In addition, THz radiation is very sensitive to polar molecules, like water. (v) That is why the strong absorption, due to water molecules (220 cm^{-1} for pure water at 1 THz and room temperature) [43], limits THz penetrability into the fresh tissues from tens to hundreds microns [42] and the capability of diagnostic only for superficial layers in vivo. (vi) The high sensitivity of water content can be used like an endogenous marker for the differentiation between fresh healthy and pathological tissues [42,43]. (vii) In particular, the time-domain spectroscopy (TDS) and imaging are insensitive to the thermal background and show high signal-to-noise ratio (SNR) [61]. (viii) TPI coherent detection ensures the record of temporal profile of the THz electric field; so, both the THz amplitude and phase can be estimated. From these, the broadband absorption coefficient and refractive index of a medium are determined, without using Kramers–Kronig relations [57,61]. The absorption is generally due to the chemical constituents of the material; thus, its spectral signatures are measured, providing useful information. As consequence, TPI has the potential to realize images with both morphological and functional information, in this frequency range. For all these reasons, THz radiation is considered very important and deserves special attention.

In this review, we introduce TPI, discussing its basic principles and performances, and the state-of-the-art applications on biomedical systems.

The review is organized as follows: in the first section, we outline the main features of pulsed THz systems and their key role in THz imaging setups, showing the equipment needed and some application ideas. Later, in the following section, after summarizing the main properties of the THz contrast image, we also focus on TPI challenges and relevant results achieved in the biomedical field.

2. THz Pulsed System/Imaging

Over the past two decades, THz generation and detection technology has considerably advanced, and several devices are commercially available now [33–37,62,63]. Their development has led to various spectroscopic and imaging techniques in the THz spectral window. According to THz sources used, the systems are divided into two general categories: the continuous wave (CW) systems, which produce a single or several discrete frequencies, and pulsed systems, characterized by a broadband frequency output, introduced for the first time as potential imaging tool in 1995 by Hu and Nuss [64]. All of them feature a large variety of properties, as well as, frequencies, power and sensitivity. In this paragraph, we briefly discuss the most common TPI systems that are the aim of our review, with a strong emphasis on THz pulsed sources and detectors.

2.1. Generation and Detection of THz Pulsed Radiation

In TPI systems, a broadband frequency emission, from tens or hundreds of GHz up to several THz, is obtained. Currently, various modalities can be listed for generating and detecting pulsed

THz radiation [65,66]: biased photoconductive antennas (PCAs) [67–69], optical rectification (OR) in nonlinear optical (NLO) crystals [70–72], THz production by plasma in air [73–75] and carrier tunneling in coupled double quantum well structures [76–78]. They exhibit different physical properties concerning powers, used materials, covered spectral range, SNR and dynamic range (DR). Among them, the most established approaches, based on PCAs and on OR, require pumping made by expensive infrared femtosecond lasers, often emitting in the near-infrared (NIR) region. For the first mentioned, a pulsed laser beam illuminates PCA gap, made of high resistive semiconductor thin film with two electrical contact pads. Thus, in the presence of the applied bias voltage and laser beam, photocurrent is generated and the static bias field accelerates the free carriers producing broadband frequency THz waves into the free space.

OR ensures the THz broadband generation through NLO centrosymmetric crystals [70–72], extended from 0.1 THz and beyond 40 THz [57,79,80], such as organic NLO crystals. 4-N,N-dimethylamino -4′-N′-methyl -stilbazolium 2,4,6-trimethylbenzenesulfonate (DSTMS) [81–84] and 4-N,N-dimethylamino-4′-N′-methyl-stilbazolium tosylate (DAST) show efficient generation from 0.3 to >16 THz in the phase-matching condition between 720 nm and 1650 nm. The NLO materials need to have high second-order NLO susceptibility, high transparency at both pump and THz frequency, and high optical damage threshold, in order to satisfy the phase-matching condition between the fundamental optical pump and generated THz waves [85,86]. When intense NIR laser beams propagate through some crystals, second order nonlinear effects occur, developing a DC or low frequency polarization [38]. This leads to the radiation of an electromagnetic single-cycle pulse with a wide frequency spectrum, ranging roughly from zero frequency to some maximum value [38].

Another way to produce THz radiation is by charges acceleration. As consequence, the charges radiate electromagnetic waves, that under particular conditions, lie in the THz region of the electromagnetic spectrum. The processes can be done either within semiconductors, in vacuum or in air [87–90]: via electrons acceleration by the intense laser pulse [87,88], by an applied bias voltage across an air gap, similarly to PCA, or by nonlinear four-wave mixing of the fundamental and the second harmonic frequencies of the laser beam, in air or in various gases [89,90].

Besides OR, photoconduction or photocurrent, surge is another primary mechanism for THz generation from semiconductor materials [57,59]. When semiconductor quantum wells (QWs) are under a bias, the mechanism for THz generation is attributed to the creation of polarized electron–hole pairs [91–93].

In its turn, two schemes are ideal to THz pulse detection, based on incoherent and coherent techniques. In terms of incoherent techniques, power-to-signal transducers, suitable for incoming energy measurements, are used. For coherent ones instead, the detectors give information concerning the power, frequency spectral range and phase of the incoming electrical signals [57]. Among THz incoherent detectors commercially available, we can mention bolometers [94]; Golay-cells, providing stable performances and high sensitivity; and pyroelectric detectors [95] and thermopiles with low sensitivity. Commercial bolometers, resistive temperature sensors, widely used in THz spectral region, are cryogenic ones. The cooling of the devices ensures high sensitivity, with noise equivalent powers (NEPs) ranging from 10^{-12} W/Hz$^{1/2}$ to 10^{-14} W/Hz$^{1/2}$. Golay-cell detectors [96] measure the thermal energy of THz propagating through a window [61]. Their spectral performances are strongly dependent on the material used for the window. In the case of diamond and crystal quartz windows, the spectral range covers from millimeters-wave to visible frequencies [61], while for HDPE (High-Density Polyethylene) window the spectral range is upper bounded at ~20 THz. Power DR can vary from hundreds nW up to 1 mW, with power DR from hundreds nW up to 1 mW. These devices can reach excellent NEPs at room temperature—around ~10^{-10} W/Hz$^{1/2}$ [61]. Instead, pyroelectric devices show lower sensitivity (NEPs ~10^{-9}–10^{-10} W/Hz$^{1/2}$) compared to bolometers and Golay performances. Table 1 summarizes some performances of the most used incoherent THz detectors.

Table 1. Main incoherent detectors suitable for THz spectral range with their performances:noise equivalent powers NEPs, THz coverage and detector bandwidth, linked to the detector response time (the higher the bandwidth, the faster the response time). Bolometer, Golay cell and Pyroelectric are pure incoherent detectors, while Schottky diode is used both for coherent and incoherent measurements.

Detector	NEP (W/ $Hz^{1/2}$)	THz Coverage (THz)	Bandwidth (Hz)
Cryogenic Bolometer	10^{-14}–10^{-12}	0.1–30	10^2–10^3
Golay cell	10^{-10}	0.01–20 [1] 0.01–700 [2]	10–10^2
Pyroelectric	10^{-10}–10^{-9}	0.1–10	10–10^8
Schottky diode	10^{-14}–10^{-11}	0.1–2	10^2–10^6

[1] High-Density Polyethylene (HDPE) window; [2] Diamond window.

THz cameras were the object of intense research activities in recent years. Actually, various THz cameras, based either on cooled and uncooled bolometers or semiconductor devices, are commercially available. Among the THz thermal cameras, the Golay cells are slow detectors and are difficult to integrate into array detectors. Consequently, the imaging scheme for them is a single-pixel detector [97]. The pyroelectric cameras, like the Pyrocam series sold by Ophir Optics, can operate with pulsed and CW sources, as well as thermopile cameras [98,99]. Generally, they are widely commercially available, and their low sensitivity makes them less attractive for THz imaging. The microbolometric cameras are arrays of bolometers; because of the relatively higher sensitivity they are considered the prime candidates for THz imaging. When cooled at cryogenic temperatures, they can achieve NEPs around 10^{-16} W/$Hz^{1/2}$. With proper design modifications, microbolometric cameras at room temperature also achieve high enough sensitivity.

Many such imaging techniques cannot be realized with existing THz cameras, but require a coherent detection scheme. Among commercial coherent or heterodyne detectors, largely used, we mention: Schottkly diodes [100,101], PCAs and NLO crystals or via electro-optical (EO) effect in nonlinear crystals [70–72]. Schottkly diodes [100,101] show limited spectral bandwidth around 1–2 THz and NEPs up to 10^{-14} W/$Hz^{1/2}$, and actually they are the object of research. PCAs (frequency range limited <10 THz and NEPs around 10–12 W/Hz1/2) can detect a THz electric field profile with a coherent detection scheme. The probe beam, derived from the same laser used to generate the THz radiation, irradiates receiver PCA, which has no external bias voltage. The laser pulses gate the detector acting as a photocurrent switch by generating charge carriers. The incoming THz radiation, on the receiver, induces the bias voltage, thus the photocarriers are accelerated [67,69]. The transient current is measured by a low-current amplifier and a lock-in technique, that are referenced against a modulation placed on the PCA emitter bias voltage.

In EO rectification NLO crystals, instead, the THz field induces birefringence in them; at a given instant, the polarization state of the probe laser pulse changes proportionally to the THz field amplitude [25,102]. Then, a balanced photodiode is used for measuring these changes in birefringence, reconstructing the THz pulse. In recent years, great interest in THz sensors and cameras is focused on the development of devices with high sensitivity, which operate at room temperature and fast operation speed [103], such as optoelectronic sensors and detectors [104–107].

2.2. *TPI Equipment*

The switch from the spectroscopy system to the imaging one is based on the possibility to collect spectroscopic information on large sample areas. For this purpose, the point-by-point signal collection requires a mechanism for scanning: moving the beam or the sample holder. The first case is realized with a beam motion or oscillating objective; using for example, galvo-mirrors and/or piezoelectric rotator or stages. In the second case, the lateral translation of the specimen stage is combined with a stationary illumination; for this purpose, a 3D linear translation stage is used, allowing sample holder motion along the THz beam (1D motion) and perpendicularly to it (2D motion).

The acquisition speed is one of the most critical challenges when using THz radiation for biomedical imaging, and it requires more attention for the implementation of clinical devices. The imaging speed depends on the specific image scheme chosen: using a point source and detector, the acquisition is slow, due to the point-by-point sample scanning. Although, the two-dimensional galvo-scanner, combined with a fast detector or sophisticated signal processing techniques [108–110], improves the scanning speed and the acquisition on a large field of view.

The object surface is illuminated by the THz beam, and it is sampled on a discrete grid, scanned in continuous or pixel-by-pixel raster modes. The information collected in each pixel is digitalized, transferred to PC from a data acquisition card (DAQ) and quantized in a finite number of bits for pre- and post-image processing.

2.2.1. TPI Far-Field Systems

A typical pulsed THz wave generation and detection system is a pump-probe setup. It can be considered as an extended THz time-domain spectroscopy (THz-TDS) method. The basic idea of THz-TDS and TPI systems is illustrated in Figure 1a.

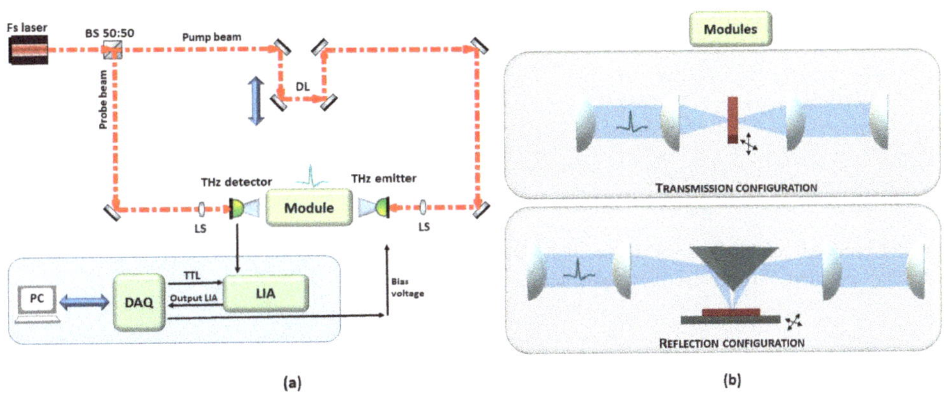

Figure 1. This is an example of TPI system switched by PCAs. (**a**) The schematic layout of common TPI system based on photoconductive switches. Here, Fs laser stands for femtosecond laser, BS for beamsplitter, DL is the delay line, LS are lenses, LIA, DAQ and PC stand for lock-in amplifier, data acquisition card and desktop computer, respectively. (**b**) The schematic layout of two potential sample orientation: transmission and reflection modes. The standard transmission imaging configuration involves the specimen placed between THz low-absorption material plates or standing in free space. The generated THz signal travels at a normal incidence to the sample surface and the transmitted one is received on the other side of the sample. Instead, for reflection mode, the THz signal illuminates one sample surface and it is reflected by the same surface.

TPI system, shown in Figure 1a, is powered by a femtosecond laser. A beamsplitter splits the initial laser beam into two parts: the pump and the probe beams. The pump beam is modulated via an optical or mechanical chopper or by modulating the THz emitter bias voltage, and then is focused on the THz emitter. The generated THz radiation is then collimated and focused onto a target sample. When a THz pulse illuminates a target, a train of THz pulses will be reflected from the various interfaces or transmitted. The transmitted (Figure 1b) and/or the reflected (Figure 1b) electric fields are, then, recollimated and refocused onto THz detector using THz transparent lenses or parabolic mirrors. The electrical signal is filtered and detected by lock-in amplifier detection. The analog output of the lock-in amplifier is collected and digitalized by a dedicated DAQ. A delay line is used in order to temporally sample the THz electric field. Thus, for each individual pulse in the detected signal, the amplitude and phase are different and are measured too.

If the pump and probe optical paths are equal, the THz electric field is measured at one fixed instant in time. In order to sample the whole THz pulse, a delay line is introduced in the optical scheme, to delay pump and probe beams. Generally, this can be performed using a mechanical translation stage, that moves at fixed steps; but rapid-scanning delay lines can also be used to speed the acquisition [111,112].

For example, we reported our home-made TPI system based on THz-TDS (see Figure 2a). It was implemented in transmission configuration.

It generates broadband THz radiation (0.1–3.5 THz) via photoconductive switches, used both for THz emission and detection. We use a mode-locked ultrafast laser (FemtoFiber NIRpro, Toptica), at 780 nm with a 100 fs temporal pulse width and a repetition rate of 80 MHz to illuminate twin G10620-11 Hamamatsu PCAs. The femtosecond laser is split into two beams by a 50:50 beamsplitter—each beam has 15 mW average power. Dielectric mirrors deflect the beams towards emitter and detector PCAs. TPX lenses are used to collimate and focalize THz radiation. The object is aligned in the focus of the THz beam, perpendicularly to the propagating THz radiation. The point-by-point spectral signal collection is obtained by scanning a large area of the object with 3D axes stage (Thorlabs). After transmission through the sample, THz pulses are collimated and refocused on the THz-detector PCA by TPX lenses. Simultaneously, the probe beam is used to gate the THz-detector PCA. The THz electric field as a function of time is measured with a delay line [112]. The output signal from the THz-detector, extracted by a lock-in amplifier (Stanford, SR830), is transferred to a National Instrument acquisition card (NI 6361-BNC connector) allowing the data collection. Our system exhibits performances comparable with the commercial systems, and being a home-made spectrometer, it can switch on the other two spectroscopic configurations using metallic flip mirrors, inserted in the THz collimated region. We can select between THz–TDS refection spectroscopy and attenuated total reflection (ATR) spectroscopy, see Figure 2a.

In Figure 2b, the temporal electric field profile, in free-space, as a function of time, is reported with relative power spectrum, for the scheme in transmission.

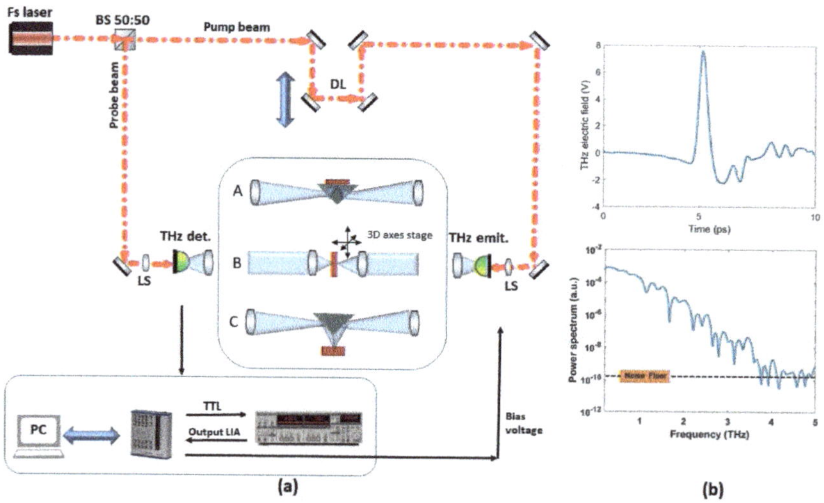

Figure 2. (**a**) Schematic layout of our TPI system. In the scheme, Fs laser stands for femtosecond laser, BS for beamsplitter, DL for delay line, LS is referred to lenses, THz det. and THz emit. are THz detector and THz emitter respectively. With two metal flip mirrors, introduced in the collimated THz beam region, one can select the scheme for THz spectroscopic imaging in transmission (B) or reflection (C) or ATR (A) spectroscopy. (**b**) The temporal electric field profile, in free-space, as a function of time, is reported with relative power spectrum.

2.2.2. TPI Near-Field Systems

The major limitation of THz far-field imaging systems is due to the problem of the diffraction-limited spatial resolution central to THz wavelengths, compared, for example, to visible radiation [113,114]. The optical resolution R in the object plane is generally defined by the Rayleigh criterion [115–118], described as:

$$R = \frac{0.61 \cdot \lambda}{NA} \approx \frac{\lambda}{2NA}$$

at the wavelength λ and numerical aperture NA. The minimal resolvable feature corresponds to the minimum size, in the object plane, of the smallest object that the optical system can resolve: as the resolution is directly proportional to the wavelength and inversely proportional to the numerical aperture of the optical components, at 1 THz the resolution turns out to be around 0.5 mm [113]. In fact, 1 THz corresponds to 0.3 mm and one must consider the loss of resolution caused mostly by lenses on the beam path. For this reason, effective and wide spreading of THz images in biomedical research and diagnostics is difficult. The wavelength-limited resolution is caused by diffraction of propagating waves, so this problem can be overcome by collecting THz pulses in near-field (NF), namely at a distance with the sample comparable to the wavelength. In this way, evanescent waves, that propagate only very close to the sample and then rapidly decay, can be detected. That is the reason why THz detector must be placed in proximity of the sample. Several techniques have been explored to enable sub-wavelength THz imaging [37,114,119,120]. With this purpose, detectors and sources of different shapes can be used; in particular, Hunsche et al. [121] proposed and demonstrated, for the first time, a method deriving from the confocal microscopy. It uses pinholes that act as spatial filter, reducing the THz beam size to sub-wavelength dimensions and blocking outside the cone of light. This results in increased lateral and depth resolutions [122,123]. However, the aperture introduces an immense loss of THz light intensity at deep sub-wavelength resolutions (<λ/10) [124], caused by the low frequency cut-off [125] and optical coupling loss. Following, it is possible to place a small detector or emitter directly behind the aperture [126–128]. The approaches without apertures, instead, are better set to achieve deep sub-wavelength resolutions, because the low-frequency cut-off is avoided. The sharp metallic tips for local field enhancement, tip apex, small detectors and/or sources elements with sub-wavelength dimensions are placed in proximity of the sample surface or in direct-contact. In this way, high resolutions are achieved (even λ/100), with the only limitations imposed by the tip diameter [129]. This means that NF resolution is not limited by diffraction so that, in this way it becomes independent of the wavelength.

The direct-contact approach, instead, is generally employed with EO sampling using THz-TDS. The sample could be then in contact with the detection crystal, leading to a direct measurement of the THz electric field in NF region. In [130], the deposition of various metal structures onto a gallium phosphite (GaP) crystal can improve the resolution by imaging at 20 μm.

3. Basics of THz Imaging

TPI can be viewed as an extension of the THz-TDS method. The classic way to obtain THz imaging from THz-TDS is illustrated above. In order to image the objects, a 2D point-by-point raster scanning of them, combined with a coherent detection, is performed; and the individual temporal profiles are recorded for each point-pixel. The raw one-dimensional time or extracted frequency data are converted into physical values [57,61,64,131,132]. The normalization is chosen to enhance the image contrast, and it can be summarized in the diagram (Figure 3).

Time domain

- Electric field amplitude at fixed time: $E_{x,y}(t_0)$
- Normalized amplitude of the main peak: $max\{|E_{x,y}(t)|\}/max\{|E_0(t)|\}$
- Time delay of the main peak with respect to the reference: $t(max\{|E_{x,y}(t)|\}) - t(max\{|E_0(t)|\})$

Frequency domain

- Spectral amplitude at fixed frequency: $E_{x,y}(\nu_0)$
- Phase : $\phi_{x,y}(\nu_0)$

Figure 3. Schematic summary concerning the image contrast in THz-TDS imaging.

In time domain, when there is interaction with the sample, the THz pulse can be attenuated, delayed or broadened, compared to the reference without the sample, generally measured in air. The electric field amplitude, at a fixed time, represents a contrast modality to normalize the image. In addition, the normalized amplitude of the main peak, as the ratio between maxima of the sample and reference electric fields, provides image information regarding the absorption, reflection or scattering losses in the object. Finally, the knowledge of time delay of the main peak with respect to the reference allows mapping the optical path changes, providing material/thickness contrast [133]. For example, the time-of-flight (TOF) technique permits to estimate the depth information about target internal dielectric profiles. A THz pulse is launched to the sample and the reflected echo is measured in amplitude or/and phase. The TOF information of the echo pulse indicates the presence of the boundaries, or inner structures along the THz propagation path, which extracts the one-dimensional depth profile. Thus, by performing a 2D scan, a 3D image of the object can be visualized, viewing into a layered structure or inside the optically opaque materials. 3D imaging potentiality is suitable for the implementation of various imaging techniques, largely used in biomedical contexts. Applying the Fourier transform on the temporal electric pulse, one gets the amplitude and phase of the spectrum. The first gives a general indication of the losses; the second one is related to the refractive index. Fixing the frequency, one can visualize the object in the frequency domain, the amplitude and the phase images [134]. This offers more contrast, because of the sufficiently different refractive indexes of the materials than the losses, which provide a meaningful contrast.

Summarizing, the THz-TDS measurements allow the extraction of both amplitude and phase of THz waveform. The complex refractive index and absorption coefficient can be obtained using the Fresnel coefficients [135,136], and from them the complex permittivity [137,138]. The knowledge of sample spectral behavior ensures to extract information concerning its morphological structure and the optical properties: the variation of refractive indexes could differentiate tissues, and could be related to the pathological status. Moreover, many materials are relatively transparent to THz radiation, and are used in the development of 3D THz imaging modalities that constitute an active research field [107,139]. Tomographic slices are employed for object/tissue 2D or 3D reconstructions, in order to obtain their internal properties and features [111]. The method tried a natural extension for the object rotation towards THz computer tomography (CT) [140,141]. In addition to THz CT, there are several other tomographic techniques based on THz radiation, including THz interaction tomography, THz tomosynthesis, time-of-flight pulsed imaging and 3D THz holography [139].

4. THz Pulsed Imaging: Uses, Advantages and Challenges for Biomedical Applications

As briefly discussed in the introduction and in the previous paragraph, the unique spectral features of THz radiation make this technology particularly interesting in the biomedical field—for ex vivo and in vivo experiments and/or analyses. During the past decades, systematic spectroscopic investigations carried out in the THz frequency region on biological targets, as biomacromolecule, cells and tissues, demonstrated differences in their optical properties. In the last five years only, the rate of imaging experiments in the THz domain is increasing.

An unavoidable issue, when talking about THz imaging, is water: THz is particularly sensitive to water content, exhibiting a strong absorption [142] and consequently limiting THz penetration depth.

In fact, the strong absorption of water limits the penetration depth in fresh tissues between tens and hundreds of microns. For example, into the human skin THz waves can penetrate only a few hundred microns [143]. In vivo clinical measurements, it limits the probing on superficial layers of the target and the reflection mode becomes the suitable choice [144,145]. The limitation could be overcome in the case of ex vivo examinations, where histopathological evaluations are required. Researchers overcome these limitations using thin and/or fixed samples and appropriate geometry. The transmitted TPI is performed on thin prepared tissues, suitable for in vitro, and reflected TPI allows the investigation of surface features, reliable in vivo and ex vivo imaging.

However, difficulties in extrapolating measurements on tissue to ex vivo include, for example, saline uptake from the sample storage environment, changes in hydration level during the measurement, temperature- and humidity-dependent and scattering effects. Considering that, the human body contains a large amount of water (~60%) and THz is heavily absorbed by water [142], fresh tissue can vastly alter THz spectroscopy measurements. As fresh tissue dehydrates when left exposed to the air, many contrast features are reduced during the data acquisition. Therefore, for biomedical studies, sample conservation is crucial.

Several research groups have investigated excised and fixed tissues considering numerous approaches; among them are dehydration [146,147], alcohol perfused [36], formalin fixed [148–150], gelatin embedding [151,152], lyophilized [153], freezing [154,155] and fixation in paraffin emulsion and embedding [156].

Formalin fixing, in order to preserve and fix excised tissues, is a common histopathological routine for diagnosis. The formalin reduces the sample variation due to the dehydration process [149,154]—the tissues' water content is replaced by formalin.

In addition, lyophilization could represent an interesting alternative to fresh tissue for THz spectroscopic measurements [153]; its pros are fast and effective water removal, and structural preservation. The problems with handling fresh tissue, such as time variability in the THz bandwidth, dynamic range and sample thickness, are therefore overcome [153]. Several studies proposed the freezing of biological sample, as an alternative technique to increase the penetration depth of THz waves in the tissues, as the absorption coefficient of ice is one order of magnitude lower than the absorption coefficient of water [143]. The discussed technique has a limitation in clinical use; the process of freezing might cause necrosis. Some suggestions are proposed by using a penetration-enhancing agent (PEA) to increase the THz radiation delivery depth. The treatment with PEAs, that are cost-effective and easy-to-find, biocompatible and nontoxic to the human body, could also ensure long preservation time for ex vivo tissues and the application in vivo. Oh et al. [157] used glycerol as PEA, easily absorbed by human skin and tissue and with an absorption coefficient much lower than that of water in the THz spectral region. They treated abdominal mouse skin with glycerol and demonstrated the enhancement of penetration depth with the reconstruction of a metal blade hidden under the tissue by the second pulse reflection of it [157].

Various alternative approaches are suggested. Oleic acid is proposed by Wang et al. [158] for trapping moisture inside thin tissue slices, in transmission acquisition mode. They realized a sandwich configuration, by placing a frozen layer of oleic acid above and below the frozen tissue sample, between tsurpica windows and waited for complete thawing, before starting the measurements at room temperature. Oleic acid is highly transparent to THz radiation and its refractive index matched tsurpica. The oleic acid layers were used to keep up the tissue hydration, showing good performances for 70 min. Fan et al. [151] used gelatin embedding for porcine skin slices reflection measurements. By looking at the THz image data and computing the optical parameters of the gelatin-embedded sample, they found a successful method to preserve the sample for at least 35 h, both for imaging and spectroscopy purposes.

Since the first demonstration of THz imaging [64], TPI and spectroscopy show their ability to differentiate human tissues [141,159], with clear differences between the tissue properties, in particular regarding absorption coefficients [160]. Fatty tissue consists of mainly hydrocarbon chains, largely nonpolar molecules, thus its absorption coefficient and refractive index are much smaller than the ones of muscle, kidney and/or liver tissues [141].

Concerning the optical characterization at the low frequency region, many spectroscopic works are present in the literature [161–167]. One significant work about the ability in distinguishing and classifying the THz response of different types of tissues is Ref. [131], which is mentioned for the chirped probe pulse technique. It offers a significant improvement concerning the slow acquisition time, and it could have great potential in imaging acquisition.

In addition, the high sensitivity of THz waves to water content in the tissues becomes the key contrast factor in many biomedical applications [168,169]; the water content evaluation allows label-free differentiation for various types of tissues as an endogenous marker, distinguishing between healthy and pathological tissues [152,168], see Figure 4.

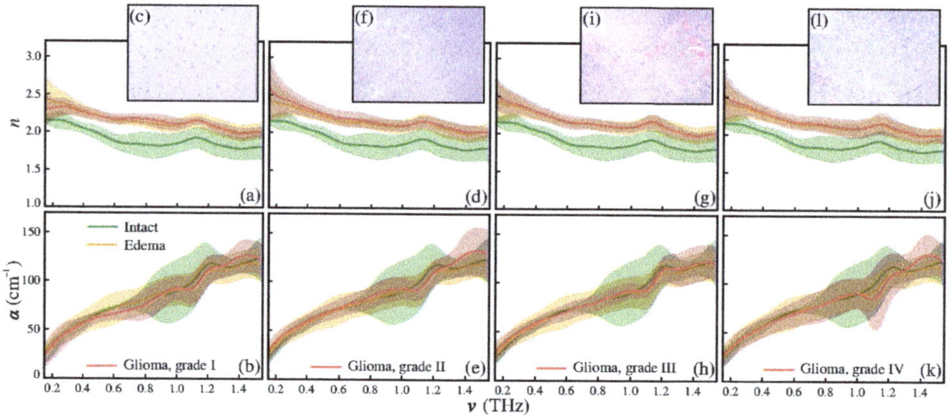

Figure 4. Refractive index n and absorption coefficient α in THz spectral region, observed by Gavdush et al. [152], and H&E-stained histology of gelatin-embedded ex vivo human tissues: intact tissue, gliomas of grade I to IV, and edema. (**a**)–(**c**) grade I; (**d**)–(**f**) grade II; (**g**)–(**i**) grade III; and (**j**)–(**l**) grade IV. The THz optical properties of gliomas are compared with equal data for the intact and edematous tissues, averages within the entire set of brain tissues specimens [152].

In the case of cancer, the contrast between healthy and malign tissue is originated from the structural variation and different water content [170]. In fact, the presence of cancer induces, in many cases, increased blood supply to the affected tissues and a local increase in tissue water content, which suggests water contrast for THz cancer imaging [171]. As a result, it involves a higher refractive index and absorption coefficient [39,169,172]. As proof, Figure 5 shows refractive indexes and absorption coefficients of normal and cancer regions of a fresh rat brain tissue sample, as a function of frequency [39].

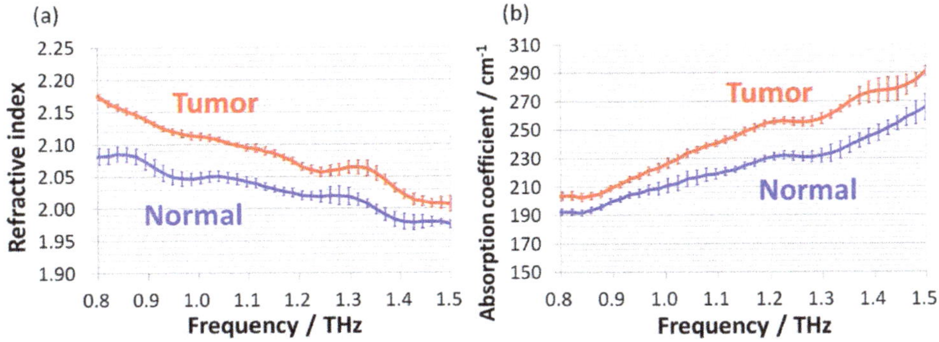

Figure 5. Optical characterization of a fresh rat brain tissue sample [39]. (**a**) Refractive index and (**b**) Absorption coefficient (cm^{-1}) of normal and tumor regions, from 0.8 to 1.5 THz. Both refractive index and absorption coefficient in the tumor region are higher than in normal tissue due to the higher water content in cancer tissue.

TPI has been successfully applied to liver [164,173,174], colon [165,175,176], intestine [161], brain [39,156,177], skin [178], ovarian [167], oral [154] and breast cancers [132,163,179].

For instance, a particular attention deserves the oral-, gastric- and intestinal- neoplasia, because an early/rapid diagnosis is required. However, it may be quite difficult due to oral-gastric-intestinal apparatus anatomy. Sim et al. [154] used a TPI reflection system in the frequency range of 0.2–1.2 THz to assess oral carcinoma. Seven oral tissues from four patients were analyzed, and then the authors compared the optical, frozen and room temperature THz, histopathological images. They found that THz images at −20 °C showed better contrast between healthy and cancer regions compared to the room temperature ones (20 °C). Additionally, frozen temperature images demonstrated better area correlation than the histopathological images [154]. The recent and detailed review of Danciu et al. [180] takes an overview of diagnostics methods and current available data on THz-based detection for digestive cancers. Some current in vitro and ex vivo progress is highlighted for identifying specific digestive neoplasia [180].

Instead, TPI has revealed the contrast between normal and neoplastic tissue regions [159,181]. One of the first applications on human ex vivo wet tissue involved imaging in excised basal cell carcinoma [182]; the regions had recognition of healthy skin and basal cell carcinoma (BBC), both in vitro [96] or in vivo and ex vivo [183,184].

According to Wallace et al. [184], TPI could differentiate between basal cell carcinoma (BCC) and normal tissue both in vivo and ex vivo, and it is under test whether TPI can facilitate the tumor margins delineation, prior to surgery. The BCC's areas have different THz properties compared to healthy tissue. The general clinical image does not identify the skin cancer distribution with accuracy in depth, under the skin surface. However, THz images, as shown in [184], clearly display the cancer distribution on the skin and also the extent of the neoplasm invasion into the skin, with a depth of 250 μm.

Comparing ex vivo and in vivo images of the same tumors, similar contrast levels can be found, however tissue deformation and shrinkage after excision leads to an exact match in the contrast pattern. Nevertheless, the contrast level, present in the THz images, was enough to map diseased tissue margins, when compared to histology. Reese et al. [185], using THz pulsed imaging for studying freshly excised colon cancer, have differentiated normal and diseased colon tissues, with a good contrast for their recognition. Moreover, Wahaia et al. [186] measured colon tissue samples both fixed in formalin and embedded in paraffin. In the first case, water is still present, and in the second one, it is eliminated. They still obtained good image contrast between healthy and cancerous regions. As a result, water is not the only factor contributing to the contrast between the two different tissue areas. Thus, the tumor boundaries in THz images can be recognized, and these are in accordance with the visible images, indicating that the THz imaging technique could be useful for diagnosing cancers. Potentially, this technique could be employed as a complementary label-free technique, allowing surgeons to determine tumor margins in real time.

While different kinds of cancer can be differentiated from healthy tissue according to the high sensitivity of THz radiation to water content, the strong potential of TPI in cancer assessment is not limited to fresh tissues, as it has been efficiently demonstrated for discrimination of pathological and healthy dehydrated tissue as well [132,174,177].

Afterwards, TPI has been used to identify tumor borders on excised breast carcinoma [187]. While THz imaging has proven potential contrast between breast cancer and healthy tissue, both in fresh and formalin-fixed, paraffin-embedded (FFPE) tumors [154,179,188,189], all published works were performed on flat sections of the tumor. The application of THz imaging to a three-dimensional sample can be exploited to produce cross section images by in-depth scanning, without slicing the tumor. Bowman et al. [190], in their work, illustrated the powerful THz pulsed imaging applications to two different carcinoma samples: infiltrating ductal carcinoma and lobular carcinoma, both embedded in paraffin blocks. THz pulsed images showed clear definition of the cancer boundaries in the block. The results were correlated in 3D with histopathology sections sliced throughout the blocks. THz images highlighted the cancer regions only when there were interfaces inside the tissue, thus near to the sample surface. In all the histopathology images, instead, the infiltrating ductal carcinoma and lobular carcinoma were clearly visible, at any section. Furthermore, the THz 3D block images could be sectioned into planar (x-y, x-z, and y-z) images in order to produce in-plane and in-depth tissue images, thus successfully identifying cancer regions, without slicing the tissue. These results demonstrate the effectiveness of THz pulsed imaging for tumor edge identification, when diseased tissue is near the surgical excision.

Reid et al. [176] conducted a study on normal, tumor and dysplastic freshly excised colon samples from 30 patients, using a conventional THz-TDS system, in reflection mode. The study showed a sensitivity of 82% and a specificity of 77% in discriminating between normal and pathological tissues, while a sensitivity of 89% and a specificity of 71% in differentiating normal and dysplastic samples [176]. Another study [191], using diverse intelligent analysis methods (neural networks, decision trees, and support vector machines) re-evaluated the data provided by [191]. This method increased the sensitivity to 90%–100% and the specificity to 86%–90%, improving the overall diagnosis precision.

With the aim of reducing patient re-operation rates and improving the ability of resection margin in breast conserving surgery, Santaolalla et al. [192] applied a multivariate Bayesian classifier to the samples waveform produced by TPI probe system. They can discriminate tumor from benign breast tissue, obtaining a sensitivity of 96% and a specificity of 95%.

Nowadays, in vivo THz detection investigates mostly superficial normal and diseased tissues, close to the epithelial layer (i.e., breast, skin). Observing the significant contrast in THz dielectric permittivity responses of healthy skin, dysplastic and non-dysplastic skin nevi, Zaytsev et al. [193,194] distinguished the precursor of melanoma, with a non-invasive in vivo analysis, proving the efficiency of THz pulsed radiation in the early diagnosis of the melanomas.

To facilitate the use of TPI to scan tumor resection margins intraoperatively, Teraview Ltd. (Cambridge, UK) has proposed and developed a handheld probe system [195]. TPI handheld probe measures tissue sample positioned in histology cassette, in reflection mode. The TPI handheld probe ensures to discriminate benign from malignant breast tissue in an ex vivo setting. The purpose of the THz device is to give valid support for rapid histopathological evaluation of excised tissues in surgery [195].

Despite this, in the last ten years, researchers began to extend THz applications to inner tissues, organs or hollow cavities of the human body. The request is for endoscopic access, achievable using highly bendable waveguides with low transmission loss [196–198]. In 2009, Ji et al. [199] fabricated and developed a miniaturized fiber-coupled THz endoscope apparatus. It emits and collects THz radiation using an optical fiber linked to a femtosecond laser. The measurement is performed close to the reflective surface of a profound tissue or an organ [199]. The researchers tested the device by looking at THz reflections from mouth, tongue and palm skin tissues and demonstrated that the moisture of internal organs is a significant confounding issue. Some years later, an innovative THz prototype, with single-channel detection based on flexible metal-coated THz waveguides [197,198] and a specific polarization exposure method, was integrated into a commercial optical endoscope and demonstrated its potentiality, successfully differentiating between normal and cancer colonic tissues [175]. This also makes it a suitable tool for investigating the water content and hydration profile of the skin. By using some basic image processing techniques like intensity windowing, histogram manipulation, edge detection and region growing [200–202], the discrimination of healthy and cancer regions by THz images can be considerably improved.

The same group also developed THz otoscope that enables the detection of otitis media [203], a disease that causes fluid and purulence in the posterior part of the tympanic membrane within the middle ear. The THz otoscope is able to measure the change in water content on the tympanic membrane; therefore, it is applicable as a medical device for otitis media diagnosis [203].

In [204] and in Figure 6, an imaging device was proposed as a potential tool for the detection of deterioration in the feet of diabetic patients. Assuming that dehydration of the feet skin of diabetics, owing to peripheral vascular disease, is a central element of their deterioration process, they take advantage of THz pulsed imaging in order to obtain a promising diagnostic tool. The results are encouraging and provide key elements that will allow the design of a clinical trial in the future.

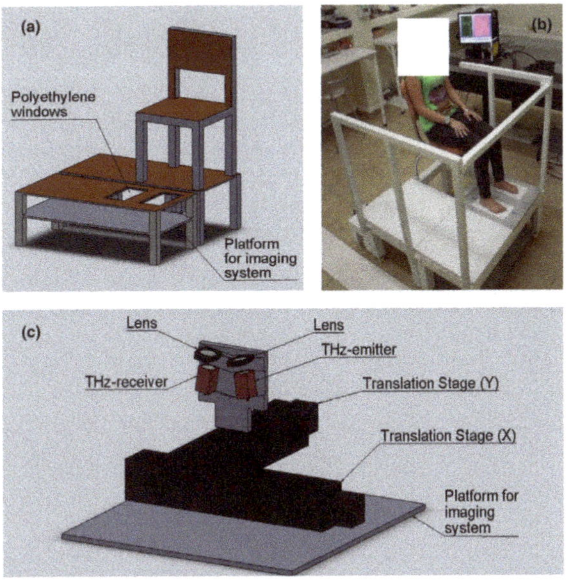

Figure 6. Platform for early screening of diabetic foot syndrome [204]. (**a**) THz setup design. It consists of an elevated surface where two high-density polyethylene windows are used to place the patient's feet. A chair is also provided in order to maintain the patient's position and avoid motion of the feet during the diagnostic image acquisition. The space under the patient sitting area is used to place the THz-TDS spectrometer and a raster scanning system [204]. (**b**) Photograph of the assembled setup during the testing. (**c**) Schematic layout of the raster-scanning imaging system, placed on the platform under the windows as indicated in (**a**).

Concerning brain tumors, Ji et al. in Figure 7 and in [205] demonstrated the effectiveness of THz reflectometry imaging (TRI) in distinguishing cancer and normal regions within a brain tissue section.

Relatively high intensity regions (depicted in red) are spatially well correlated to the tumor areas found with green fluorescent protein (GFP) and Hematoxylin and eosin (H&E) techniques.

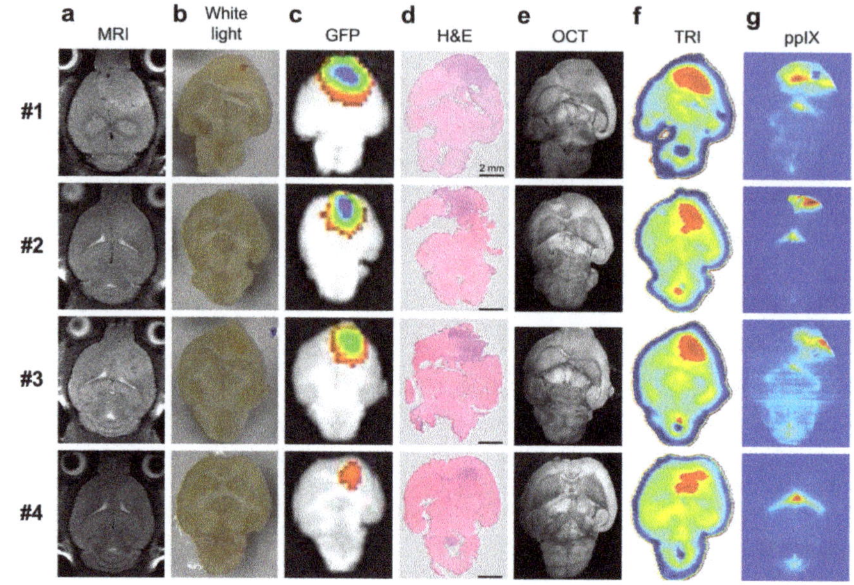

Figure 7. Brain tumor discrimination with different imaging techniques [205]. (**a**) MRI (magnetic resonance imaging). (**b**) White light imaging. (**c**) GFP (green fluorescent protein) imaging. (**d**) H&E (Hematoxylin and eosin) stained imaging. (**e**) OCT (optical coherence tomography) imaging. (**f**) THz reflectometry imaging (TRI). (**g**) 5-ALA-induced ppIX fluorescence imaging. TRI images show red regions (relatively high intensity) that are in agreement with tumor regions images obtained with GFP and H&E modalities, while in white light images, cancer regions are not identified.

Yamaguchi et al. [39], instead, take advantage of different optical properties (like refractive indexes and absorption coefficients) in normal and tumor regions to produce a THz image of a fresh rat brain tissue sample, see Figure 8.

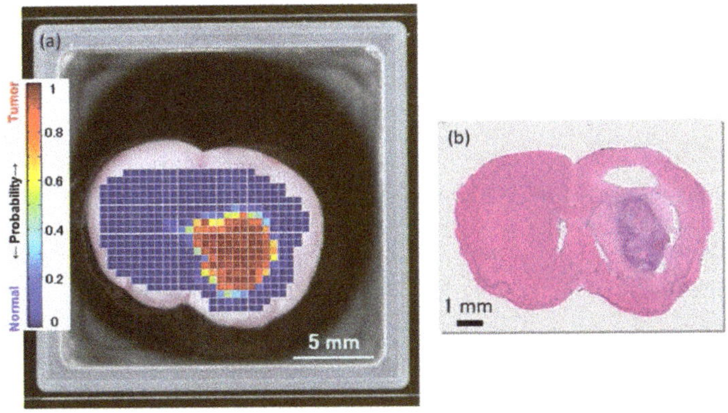

Figure 8. Brain tumor images of rat fresh tissue using different techniques: (**a**) THz spectroscopy and (**b**) HE-stained (hematoxylin and eosin) image of the same tissue section [39]. The THz image has been realized by computing a Tumor probability, where zero value corresponds to healthy tissue, based on the different refractive indexes of tumor and normal tissue regions. The red area in (**a**) shows the tumor region and is well in agreement with the dark purple area in HE-stained image (**b**).

Figure 8a shows the THz image based on the refractive index information (expressed as tumor probability), that results higher in the tumor than in the normal region (see also Figure 5). The red area locates the diseased tissue with great accordance to the HE-stained (hematoxylin and eosin) image of the same tissue section, where a darker purple region is visible (Figure 8b) [39].

Table 2 summarizes the main kind of tumor studied with TPI, with their relative references. Most works are ex vivo, while some papers for skin cancer are available in vivo and in vitro too.

Table 2. Different kind of tumors investigated with TPI and their references.

Tumor	Sample Status	References
Liver	*Ex vivo*	[173,174]
Brain-cervical	*Ex vivo*	[39,156,157,177,205,206]
Breast	*Ex vivo*	[132,163,172,179,181,187–190,192,195]
Oral-gastric-intestinal	*Ex vivo*	[154,161,162,164–166,176,180,185,186,207–210]
Skin	*Ex vivo* *In vivo* *In vitro*	[159,182–184,193,194]
Ovarian	*Ex vivo*	[167]

Additionally, THz radiation is [211] a non-ionizing one, such that it is biologically safe for in vivo applications [212]. Moreover, the limited penetration depth of THz radiation has focused medical imaging research into the dentistry area [213]. TPI has been found to be an interesting technique for dental tissue (enamel, dentine and pulp) discrimination because refractive index differences enable the three tissue regions to be identified [66,148,214–217]. TPI is not the only possible technique for dental disease monitoring [218]: (i) visual caries examination, for example, loss of enamel translucency in the region between the contacting proximal surfaces of two adjacent teeth, (ii) X-ray imaging, (iii) electrical impedance measurements, (iv) ultrasound and (v) fluorescence-based methods [218,219] can also be adopted. However, many disadvantages can be assessed to some of these techniques: (i) is not possible at early detection stages, (i) and (ii) are difficult for posterior teeth, (ii) is potentially harmful for patients' health because it uses ionizing radiation. For these reasons, additional strategies for dental health monitoring, with a particular emphasis on diagnosis at an earlier stage of formation and the use of non-ionizing radiations, are widely requested [218].

Arnone et al. used TPI in order to distinguish different animal and human tissues, in particular enamel, dentine and pulp, in human teeth. They realized two different kind of images with the same data set: TOF and absorption images. Being enamel ~99% mineral and dentine ~70% mineral, they show different refractive indexes, that result in different THz TOFs. The produced TOF image shows enamel only and enamel and dentine regions. In the absorption image, instead, the inner pulp region of tooth is visible, because it shows a strong absorption, thanks to the additional material in that tooth area [214]. Zinov'eva et al. [215] did some transmission images of human permanent molar tooth slices, at different THz frequencies, ranging from 0.2 to 1.5 THz. The teeth used in the experiments were processed to produce artificial lesions by chemical demineralization. The images clearly resolved enamel and dentine areas. In addition, demineralized tooth regions have shown an increase of THz transmission signal in comparison with healthy tooth tissue. This difference can be used to trace demineralization development in dental tissues [215]. Artificial demineralization detection has been the first step towards dental caries detection because caries regions are the result of mineral loss from enamel, causing a change in enamel refractive indexes and absorption coefficients. These changes can be exploited to identify lesions not visible to the naked eye. Early detection is important because initial stages of demineralization are reversible [66,148]. Crawley et al. [148] calculated enamel, dentine and caries absorption coefficients and refractive indexes in the range between 0.3 and 2.0 THz, in THz transmission geometry. Like [214], they produced a THz absorption map in a 210 μm thick tooth slice and a TOF image, but with more accurate spatial resolution. Caries are correctly detected by the absorption image because average absorption coefficient of carious enamel is typically 35% larger than the one associated with healthy enamel, in that frequency range, using TPI. Enamel and dentine are precisely identified using TOF data because they are related to different refractive indexes. In reflection geometry, instead, some images have been created by plotting the change in THz pulse height at a specific time delay after reflection from the sample surface. In these images, both caries are detected, and dentine and enamel are differentiated. Moreover, they investigated a hypomineralization region of a 200 μm thick tooth slice and demonstrated that TPI images can distinguish hypomineralization from enamel caries (demineralization) [216]. Finally, they demonstrated that even a TPI image of a tooth hemisection (much thicker with respect to previous slices) can discriminate caries inside it.

A 3D study of dental tissue has been reported in [217]. The enamel-dentine junction in 12 human incisor teeth has been detected in 91% of the cases. A series of ~100 μm deep steps were chemically produced in order to alter enamel thickness. They were imaged and the authors demonstrated that they accurately and reliably make direct measurements of enamel thickness; this is necessary for monitoring enamel erosion, a common dental disorder.

Finally, in order to verify TPI validity and effectiveness in dental caries detection, Kamburoğlu et al. [218] compared TPI (static images and movie video) with common radiological techniques: intraoral photostimulable phosphor late (PSP) and cone beam CT (CBCT) for the detection of dental caries ex vivo [218–220]. They demonstrated that TPI shows good performances for caries identification, compared to the most used techniques, see Figure 9.

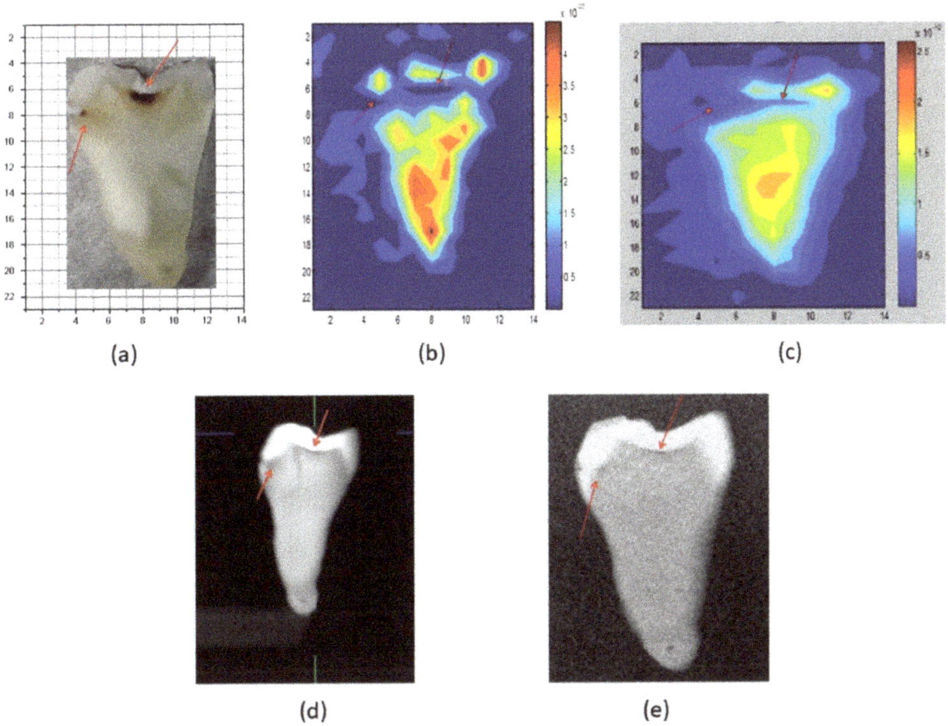

Figure 9. Tooth caries identification with different imaging techniques [218]. (**a**) White light image. (**b**) THz static image at 0.35 THz. (**c**) THz image from video movie. (**d**) CBCT (cone beam CT) image. (**e**) PSP (photostimulable phosphor late) image. Red arrows indicate dental caries in tooth tissue. The five images are in agreement in identifying caries regions.

5. Conclusions

In summary, the rapid THz technological advances have led THz radiation to emerge as a promising and useful tool in medicine. We have outlined the detection techniques of THz pulsed radiation and biomedical applications, based on the use of TPI systems.

In this perspective, we have summarized the prominent advantages in the use of THz radiation, with particular attention to THz pulsed radiation. Briefly, we have illustrated THz pulsed sources, and subsequently incoherent and coherent detectors, that are commercially available and/or of interest for technological research. The common schematic layouts for reflection and transmission imaging modes are discussed too.

Concerning the THz radiation properties, we have discussed the major application of TPI in the field of biomedicine, reporting many examples of ex vivo and in vivo studies, including cases of histopathological imaging of cancers, which are suitable targets. The enhancement of the sensing capabilities and penetration depth within tissues, in medical applications, were addressed. This represents a delicate point to question, and in particular the use of PEAs should require more attention, and many extended trials should be carried out to allow the application of TPI systems in clinical fields and hospitals.

Although THz technology is rapidly increasing as a medical imaging modality, and THz imaging applications in biomedical fields have drawn extensive interest, the technologies are maturing and the principles have been demonstrated in order to direct efforts towards the realization that THz clinical applications are viable in the real world.

Author Contributions: Writing-Original Draft Preparation, A.D. and M.D.F.; Writing-Review & Editing, all authors; Formal Analysis, M.D.F. and V.D.; Supervision, S.L. and M.P.; Project Administration, S.L.; Funding Acquisition, S.L. All authors have read and agreed to the published version of the manuscript.

Funding: We acknowledge the financial support of the Bilateral Cooperation Agreement between Italy and Japan of the Italian Ministry of Foreign Affairs and of the International Cooperation (MAECI), in the framework of the project of major relevance N. PGR00728.

Conflicts of Interest: The authors declare no conflict of interest.

References

1. Cao, Q.; Zhegalova, N.; Wang, S.; Akers, W.; Berezin, M. Multispectral imaging in the extended near-infrared window based on endogenous chromophores. *J. Biomed. Opt.* **2013**, *18*, 101318. [CrossRef] [PubMed]
2. Wilson, R.; Nadeau, K.; Jaworski, F.; Rowland, R.; Nguyen, J.; Crouzet, C.; Saager, R.; Choi, B.; Tromberg, B.; Durkin, A. Quantitative short-wave infrared multispectral imaging of in vivo tissue optical properties. *J. Biomed. Opt.* **2014**, *19*, 086011. [CrossRef] [PubMed]
3. Bazylewski, P.; Ezugwu, S.; Fanchini, G. A review of Three-Dimensional Scanning Near-Field Optical Microscopy (3D-SNOM) and Its Applications in Nanoscale Ligth Management. *Appl. Sci.* **2017**, *7*, 973. [CrossRef]
4. Weng, Q.; Panchai, V.; Lin, K.T.; Sun, L.; Kajihara, Y.; Tzalenchuk, A.; Komiyama, S. Comparison of active and passive methods for the infrared scanning near-field microscopy. *Appl. Phys. Lett.* **2019**, *114*, 153101. [CrossRef]
5. Jeon, S.; Kim, J.; Lee, D.; Baik, J.W.; Kim, C. Review on pratical photoacoustic microscopy. *Photoacoustics* **2019**, *15*, 100141. [CrossRef]
6. Guo, R.; Lu, G.; Qin, B.; Fe, B. Ultrasound Imaging Technologies for Breast Cancer Detection and Management: A Review. *Ultrasound Med. Biol.* **2018**, *44*, 37–70. [CrossRef]
7. Xu, C.; Carney, P.S.; Boppart, S.A. Wavelength-dependent scattering in spectroscopic optical coherence tomography. *Opt. Express* **2005**, *13*, 450–462. [CrossRef]
8. Xu, C.; Vinegoni, C.; Ralston, T.S.; Luo, W.; Tan, W.; Boppart, S.A. Spectroscopic spectral-domain optical coherence microscopy. *Opt. Lett.* **2006**, *31*, 1079–1081. [CrossRef]
9. Kim, M.K. Principles and techniques of digital holographic microscopy. *SPIE Rev.* **2010**, *1*, 018005. [CrossRef]
10. Marquet, P.; Depeursinge, C.; Magistretti, P.J. Review of quantitative phase-digital holographic microscopy: Promising novel imaging technique to resolve neuronal network activity and identify cellular biomarkers of psychiatric disorders. *Neurophotonics* **2014**, *1*, 020901. [CrossRef]
11. Pitkäaho, T.; Manninen, A.; Naughton, T.J. Focus prediction in digital holographic microscopy using deep convolutional neural networks. *Appl. Opt.* **2019**, *58*, A202–A208. [CrossRef] [PubMed]
12. Jermyn, M.; Mok, K.; Mercier, J.; Desroches, J.; Pichette, J.; Saint-Arnaud, K.; Bernstein, L.; Guiot, M.C.; Petrecca, K.; Leblond, F. Intraoperative brain cancer detection with Raman spectroscopy in humans. *Sci. Transl. Med.* **2015**, *7*, 274ra19. [CrossRef] [PubMed]
13. Min, W.; Freudiger, C.W.; Lu, S.; Xie, X.S. Coherent Nonlinear Optical Imaging: Beyond Fluorescence Microscopy. *Annu. Rev. Phys. Chem.* **2011**, *62*, 507–530. [CrossRef] [PubMed]
14. Zumbusch, A.; Langbein, W.; Borri, P. Nonlinear vibrational microscopy applied to lipid biology. *Prog. Lipid. Res.* **2013**, *52*, 615–632. [CrossRef] [PubMed]
15. Zhang, C.; Zhang, D.; Cheng, J.X. Coherent Raman Scattering Microscopy in Biology and Medicine. *Annu. Rev. Biomed. Eng.* **2015**, *17*, 415–445. [CrossRef] [PubMed]
16. D'Arco, A.; Brancati, N.; Ferrara, M.A.; Indolfi, M.; Frucci, M.; Sirleto, L. Subcellular chemical and morphological analysis by stimulated Raman scattering microscopy and image analysis techniques. *BOE* **2016**, *7*, 1853–1864.
17. Cheng, J.X.; Xie, X.S. Vibrational spectroscopic imaging of living systems: An emerging platform for biology and medicine. *Science* **2015**, *350*, aaa8870. [CrossRef]
18. Lu, F.K.; Calligaris, D.; Olibiyi, O.I.; Norton, I.; Yang, W.; Santagata, S.; Xie, X.S.; Golby, A.J.; Agar, N.Y.R. Label-Free Neurosurgical pathology with Stimulated Raman Imaging. *Cancer Res.* **2016**, *76*, OF1–OF12. [CrossRef] [PubMed]

19. Oheim, M.; Michael, D.J.; Geisbauer, M.; Madsen, D.; Chow, R.H. Principles of two-photon excitation fluorescence microscopy and other nonlinear imaging approaches. *Adv. Drug Deliv. Rev.* **2006**, *58*, 788–808. [CrossRef]
20. Müller, M.; Zumbusch, A. Coherent anti-Stokes Raman Scattering Microscopy. *Chem. Phys. Chem.* **2007**, *8*, 2156–2170. [CrossRef]
21. Li, J.; Chen, Q.; Sun, J.; Zhang, J.; Ding, J.; Zuo, C. Three-dimensional tomographic microscopy technique with multi-frequency combination with partially coherent illuminations. *BOE* **2018**, *9*, 2526–2542. [CrossRef] [PubMed]
22. Li, J.; Lin, P.; Tan, Y.; Cheng, J.X. Volumetric stimulated Raman scattering imaging of cleared tissues towards three-dimensional chemical histopathology. *BOE* **2019**, *10*, 4329. [CrossRef] [PubMed]
23. Orringer, D.A.; Pandian, B.; Niknafs, Y.S.; Hollon, T.C.; Boyle, J.; Lewis, S.; Garrard, M.; Hervey-Jumper, S.L.; Garton, H.J.L.; Maher, C.O.; et al. Rapid intraoperative histology of unprocessed surgical specimens via fibre-laser-based stimulated Raman scattering microscopy. *Nat. Biomed. Eng.* **2017**, *1*, 0027. [CrossRef]
24. Denk, W.; Strickler, J.H.; Webb, W.W. Two-photon laser scanning fluorescence microscopy. *Science* **1990**, *248*, 73–76. [CrossRef] [PubMed]
25. Cicchi, R.; Kapsokalyvas, D.; Pavone, F.S. Clinical Nonlinear Laser Imaging of Human Skin: A Review. *Biomed. Res. Int.* **2014**, *2014*, 903589. [CrossRef] [PubMed]
26. Campagnola, P.J.; Loew, L.M. Second-harmonic imaging microscopy for visualizing biomolecular arrays in cells, tissues and organisms. *Nat. Biotechnol.* **2003**, *21*, 1356–1360. [CrossRef]
27. Miller, D.R.; Jarrett, J.W.; Hassan, A.M.; Dunn, A.K. Deep tissue imaging with multiphoton fluorescence microscopy. *Curr. Opin. Biomed. Eng.* **2017**, *4*, 32–39. [CrossRef]
28. Castello, M.; Tortarolo, G.; Buttafava, M.; Deguchi, T.; Villa, F.; Koho, S.; Pesce, L.; Oneto, M.; Pelicci, S.; Lanzanò, L.; et al. A robust and versatile platform for imaging scanning microscopy enabling super-resolution FLIM. *Nat. Methods* **2019**, *16*, 175–178. [CrossRef]
29. Müller, T.; Schumann, C.; Kraegeloh, A. STED Microscopy and its Applications: New Insights into Cellular Processes on the Nanoscale. *Chem. Phys. Chem.* **2012**, *13*, 1986–2000. [CrossRef]
30. Huang, B.; Babcock, H.; Zhuang, X. Breaking the Diffraction Barrier: Super-Resolution Imaging of Cells. *Cell* **2010**, *143*, 1047–1058. [CrossRef]
31. Pietraszewska-Bogiel, A.; Gadella, T.W.J. FRET microscopy: From principle to routine technology in cell biology. *J. Microsc.* **2011**, *241*, 111–118. [CrossRef] [PubMed]
32. Auston, D.H.; Nuss, M.C. Electrooptical generation and detection of femtosecond electrical transients. *IEEE J. Quantum Electron.* **1988**, *24*, 184–197. [CrossRef]
33. Mittleman, D.; Gupta, M.; Neelamani, R.; Baraniuk, R.G.; Rudd, J.V.; Koc, M. Recent advances in terahertz imaging. *Appl. Phy. B* **1999**, *68*, 1085–1094. [CrossRef]
34. Siegel, P.H. Terahertz technology in biology and medicine. *IEEE Trans. Microw. Theory* **2004**, *52*, 2438–2447. [CrossRef]
35. Zhang, X.C. Terahertz wave imaging: Horizons and hurdles. *Phys. Med. Biol.* **2002**, *47*, 3667–3677. [CrossRef]
36. Wallace, V.P.; Taday, P.F.; Fitzgerald, A.J.; Woodward, R.M.; Cluff, J.; Pye, R.J.; Arnone, D.D. Terahertz pulsed imaging and spectroscopy for biomedical and pharmaceutical applications. *Faraday Discuss* **2004**, *126*, 255–263. [CrossRef]
37. Withayachumnankul, W.; Png, G.M.; Yin, X.; Atakaramians, S.; Jones, I.; Lin, H.; Ung, B.S.Y.; Balakrishnan, J.; Ng, B.W.-H.; Ferguson, B.; et al. T-ray sensing and imaging. *Proc. IEEE* **2007**, *95*, 1528–1558. [CrossRef]
38. Mickan, S.; Abbott, D.; Munchb, J.; Zhang, X.C.; van Doornd, T. Analysis of system trade-offs for terahertz imaging. *Microelectron. J.* **2000**, *31*, 503–514. [CrossRef]
39. Yamaguchi, S.; Yamaguchi, S.; Fukushi, Y.; Kubota, O.; Itsuji, T.; Ouchi, T.; Yamamoto, S. Brain tumor imaging of rat fresh tissue using terahertz spectroscopy. *Sci. Rep.* **2016**, *6*, 30124. [CrossRef]
40. Bajwa, N.; Au, J.; Jarrahy, R.; Sung, S.; Fishbein, M.C.; Riopelle, D.; Ennis, D.B.; Aghaloo, T.; John, M.A.; Grundfest, W.S.; et al. Non-invasive terahertz imaging of tissue water content for flap viability assessment. *BOE* **2017**, *8*, 460–474. [CrossRef]
41. Karagoz, B.; Altan, H.; Kamburoglu, K. Terahertz pulsed imaging study of dental caries. In *Medical Laser Applications and Laser-tissue Interactions VII. Proceedings of the European Conference on Biomedical Optics, Optical Society of America, Munich, Germany, 21–25 June 2015*; Lilge, L., Sroka, R., Eds.; SPIE: Washington, NY, USA, 2015; Volume 9542, 95420N.

42. Zaytsev, I.; Dolganova, I.N.; Chernomyrdin, N.V.; Katyba, G.M.; Gavdush, A.A.; Cherkasova, O.P.; Komandin, G.A.; Shchedrina, M.A.; Khodan, A.N.; Ponomarev, D.S.; et al. The progress and perspectives of terahertz technology for diagnosis of neoplasms: A review. *J. Opt.* **2020**, *22*, 013001. [CrossRef]
43. Yang, X.; Zhao, X.; Yang, K.; Liu, Y.; Liu, Y.; Fu, W.; Luo, Y. Biomedical Applications of Terahertz Spectroscopy and Imaging. *Trends Biotechnol.* **2016**, *34*, 810–824. [CrossRef] [PubMed]
44. Son, J.H.; Oh, S.J.; Cheon, H. Potential clinical applications of terahertz radiation. *J. Appl. Phys.* **2019**, *125*, 190901. [CrossRef]
45. Leahy-Hoppa, M.R.; Miragliotta, J.; Osiander, R.; Burnett, J.; Dikmelik, Y.; McEnnis, C.; Spicer, J.B. Ultrafast Laser-Based Spectroscopy and Sensing: Applications in LIBS, CARS, and THz Spectroscopy. *Sensors* **2010**, *10*, 4342–4372. [CrossRef] [PubMed]
46. Fischer, B.; Hoffmann, M.; Helm, H.; Modjesch, G.; Uhd Jepsen, P. Chemical recognition in terahertz time-domain spectroscopy and imaging. *Semicond. Sci. Technol.* **2005**, *20*, S246. [CrossRef]
47. D'Arco, A.; Di Fabrizio, M.; Dolci, V.; Marcelli, A.; Petrarca, M.; Della Ventura, G.; Lupi, S. Characterization of volatile organic compounds (VOCs) in their liquid-phase by terahertz time-domain spectroscopy. *BOE* **2020**, *11*, 1–7. [CrossRef] [PubMed]
48. Stoik, C.D.; Bohn, M.J.; Blackshire, J.L. Nondestructive evaluation of aircraft composites using transmissive terahertz time domain spectroscopy. *Opt. Express* **2008**, *16*, 17039–17051. [CrossRef] [PubMed]
49. Heimbeck, M.S.; Ng, W.R.; Golish, D.R.; Gehm, M.E.; Everitt, H.O. Terahertz digital holographic imaging of voids within visibly opaque dielectrics. *IEEE Trans. Terahertz Sci. Technol.* **2015**, *5*, 110–116. [CrossRef]
50. Giuliano, B.M.; Gavdush, A.A.; Müller, B.; Zaytsev, K.I.; Grassi, T.; Ivlev, A.V.; Palumbo, M.E.; Baratta, G.A.; Scirè, C.; Komandin, G.A.; et al. Broadband spectrsocpy of astrophysical ice analogues I. Direct measurement of the complex refractive index of CO ice using terahertz time-domain spectroscopy. *Astron. Astrophys.* **2019**, *629*, A112. [CrossRef]
51. Federici, J.F.; Schulkin, B.; Huang, F.; Gary, D.; Barat, R.; Oliveira, F.; Zimdars, D. THz imaging and sensing for security applications—Explosives, weapons and drugs. *Semicond. Sci. Technol.* **2005**, *20*, S266. [CrossRef]
52. Ergün, S.; Sönmez, S. Terahertz Technology for Military Applications. *J. Assoc. Inf. Sci. Technol.* **2015**, *3*, 13–16. [CrossRef]
53. Liu, H.B.; Zhong, H.; Karpowicz, N.; Chen, Y.; Zhang, X.C. Terahertz spectroscopy and imaging for defense and security applications. *Proc. IEEE* **2007**, *95*, 1514–1527. [CrossRef]
54. Wang, K.; Sun, D.W.; Pu, H. Emerging non-destructive terahertz spectroscopic imaging technique: Principle and applications in the agri-food industry. *Trend Food Sci. Technol.* **2017**, *67*, 93–105. [CrossRef]
55. Cosentino, A. Terahertz and cultural heritage science: Examination of art and archeology. *Technologies* **2016**, *4*, 6. [CrossRef]
56. Lee, Y.S. *Principles of Terahertz Science and Technology*; Springer: Berlin/Heidelberg, Germany, 2009.
57. Zhang, X.C.; Xu, J. *Introduction to THz Wave Photonics*; Springer: Berlin/Heidelberg, Germany, 2010.
58. Manti, L.; D'Arco, A. Cooperative biological effects between ionizing radiation and other physical and chemical agents. *Mutat. Res.* **2010**, *704*, 115–122. [CrossRef]
59. Titova, L.V.; Rodriguez-Juarez, R.; Woycicki, R.; Hegmann, F.A.; Kovalchuk, O. Intense THz pulses down-regulate genes associated with skin cancer and psoriasis: A new therapeutic avenue? *Sci. Rep.* **2013**, *3*, 2363. [CrossRef] [PubMed]
60. Fëdorov, V.I.; Serdyukov, D.S.; Cherkasova, O.P.; Popova, S.S.; Nemova, E.F. The influence of terahertz radiation on the cell's genetic apparatus. *J. Opt. Technol.* **2017**, *84*, 509–514. [CrossRef]
61. Naftaly, M. (Ed.) *Terahertz Metrology*; Artech House: London, UK, 2015.
62. Chevalier, P.; Amirzhan, A.; Wang, F.; Piccardo, M.; Johnson, S.G.; Capasso, F.; Everitt, H.O. Widely tunable compact terahertz gas lasers. *Science* **2019**, *366*, 856–860. [CrossRef]
63. Otsuji, T. Trends in the research of modern terahertz detectors: Plasmon detectors. *IEEE Trans. Terahertz Sci. Technol.* **2015**, *5*, 1110–1120.
64. Hu, B.B.; Nuss, M.C. Imaging with terahertz waves. *Opt. Lett.* **1995**, *20*, 1716–1718. [CrossRef]
65. Walther, M.; Fischer, B.M.; Ortner, A.; Bitzer, A.; Thoman, A.; Helm, H. Chemical sensing and imaging with pulsed terahertz radiation. *Anal. Bioanal. Chem.* **2010**, *397*, 1009–1017. [CrossRef] [PubMed]
66. Pickwell, E.; Wallace, V.P. Biomedical applications of terahertz technology. *J. Phys. D Appl. Phys.* **2006**, *39*, R301–R310. [CrossRef]

67. Auston, D.H. Picosecond optolectronic switching and gating in silicon. *Appl. Phys. Lett.* **1975**, *26*, 101–103. [CrossRef]
68. Grischkowsky, D.; Keiding, S.; van Exter, M.; Fattinger, C.H. Far-infrared time-domain spectroscopy with terahertz beams of dielectrics and semiconductors. *J. Opt. Soc. Am. B* **1990**, *7*, 2006–2015. [CrossRef]
69. Tani, M.; Herrmann, M.; Sakai, K. Generation and detection of terahertz pulsed radiation with photoconductive antennas and its application to imaging. *Meas. Sci. Technol.* **2002**, *13*, 1739–1745. [CrossRef]
70. Zhang, X.C.; Ma, X.F.; Jin, Y.; Lu, T.M.; Boden, E.P.; Phelp, P.D.; Stewart, K.R.; Yakymyshyn, C.P. Terahertz optical rectification from a nonlinear organic crystal. *Appl. Phys. Lett.* **1992**, *61*, 3080–3082. [CrossRef]
71. Wu, Q.; Zhang, X.C. Terahertz broadband GaP electro-optic sensor. *Appl. Phys. Lett.* **1997**, *70*, 1784. [CrossRef]
72. Winnewisser, C.; Jepsen, P.; Schall, M.; Schyja, V.; Helm, H. Electro-optic detection of THz radiation in $LiaTO_3$, $LiNbO_3$ and ZnTe. *Appl. Phys. Lett.* **1997**, *70*, 3069. [CrossRef]
73. Kim, K.Y.; Taylor, A.J.; Glownia, J.H.; Rodriguez, G. Coherent control of terahertz supercontinuum generation in ultrafast laser-gas interactions. *Nat. Photonics* **2008**, *2*, 605–609. [CrossRef]
74. Thomson, M.D.; Blank, V.; Roskos, H.G. Terahertz white-ligth pulses from an air plasma photo-induced by incommensurate two-color optical fields. *Opt. Express* **2010**, *18*, 23173–23182. [CrossRef]
75. Dai, J.; Liu, J.; Zhang, X.C. Terahertz wave air photonics: Terahertz wave generation and detection with laser-induced gas plasma. *IEEE J. Sel. Top. Quantum Electron.* **2011**, *17*, 183–190. [CrossRef]
76. Planken, P.C.M.; Nuss, M.C.; Knox, W.H.; Miller, D.A.B.; Goossen, K.W. THz pulses from the creation of polarized electron-hole pairs in biased quantum wells. *Appl. Phys. Lett.* **1992**, *61*, 2009–2011. [CrossRef]
77. Sun, G.; Xu, G.; Ding, Y.J.; Zhao, H.; Liu, G.; Zhang, J.; Tansu, N. Efficient Terahertz Generation Within InGaN/GaN Multiple Quantum Wells. *IEEE J. Sel. Top. Quantum Electron.* **2011**, *17*, 48–53. [CrossRef]
78. Roskos, H.G.; Nuss, M.C.; Shah, J.; Leo, K.D.; Miller, A.; Fox, A.M.; Schmitt-Rink, S.; Köhler, K. Coherent submillimeter-wave emission from change oscillations in a double-well potential. *Appl. Phys. Lett.* **1992**, *68*, 2216–2219. [CrossRef] [PubMed]
79. Huber, R.; Brodshelm, A.; Tauser, F.; Leitenstorfer, A. Generation and field-resolved detection of femtosecond electromagnetic pulses tunable up to 41 THz. *Appl. Phys. Lett.* **2000**, *76*, 3191–3193. [CrossRef]
80. Kono, S.; Tani, M.; Sakai, K. Ultrabroadband Photoconductive Detection: Comparison with Free-Space Electro-Optic Sampling. *Appl. Phys. Lett.* **2001**, *79*, 898–900. [CrossRef]
81. Houver, S.; Huber, L.; Savoini, M.; Abreu, E.; Johnson, S.L. 2D THz spectroscopic investigation of ballistic conduction-band electron dynamics in InSb. *Opt. Express* **2019**, *27*, 10854. [CrossRef]
82. Curcio, A.; Dolci, V.; Lupi, S.; Petrarca, M. Terahertz-based retrieval of the spectral phase and amplitude of ultrashort laser pulses. *Opt. Lett.* **2018**, *43*, 783. [CrossRef]
83. Liu, B.; Bromberger, H.; Cartella, A.; Gebert, T.; Först, M.; Cavalleri, A. Generation of narrowband, high-intensity,carrier-envelope phase-stable pulses tunable between 4 and 18 THz. *Opt. Lett.* **2017**, *42*, 129. [CrossRef]
84. Zhang, Y.; Zhang, X.; Li, S.; Gu, J.; Li, Y.; Tian, Z.; Ouyang, C.; He, M.; Hanand, J.; Zhang, W. Broadband THz-TDS System based on DSTMS emitter and LTGInGaAs/InAlAs Photoconductive Antenna Detector. *Sci. Rep.* **2016**, *6*, 26949. [CrossRef]
85. Jazbinšek, M.; Puc, U.; Abina, A.; Zidansek, A. Organic crystals for THz photonics. *Appl. Sci.* **2019**, *9*, 882. [CrossRef]
86. Hebling, J.; Yeh, K.L.; Hoffmann, M.C.; Bartal, B.; Nelson, K.A. Generation of High-Power Terahertz Pulses by Tilted-Pulse-Front Excitation and Their Application Possibilities. *J. Opt. Soc. Am. B* **2008**, *25*, B6–B19. [CrossRef]
87. Hamster, H.; Sullivan, A.; Gordon, S.; White, W.; Falcone, R.W. Subpicosecond Electromagnetic Pulses from Intense Laser-Plasma Interaction. *Phys. Rev. Lett.* **1993**, *71*, 2725. [CrossRef] [PubMed]
88. Hamster, H.; Sullivan, A.; Gordon, S.; Falcone, R.W. Short-Pulse Terahertz Radiation from High-Intensity-Laser- Produced Plasmas. *Phys. Rev. E* **1994**, *49*, 671. [CrossRef]
89. Löffler, T.; Jacob, F.; Roskos, H.G. Generation of terahertz pulses by photoionization of electrically biased air. *Appl. Phys. Lett.* **2000**, *77*, 453–455. [CrossRef]
90. Cook, D.J.; Hochstrasser, R.M. Intense terahertz pulses by four-wave rectification in air. *Opt. Lett.* **2000**, *25*, 1210–1212. [CrossRef]

91. Schmuttenmaer, C.A. Exploring dynamics in the far-infrared with terahertz spectroscopy. *Chem. Rev.* **2004**, *104*, 1759–1780. [CrossRef]
92. Greene, B.I.; Federici, J.F.; Dykaar, D.R.; Levi, A.F.J.; Pfeiffer, L. Picosecond pump and probe spectroscopy utilizing freely propagating terahertz radiation. *Opt. Lett.* **1991**, *16*, 48–49. [CrossRef]
93. Gouider, F.; Vasilyev, Y.B.; Bugár, M.; Könemann, J.; Buckle, P.D.; Nachtwei, G. Terahertz photoresponse of AlInSb/InSb/AlInSb quantum well structures. *Phys. Rev. B* **2010**, *81*, 155304. [CrossRef]
94. Oda, N. Uncooled bolometer-type terahertz focal plane array and camera for real-time imaging. *C. R. Phys.* **2010**, *11*, 496–509. [CrossRef]
95. Dean, P.; Shaukat, M.U.; Khanna, S.P.; Lachab, M.; Burnett, A.; Davies, A.G.; Linfield, E.H.; Chakraborty, S. Absorption–sensitive diffuse reflection imaging of concealed powders using a terahertz quantum cascade laser. *Opt. Express* **2008**, *16*, 5997–6007. [CrossRef] [PubMed]
96. Golay, M.J.E. The theoretical and pratical sensitivity of the pneumatic infrared detector. *Rev. Sci. Instrum.* **1949**, *20*, 816–820. [CrossRef] [PubMed]
97. Karpowicz, N.; Zhong, H.; Xu, J.; Lin, K.I.; Hwang, J.S.; Zhang, X.C. Comparison between pulsed terahertz time-domain imaging and continuous wave terahertz imaging. *Semicond. Sci. Technol.* **2005**, *20*, S293–S299. [CrossRef]
98. Cox, J.A.; Higashi, R.; Nusseibeh, F.; Newstrom-Peitso, K.; Zins, C. Uncooled MEMS-based detector arrays for THz imaging applications. In *Terahertz Physics, Devices, and Systems III: Advanced Applications in Industry and Defense, Proceedings of the SPIE Defense, Security, and Sensing, Orlando, FL, USA, 13–17 April 2009*; SPIE: Washington, NY, USA; Volume 7311, 73110R.
99. Kašalynas, I.; Adam, A.J.L.; Klaassen, O.; Hovenier, N.J.; Pandraud, G.; Iordanov, V.P.; Sarro, P.M. Some properties of a room temperature THz detection array. In *Advanced Optical Materials, Technologies, and Devices, Proceedings of the SPIE Advanced Optical Materials, Technologies, and Devices, Vilnius, Lithuania, 27–30 August 2006*; SPIE: Washington, NY, USA; Volume 6596, 65960J.
100. Semenov, A.; Cojocari, O.; Hübers, H.-W.; Song, F.; Klushin, A.; Müller, A.-S. Application of Zero-Bias Optical Schottky-Diode Detectors for Monitoring Short-Pulse and Weak Terahertz Radiation. *IEEE Electron Device Lett.* **2010**, *31*, 674–676. [CrossRef]
101. Maestrini, A.; Thomas, B.; Wang, H.; Jung, C.; Treuttel, J.; Jin, Y.; Chattopadhyay, G.; Mehdi, I.; Beaudin, G. Schottky diode-based terahertz frequency multipliers and mixers. *C. R. Phys.* **2010**, *11*, 480–495. [CrossRef]
102. Nazarov, M.M.; Makarova, S.A.; Shkurinov, A.P.; Okhotnikov, O.G. The use of combination of nonlinear optical materials to control terahertz pulse generation and detection. *Appl. Phys. Lett.* **2008**, *92*, 021114–021117. [CrossRef]
103. Zhang, Y.; Hosono, S.; Nagai, N.; Song, S.-H.; Hirakawa, K. Fast and sensitive bolometric terahertz detection at room temperature through thermomechanical transduction. *J. Appl. Phys.* **2019**, *125*, 151602. [CrossRef]
104. Rezvani, J.; Di Gioacchino, D.; Gatti, C.; Poccia, N.; Ligi, C.; Tocci, S.; Cestelli Guidi, M.; Cibella, S.; Lupi, S.; Marcelli, A. Tunable Vortex Dynamics in Proximity Junction Arrays: A Possible Accurate and Sensitive 2D THz Detector. In Proceedings of the LIV Zakopane School of Physics, Breaking Frontiers, Zakopane, Poland, 21–25 May 2019; Volume 137, pp. 17–20.
105. Knap, W.; Dyakonov, M.I. Field effect transistors for terahertz applications. In *Handbook of Terahertz Technologiey for Imaging, Sensing and Communications*; Volume in Woodhead Publishing Series in Electronics and Optical Materials; Saeedkia, D., Ed.; Elsevier: Amsterdam, The Nederland, 2013; pp. 121–155.
106. Fan, K.; Suen, J.Y.; Liu, X.; Padilla, W.J. All-dielectric metasurface absorbers for uncooled terahertz imaging. *Optica* **2017**, *4*, 601–604. [CrossRef]
107. Guerboukha, H.; Nallappan, K.; Skorobogatiy, M. Toward real-time terahertz imaging. *Adv. Opt. Phot.* **2018**, *10*, 843–938. [CrossRef]
108. Cho, H.; Lee, S.H.; Nam-Gung, C.; Oh, S.J.; Oh, J.H.; Park, H.; Ahn, C.B. Fast terahertz reflection tomography using block-based compressed sensing. *Opt. Express* **2011**, *19*, 16401–16409. [CrossRef]
109. Hwang, B.M.; Lee, S.H.; Lim, W.T.; Ahn, C.B.; Son, J.H.; Park, H. A fast spatial-domain terahertz imaging using block-based compressed sensing. *J. Infrared Millim. Terahertz Waves* **2011**, *32*, 1328–1336. [CrossRef]
110. Hong, H.J.; Park, J.; Park, H.; Son, J.H.; Ahn, C.B. Pre- and post-processing for tomographic reconstruction of terahertz time-domain spectroscopy. *Opt. Express* **2013**, *21*, 19943–19950. [CrossRef] [PubMed]

111. Wallace, V.P.; Macpherson, E.; Zeitler, J.A.; Reid, C. Three-dimensional imaging of optically opaque materials using nonionizing terahertz radiation. *J. Opt. Soc. Am. Opt. Image Sci. Vis.* **2008**, *25*, 3120–3133. [CrossRef] [PubMed]
112. Castro-Camus, E.; Alfaro, M. Photoconductive devices for terahertz pulsed spectroscopy: A review. *Photon. Res.* **2016**, *4*, A36. [CrossRef]
113. Chan, W.L.; Deibel, J.; Mittleman, D.M. Imaging with terahertz radiation. *Rep. Prog. Phys.* **2007**, *70*, 1325–1379. [CrossRef]
114. Adam, A.J.L. Review of Near-Field Terahertz Measurement Methods and Their Applications. *J. Infrared Millim. Terahertz Waves* **2011**, *32*, 976–1019. [CrossRef]
115. Goodman, J.W. *Introduction to Fourier Optics*, 2nd ed.; McGraw-Hill: New York, NY, USA, 1996.
116. Abbe, E. Beiträge zur theroie des mikroskops und der mikroskopischen wahrnehmung. *Arch. Mikrosk. Anat.* **1873**, *9*, 413–468. [CrossRef]
117. Born, M.; Wolf, E. *Principles of Optics*, 7th ed.; Cambridge University Press: Cambridge, UK, 1999.
118. Pawley, J.B. *Handbook of Biological Confocal Microscopy*; Springer: Berlin/Heidelberg, Germany, 2006.
119. Yuan, T.; Xu, Z.; Zhang, X.C. Development of terahertz wave microscopes. *Infrared Phys. Technol.* **2004**, *45*, 417–425. [CrossRef]
120. Blanchard, F.; Doi, A.; Tanaka, T.; Tanaka, K. Real-time, subwavelength terahertz imaging. *Annu. Rev. Mater. Res.* **2013**, *43*, 237–259. [CrossRef]
121. Hunsche, S.; Koch, M.; Brener, I.; Nuss, M.C. THz near-field imaging. *Opt. Commun.* **1998**, *150*, 22–26. [CrossRef]
122. Flammini, M.; Bonsi, C.; Ciano, C.; Giliberti, V.; Pontecorvo, E.; Italia, P.; DelRe, E.; Ortolani, M. Confocal terahertz imaging of ancient manuscripts. *J. Infrared Millim. Terahertz Waves* **2017**, *38*, 435–442. [CrossRef]
123. Salhi, M.A.; Pupeza, I.; Koch, M. Confocal THz laser microscope. *J. Infrared Millm. Terahertz Waves* **2010**, *31*, 358–366. [CrossRef]
124. Saeedkia, D. *Handbook of Terahertz Technology for Imaging, Sensing and Communications*; Volume in Woodhead Publishing Series in Electronics and Optical Materials; Saeedkia, D., Ed.; Elsevier: Amsterdam, The Nederland, 2013.
125. Adam, A.J.; Brok, J.M.; Seo, M.A.; Ahn, K.J.; Kim, D.S.; Kang, J.H.; Park, Q.H.; Nagel, M.; Planken, P.C. Advanced terahertz electric near-field measurements at sub-wavelength diameter metallic apertures. *Opt. Express* **2008**, *16*, 7407–7417. [CrossRef] [PubMed]
126. Mitrofanov, O.; Brener, I.; Harel, R.; Wynn, J.D.; Pfeiffer, L.N.; West, K.W.; Federici, J. Terahertz near-field microscopy based on a collection mode detector. *Appl. Phys. Lett.* **2000**, *77*, 3496. [CrossRef]
127. Iwami, K.; Ono, T.; Esashi, M. A New Approach to Terahertz Local Spectroscopy Using Microfabricated Scanning Near-Field Probe. *Jpn. J. Appl. Phys.* **2008**, *47*, 8095–8097. [CrossRef]
128. Cao, H.; Agrawal, A.; Nahata, A. Controlling the transmission resonance lineshape of a single subwavelength aperture. *Opt. Express* **2005**, *13*, 763–769. [CrossRef]
129. Keilmann, F.; Hillenbrand, R. Near-field microscopy by elastic light scattering from a tip. *Philos. Trans. Math. Phys. Eng. Sci.* **2004**, *362*, 787–805. [CrossRef]
130. Adam, A.J.L.; Brok, J.M.; Planken, P.C.M.; Seo, M.A.; Kim, D.S. THz near-field measurements of metal structure. *C. R. Phys.* **2008**, *9*, 161–168. [CrossRef]
131. Löffler, T.; Siebert, K.; Czasch, S.; Bauer, T.; Roskos, H.G. Visualization and classification in biomedical terahertz pulsed imaging. *Phys. Med. Biol.* **2002**, *47*, 3847–3852. [CrossRef]
132. Fitzgerald, A.J.; Wallace, V.P.; Pinder, S.E.; Purushotham, A.D.; O'Kelly, P.; Ashworth, P.C. Classification of terahertz-pulsed imaging data from excised breast tissue. *J. Biomed. Opt.* **2012**, *17*, 016005. [CrossRef]
133. Ushakov, A.; Chizhov, P.; Bukin, V.; Savel'ev, A.; Garnov, S. Broadband in-line terahertz 2D imaging: Comparative study with time-of-ligth, cross-correlation, and Fourier transform data processing. *J. Opt. Soc. Am. B* **2018**, *35*, 1159–1164. [CrossRef]
134. Wan, M.; Healy, J.J.; Sheridan, J.T. Terahertz phase imaging and biomedical applications. *Opt. Laser Technol.* **2020**, *122*, 105859. [CrossRef]
135. Dressel, M.; Grüner, G. *Electrodynamics of Solids*; Cambridge University Press: Cambridge, UK, 2002.
136. Kužel, P.; Němec, H.; Kadlec, F. Gouy shift correction for highly accurate refractive index retrieval in time-domain terahertz spectroscopy. *Opt. Express* **2010**, *18*, 15338–15348. [CrossRef] [PubMed]

137. Hegmann, F.A.; Ostroverkhova, O.; Cooke, D.G. Probing organic semiconductors with terahertz pulses. In *Photophysics of Molecular Materials*; Wiley: Hoboken, NJ, USA, 2000; pp. 367–428.
138. Llioyd-Hughes, J.; Jeon, T.I. A review of the terahertz conductivity of bulk and nano-materials. *J. Infrared Millim. Terahertz Waves* **2012**, *33*, 871–925. [CrossRef]
139. Guillet, J.P.; Recur, B.; Frederique, L.; Bousquet, B.; Canioni, L.; Manek-Hönninger, I.; Desbarats, P.; Mounaix, P. Review of Terahertz Tomography Techniques. *J. Infrared. Millim. Terahertz Waves* **2014**, *35*, 382–411. [CrossRef]
140. Mittleman, D.M. Twenty years of terahertz imaging. *Opt. Express* **2018**, *26*, 9417–9431. [CrossRef]
141. Ferguson, B.; Wang, S.; Gray, D.; Abbot, D.; Zhang, X.-C. T-ray computed tomography. *Opt. Lett.* **2002**, *27*, 1312–1314. [CrossRef]
142. Thrane, L.; Jacobsen, R.H.; Jepsen, P.U.; Keiding, S.R. THz reflection spectroscopy of liquid water. *Chem. Phys. Lett.* **1995**, *240*, 330–333. [CrossRef]
143. Cheon, H.; Yang, H.J.; Son, J.-H. Toward Clinical Cancer Imaging Using Terahertz Spectroscopy. *IEEE J. Sel. Top. Quantum Electron.* **2017**, *23*, 8600109. [CrossRef]
144. Zaytsev, K.I.; Gavdush, A.; Chernomyrdin, N.; Yurchenko, S. Highly accurate in vivo terahertz spectroscopy of healthy skin: Variation of refractive index and absorption coefficient along the human body. *IEEE Trans. Terahertz Sci. Technol.* **2015**, *5*, 817–827. [CrossRef]
145. Parrott, E.; Sy, S.; Blu, T.; Wallace, V.; Pickwell-MacPherson, E. Terahertz pulsed imaging in vivo: Measurements and processing methods. *J. Biomed. Opt.* **2011**, *16*, 106010. [CrossRef] [PubMed]
146. Kolesniko, A.; Kolesnikova, E.; Popov, A.; Nazarov, M.; Shkurinov, A.; Tuchin, V. In vitro terahertz monitoring of muscle tissue dehydration under the action of hyperosmotic agents. *Quantum Electron.* **2014**, *44*, 633. [CrossRef]
147. Smolyanskaya, O.; Schelkanova, I.; Kulya, M.; Odlyanitskiy, E.; Goryachev, I.; Tcypkin, A.; Grachev, Y.; Toropova, Y.; Tuchin, V. Glycerol dehydration of native and diabetic animal tissues studied by THz-TDS and NMR methods. *BOE* **2018**, *9*, 1198–1215. [CrossRef] [PubMed]
148. Fitzgerald, A.J.; Berry, E.; Zinov'ev, N.N.; Homer-Vanniasinkam, S.; Miles, R.E.; Chamberlain, J.M.; Smith, M.A. Catalogue of human tissue optical properties at terahertz frequencies. *J. Biol. Phys.* **2003**, *29*, 123–128. [CrossRef] [PubMed]
149. Kan, W.C.; Lee, W.S.; Cheung, W.H.; Wallace, V.P.; Pickwell-Macpherson, E. Terahertz pulsed imaging of knee cartilage. *BOE* **2010**, *1*, 967–974. [CrossRef] [PubMed]
150. Sun, Y.; Fischer, B.; Pickwell-MacPherson, E. Effects of formalin fixing on the terahertz properties of biological tissues. *J. Biomed. Opt.* **2009**, *14*, 064017. [CrossRef]
151. Fan, S.; Ung, B.; Parrott, E.P.J.; Pickwell-MacPherson, E. Gelatin embedding: A novel way to preserve biological samples for terahertz imaging and spectroscopy. *Phys. Med. Biol.* **2015**, *60*, 2703–2713. [CrossRef]
152. Gavdush, A.A.; Chernomyrdin, N.V.; Malakhov, K.M.; Beshplav, S.-I.T.; Dolganova, I.N.; Kosyrkova, A.V.; Nikitin, P.V.; Musina, G.R.; Katyba, G.M.; Reshetov, I.V.; et al. Terahertz spectroscopy of gelatin-embedded human brain gliomas of different grades: A road toward intraoperative THz diagnosis. *J. Biomed. Opt.* **2019**, *24*, 027001. [CrossRef]
153. Png, G.M.; Choi, J.W.; Ng, B.W.; Mickan, S.P.; Abbott, D.; Zhang, X.C. The impact of hydration changes in fresh bio-tissue on THz spectroscopic measurements. *Phys. Med. Biol.* **2008**, *53*, 3501. [CrossRef]
154. Sim, Y.; Park, J.; Ahn, K.-M.; Park, C.; Son, J.-H. Terahertz imaging of excised oral cancer at frozen temperature. *BOE* **2013**, *4*, 1413–1421. [CrossRef]
155. He, Y.; Ung, B.-Y.; Parrott, E.; Ahuja, A.; Pickwell-MacPherson, E. Freeze-thaw hysteresis effects in terahertz imaging of biomedical tissues. *BOE* **2016**, *7*, 4711–4717. [CrossRef] [PubMed]
156. Meng, K.; Chen, T.N.; Chen, T.; Zhu, T.G.; Liu, Q.; Li, Z.; Li, F.; Zhong, S.-C.; Li, Z.-R.; Feng, H.; et al. Terahertz pulsed spectroscopy of paraffin embedded brain glioma. *J. Biomed. Opt.* **2014**, *19*, 077001. [CrossRef] [PubMed]
157. Oh, S.J.; Kim, S.H.; Jeong, K.; Park, Y.; Huh, Y.M.; Son, J.H.; Suh, J.S. Measurement depth enhancement in terahertz imaging of biological tissues. *Opt. Express* **2013**, *21*, 21299–21305. [CrossRef] [PubMed]
158. Wang, Y.; Notake, T.; Tang, M.; Nawata, K.; Ito, H.; Minamide, H. Terahertz-wave water concentration and distribution measurement in thin biotissue based on a novel sample preparation. *Phys. Med. Biol.* **2011**, *56*, 4517. [CrossRef] [PubMed]
159. Wallace, V.P.; Fitzgerald, A.J.; Pickwell, E.; Pye, R.J.; Taday, P.F.; Flanagan, N.; Ha, T. Terahertz pulsed spectroscopy of human basal cell carcinoma. *Appl. Spectrosc.* **2006**, *60*, 1127–1133. [CrossRef] [PubMed]

160. Sun, Y.; Sy, M.Y.; Wang, Y.X.J.; Ahuja, A.T.; Zhang, Y.T.; Pickwell-MacPherson, E. A promising diagnostic method: Terahertz pulsed imaging and spectroscopy. *World J. Radiol.* **2011**, *3*, 55–65. [CrossRef]
161. Yeo, W.G.; Gurel, O.; Hitchcock, C.L.; Park, S.; Sertel, K.; Nahar, N.K. Evaluation of cancer tissue morphology via THz spectroscopic imaging: Human lung and small intestine malignancies. *Infrared Phys. Technol.* **2019**, *97*, 411–416. [CrossRef]
162. Wahaia, F.; Kasălynas, I.; Minkevičius, L.; Carvalho Silva, C.D.; Urbanowicz, A.; Valušis, G. Terahertz spectroscopy and imaging for gastic cancer diagnosis. *J. Spectr. Imaging* **2020**, *9*, 1–8.
163. El-Shenawee, M.; Vohra, N.; Bowman, T.; Bailey, K. Cancer detection in excised breast tumors using terahertz imaging and spectroscopy. *Biomed. Spectrosc. Imaging* **2019**, *8*, 1–9. [CrossRef]
164. Cao, Y.; Huang, P.; Chen, J.; Ge, W.; Hou, D.; Zhang, G. Qualitative and quantitative detection of liver injury with terahertz time-domain spectroscopy. *BOE* **2020**, *11*, 982–993. [CrossRef]
165. Cao, Y.; Chen, J.; Huang, P.; Ge, W.; Hou, D.; Zhang, G. Inspecting human colon adenocarcinoma cell lines by using terahertz time-domain reflection spectroscopy. *Spectrochim. Acta* **2019**, *211*, 356–363. [CrossRef] [PubMed]
166. Kuzikova, A.V.; Grigorev, R.O.; Kurasova, A.P.; Demchenko, P.S.; Senyuk, A.V.; Zakharenko, A.A.; Belolipetskaya, J.R.; Khamid, A.H.; Khodzitsky, M.K. Study of refractive index of human stomach cancer tissue in THz frequency range. In Proceedings of the 6th International School and Conference "Saint Petersburg OPEN 2019": Optoelectronics, Photonics, Engineering and Nanostructure, Saint Petersburg, Russia, 22–25 April 2019; Volume 1410, 012070.
167. Globus, T.; Moskaluk, C.; Pramoonjago, P.; Gelmont, B.; Moyer, A.; Bykhovski, A.; Ferrance, J. Sub-terahertz vibrational spectroscopy of ovarian cancer and normal control tissue for molecular diagnostic technology. *Cancer Biomark.* **2019**, *24*, 405–419. [CrossRef] [PubMed]
168. Yamaguchi, S.; Fukushi, Y.; Kubota, O.; Itsuji, T.; Ouchi, T.; Yamamoto, S. Origin and quantification of differences between normal and tumor tissues observed by terahertz spectroscopy. *Phys. Med. Biol.* **2016**, *68*, 6808–6820. [CrossRef] [PubMed]
169. Chau, D.Y.S.; Dennis, A.R.; Lin, H.; Zeitler, J.A.; Tunnacliffe, A. Determination of Water Content in Dehydrated Mammalian Cells Using Terahertz Pulsed Imaging: A Feasibility Study. *Curr. Pharm. Biotechnol.* **2015**, *17*, 200–207. [CrossRef]
170. Smolyanskaya, O.A.; Chernomyrdin, N.V.; Konovko, A.A.; Zaytsev, K.I.; Ozheredov, I.A.; Cherkasova, O.P.; Nazarov, M.M.; Guillet, J.P.; Kozlov, S.A.; Kistenev, Y.V.; et al. Terahertz biophotonics as a tool for studies of dielectric and spectral properties of biological tissues and liquids. *Prog. Quantum Electron.* **2018**, *62*, 1–77. [CrossRef]
171. Brun, M.A.; Formanek, F.; Yasuda, A.; Sekine, M.; Ando, N.; Eishii, Y. Terahertz imaging applied to cancer diagnosis. *Phys. Med. Biol.* **2010**, *55*, 4615–4623. [CrossRef]
172. Chen, H.; Chen, T.H.; Tseng, T.F.; Lu, J.T.; Kuo, C.C.; Fu, S.C.; Lee, W.J.; Tsai, Y.F.; Huang, Y.Y.; Chuang, E.Y.; et al. High-sensitivity in vivo THz transmission imaging of early human breast cancer in a subcutaneous xenograft mouse model. *Opt. Express* **2011**, *19*, 21552–21562. [CrossRef]
173. Miura, Y.; Kamataki, A.; Uzuki, M.; Sasaki, T.; Nishizawa, J.I.; Sawai, T. Terahertz-wave spectroscopy for precise histopathological imaging of tumor and non-tumor lesions in paraffin sections. *J. Exp. Med.* **2011**, *223*, 291–296. [CrossRef]
174. Knobloch, P.; Schildknecht, C.; Kleine-Ostmann, T.; Koch, M.; Hoffmann, S.; Hofmann, M.; Rehberg, E.; Sperling, M.; Donhuijsen, K.; Hein, G.; et al. Medical THz imaging: An investigation of histo-pathological samples. *Phys. Med. Biol.* **2002**, *47*, 3875. [CrossRef]
175. Doradla, P.; Alavi, K.; Joseph, C.; Giles, R. Single-channel prototype terahertz endoscopic system. *J. Biomed. Opt.* **2014**, *19*, 080501. [CrossRef]
176. Reid, C.B.; Fitzgerald, A.; Reese, G.; Goldin, R.; Tekkis, P.; O'Kelly, P.S.; Pickwell-MacPherson, E.; Gibson, A.P.; Wallace, V.P. Terahertz pulsed imaging of freshly excised human colonic tissues. *Phys. Med. Biol.* **2011**, *56*, 4333–4353. [CrossRef] [PubMed]
177. Oh, S.J.; Kim, S.H.; Ji, Y.B.; Jeong, K.; Park, Y.; Yang, J.; Park, D.W.; Noh, S.K.; Kang, S.G.; Huh, Y.M.; et al. Study of freshly excised brain tissues using terahertz imaging. *BOE* **2014**, *5*, 2837–2842. [CrossRef] [PubMed]
178. Zhong, S. Progress in terahertz nondestructive testing: A review. *Front. Mech. Eng.* **2019**, *14*, 273–281. [CrossRef]

179. Bowman, T.C.; El-Shenawee, M.; Campbell, L.K. Terahertz Imaging of Excised Breast Tumor Tissue on Paraffin Sections. *IEEE Trans. Antennas Propag.* **2015**, *63*, 2088–2097. [CrossRef]
180. Danciu, M.; Alexa-Stratulat, T.; Stefanescu, C.; Dodi, G.; Tamba, B.I.; Mihai, C.T.; Stanciu, G.D.; Luca, A.; Spiridon, I.A.; Ungureanu, L.B.; et al. Terahertz Spectroscopy and Imaging: A Cutting-Edge Method for Diagnosing Digestive Cancers. *Materials* **2019**, *12*, 1519. [CrossRef] [PubMed]
181. Wallace, V.P.; MacPherson, E.; Fitzgerald, A.J.; Lo, T.; Provenzano, E.; Pinder, S.; Purushotham, A. Terahertz pulsed imaging and spectroscopy of breast tumors. In Proceedings of the Optical Methods in the Life Sciences, Boston, MA, USA, 1–3 October 2006.
182. Woodward, R.M.; Wallace, V.P.; Pye, R.J.; Cole, B.E.; Arnone, D.D.; Linfield, E.H.; Pepper, M. Terahertz pulse imaging of ex vivo basal cell carcinoma. *J. Investig. Dermatol.* **2003**, *120*, 72–78. [CrossRef]
183. Woodward, R.M.; Cole, B.E.; Wallace, V.P.; Pye, R.J.; Arnone, D.D.; Linfield, E.H.; Pepper, M. Terahertz pulse imaging in reflection geometry of human skin cancer and skin tissue. *Phys. Med. Biol.* **2002**, *47*, 3853. [CrossRef]
184. Wallace, V.P.; Fitzegerald, A.J.; Shankar, S.; Flanagan, N.; Pye, R.; Cluff, J.; Arnone, D.D. Dermatological Surgery and Lasers. Terahertz pulsed imaging of basal cell carcinoma ex vivo and in vivo. *Br. J. Dermatol.* **2004**, *151*, 424–432. [CrossRef]
185. Reese, G.; Reid, C.; Goldin, R.; Tran-Dang, M.A.; Fitzgerald, A.; Tekkis, P.; Wallace, V.P. Using terahertz pulsed imaging (TPI) to identify colonic pathology. In Proceedings of the 33rd International Conference on Infrared, Millimeter and Terahertz Waves, IEEE, Pasadena, CA, USA, 15–19 September 2008.
186. Wahaia, F.; Valusis, G.; Bernardo, L.M.; Almeida, A.; Moreira, J.A.; Lopes, P.C.; Macutkevic, J.; Kasalynas, I.; Seliuta, D.; Adomavicius, R.; et al. Detection of colon cancer by terahertz techniques. *J. Mol. Struct.* **2011**, *1006*, 77–82. [CrossRef]
187. Fitzgerald, A.J.; Wallace, V.P.; Jimenez-Linan, M.; Bobrow, L.; Pye, R.J.; Purushotham, A.D.; Arnone, D.D. Terahertz pulsed imaging of human breast tumors. *Radiology* **2006**, *239*, 533–540. [CrossRef]
188. Bowman, T.C. *Experimental Terahertz Imaging and Spectroscopy for Ex-vivo Breast Cancer Tissue*; Master of Science in Electrical Engineering (Graduate), University of Arkansas: Fayetteville, AR, USA, August 2014.
189. Ashworth, P.C.; Pickwell-MacPherson, E.; Provenzano, E.; Pinder, S.E.; Purushotham, A.D.; Pepper, M.; Wallace, V.P. Terahertz pulsed spectroscopy of freshly excised human breast cancer. *Opt. Express* **2009**, *17*, 12444–12454. [CrossRef] [PubMed]
190. Bowman, T.; Wu, Y.; Gauch, J.; Campbell, L.K.; El-Shenawee, M. Terahertz Imaging of Three-Dimensional Dehydrated Breast Cancer Tumors. *J. Infrared Millim. Terahertz Waves* **2017**, *38*, 766–786. [CrossRef]
191. Eadie, L.H.; Reid, C.B.; Fitzgerald, A.J.; Wallace, V.P. Optimizing multi-dimensional terahertz imaging analysis for colon cancer diagnosis. *Expert Syst. Appl.* **2013**, *40*, 2043–2205. [CrossRef]
192. Santaolalla, A.; Sheikh, M.; Van Hemelrijck, M.; Portieri, A.; Coolen, A.C.C. Improved resection margins in breast-conserving surgery using Terahertz Pulsed imaging data. *arXiv* **2018**, arXiv:1805.01349.
193. Zaytsev, K.I.; Chernomyrdin, N.V.; Kudrin, K.C.; Gavdush, A.A.; Nosova, P.A.; Yurchenko, S.O.; Reshetov, I.V. In vivo terahertz pulsed spectroscopy of dysplastic and non-dysplastic skin nevi. *J. Phys. Conf. Ser.* **2016**, *735*, 012076. [CrossRef]
194. Zaytsev, K.I.; Kudrin, K.G.; Karasik, V.E.; Reshetov, I.V.; Yurchenko, S.O. In vivo terahertz spectroscopy of pigmentary skin nevi: Pilot study of non-invasive early diagnosis of dysplasia. *Appl. Phys. Lett.* **2015**, *106*, 053702. [CrossRef]
195. Grootendorst, M.R.; Fitzgerald, A.J.; Brouwer, S.G.; de Koning, S.A.; Portieri, A.; Van Hemelrijck, M.; Young, M.R.; Owen, J.; Cariati, M.; Pepper, M.; et al. Use of a handheld terahertz pulsed imaging device to differentiate benign and malignant breast tissue. *BOE* **2017**, *8*, 2932–2945. [CrossRef]
196. Matsuura, Y.; Takeda, E. Hollow optical fibers loaded with an inner dielectric film for terahertz broadband spectroscopy. *JOSA* **2008**, *25*, 1949–1954. [CrossRef]
197. Doradla, P.; Joseph, C.S.; Kumar, J.; Giles, R.H. Characterization of bending loss in hollow flexible terahertz waveguides. *Opt. Express* **2012**, *20*, 19176–19184. [CrossRef]
198. Doradla, P.; Joseph, C.S.; Kumar, J.; Giles, R.H. Propagation loss optimization in metal/dielectric coated hollow flexible terahertz waveguides. *Proc. SPIE* **2012**, *8261*, 82610P1–82610P10.

199. Ji, Y.B.; Lee, E.S.; Kim, S.-H.; Son, J.H.; Jeon, T.I. A miniaturized fiber-coupled terahertz endoscope system. *Opt. Express* **2009**, *17*, 17082–17087. [CrossRef] [PubMed]
200. Gonzalez, R.C.; Woods, R.E. *Digital Image Processing*, 3rd ed.; Pearson Prentice Hall: Upper Saddle River, NJ, USA, 2008; pp. 120–144.
201. Pratt, W.K. *Digital Image Processing*, 4th ed.; John Wiley & Sons: Hoboken, NJ, USA, 2007; pp. 288–291.
202. Canny, J.F. A Computational Approach to Edge Detection. *IEEE Trans. Pattern Anal. Mach. Intell.* **1986**, *8*, 679–698. [CrossRef] [PubMed]
203. Ji, Y.B.; Moon, I.-S.; Bark, H.S.; Kim, S.H.; Park, D.W.; Noh, S.K.; Huh, Y.-M.; Suh, J.-S.; Oh, S.J.; Jeon, T.-I. Terahertz otoscope and potential for diagnosis otitis media. *BOE* **2016**, *7*, 1201. [PubMed]
204. Hernandez-Cardoso, G.G.; Rojas-Landeros, S.C.; Alfaro-Gomez, M.; Hernandez-Serrano, A.I.; Salas-Gutierrez, I.; Lemus-Bedolla, E.; Castillo-Guzman, A.R.; Lopez-Lemus, H.L.; Castro-Camus, E. Terahertz imaging for early screening of diabetic foot syndrome: A proof of concept. *Sci. Rep.* **2017**, *7*, 42124. [CrossRef] [PubMed]
205. Jung, E.; Lim, M.H.; Moon, K.W.; Do, Y.W.; Lee, S.S.; Han, H.W.; Choi, H.J.; Cho, K.S.; Kim, K.R. Terahertz pulse imaging of micro-metastatic lymph nodes in early-stage cervical cancer patients. *J. Opt. Soc.* **2011**, *15*, 155–160. [CrossRef]
206. Ji, Y.B.; Oh, S.J.; Kang, S.G.; Heo, J.; Kim, S.H.; Choi, Y.; Song, S.; Son, S.H.; Lee, J.H.; Haam, S.J.; et al. Terahertz reflectometry imaging for low and high grade gliomas. *Sci. Rep.* **2016**, *6*, 36040. [CrossRef]
207. Kashanian, H.A.; Ghaary, H.B.; Bagherzadeh, N.C. Gastric Cancer Diagnosis Using Terahertz Imaging. *Majlesi J. Multimed. Process.* **2015**, *4*, 1–7.
208. Ji, Y.B.; Kim, S.-H.; Jeong, K.; Choi, Y.; Son, J.-H.; Park, D.W.; Noh, S.K.; Jeon, T.-I.; Huh, Y.-M.; Seungjoo, H.; et al. Terahertz spectroscopic imaging and properties of gastrointestinal tract in a rat model. *BOE* **2014**, *5*, 4162–4170. [CrossRef]
209. Ji, Y.B.; Park, C.H.; Kim, H.; Kim, S.H.; Lee, G.M.; Noh, S.K.; Jeon, T.-I.; Son, J.-H.; Huh, Y.-M.; Seungjoo, H.; et al. Feasibility of terahertz reflectometry for discrimination of human early gastric cancers. *BOE* **2015**, *6*, 1398–1406. [CrossRef]
210. Sim, Y.C.; Ahn, K.-M.; Park, J.Y.; Park, C.; Son, J.-H. Temperature-Dependent Terahertz Imaging of Excised Oral Malignant Melanoma. *IEEE J. Biomed. Health Inform.* **2013**, *17*, 779. [CrossRef]
211. Taylor, Z.D.; Singh, R.S.; Bennett, D.B.; Tewari, P.; Kealey, C.P.; Bajwa, N.; Culjat, M.O.; Hubschman, J.; Brown, E.R.; Grundfest, W.S. THz Medical Imaging: In vivo Hydration Sensing. *IEEE Trans. Terahertz Sci. Technol.* **2011**, *1*, 201–219. [CrossRef]
212. Wilmink, G.J.; Grundt, J.E. Current State of Research on Biological Effects of Terahertz Radiation. *J. Infrared Millim. Terahertz Waves* **2011**, *32*, 1074–1122. [CrossRef]
213. Humphreys, K.; Loughran, J.P.; Gradziel, M.; Lanigan, W.; Ward, T.; Murphy, J.A.; O'Sullivan, C. Medical applications of terahertz imaging: A review of current technology and potential applications in biomedical engineering. In Proceedings of the 26th Annual International Conference of the IEEE Engineering in Medicine and Biology Society, San Francisco, CA, USA, 1–5 September 2004.
214. Arnone, D.D.; Ciesla, C.M.; Corchia, A.; Egusa, S.; Pepper, M.; Chamberlain, J.M.; Bezant, C.; Linfield, E.H.; Clothier, R.; Khammo, N. Applications of terahertz (THz) technology to medical imaging. In Proceedings of the Terahertz Spectroscopy and Applications II, Munich, Germany, 16–18 June 1999; Volume 3828.
215. Zinov'ev, N.N.; Fitzgerald, A.F.; Strafford, S.M.; Wood, D.J.; Carmichael, F.A.; Miles, R.E.; Smith, M.A.; Chamberlain, J.M. Identification of tooth decay using terahertz imaging and spectroscopy. In Proceedings of the Twenty Seventh International Conference on Infrared and Millimeter Waves, San Diego, CA, USA, 22–26 September 2002.
216. Crawley, D.A.; Longbottom, C.; Cole, B.E.; Ciesla, C.M.; Arnone, D.; Wallace, V.P.; Pepper, M. Terahertz pulse imaging: A pilot study of potential applications in dentistry. *Caries Res.* **2003**, *37*, 352–359. [CrossRef]
217. Crawley, D.; Longbottom, C.; Wallace, V.P.; Cole, B.; Arnone, D.; Pepper, M. Three-dimensional terahertz pulse imaging of dental tissue. *J. Biomed. Opt.* **2003**, *8*, 303–307. [CrossRef]
218. Kamburoğlu, K.; Karagöz, B.; Altan, H.; Özen, D. An ex vivo comparative study of occlusal and proximal caries using terahertz and X-ray imaging. *Dentomaxillofac. Radiol.* **2019**, *48*, 20180250. [CrossRef]

219. Kamburoğlu, K.; Kurt, H.; Kolsuz, E.; Öztaş, B.; Tatar, I.; Çelik, H.H. Occlusal caries depth measurements obtained by five different imaging modalities. *J. Digit. Imaging* **2011**, *24*, 804–813. [CrossRef]
220. Torres, M.G.; Santos, A.S.; Neves, F.S.; Arriaga, M.L.; Campos, P.S.F.; Crusoé-Rebello, I. Assessment of enamel-dentin caries lesions detection using bitewing PSP digital images. *J. Appl. Oral Sci.* **2011**, *19*, 462–468. [CrossRef]

condensed matter

MDPI

Review

Fabry-Perot Cavity Leaky Wave Antennas with Tunable Features for Terahertz Applications

Silvia Tofani * and Walter Fuscaldo

Department of Information Engineering, Electronics and Telecommunications, "Sapienza" University of Rome, via Eudossiana 18, 00184 Rome, Italy; walter.fuscaldo@uniroma1.it

* Correspondence: silvia.tofani@uniroma1.it

Received: 13 January 2020; Accepted: 22 January 2020; Published: 25 January 2020

Abstract: Terahertz (THz) radiation is a very appealing band of the electromagnetic spectrum due to its practical applications. In this context, the THz generation and manipulation is an essential part of the technological development. The demand of THz antennas is still high because it is already difficult to obtain directive, efficient, planar, low-cost, and easy-to-fabricate THz radiating systems. In this regard, Fabry-Perot cavity leaky-wave antennas are gaining increasing attention at THz, due to their very interesting radiating features: the combination of planar designs with metamaterials and metasurfaces could offer a promising platform for future THz manipulation technologies. In this short review, we focus on different classes of leaky-wave antennas, based on materials with tunable quasi-optical parameters. The possibility of producing directive patterns with particularly good efficiencies, as well as the capability of dynamically reconfiguring their radiating features, are discussed by taking into account the risk of increasing costs and fabrication complexity.

Keywords: terahertz; leaky-wave antennas; metasurface; liquid crystals; graphene

1. Introduction

The terahertz (THz) band conventionally occupies the region of the electromagnetic spectrum between the high-frequency limit of the microwaves, often indicated as 300 GHz, and the low-frequency edge of the infrared frequency range, usually fixed at 10 THz. This corresponds to wavelengths in the range between 300 μm to 1 mm. Historically, THz detection was conceived for interstellar dust sensing between 100 GHz and 3 THz. Nowadays, several applications have been exploited, such as molecular spectroscopy and high-resolution imaging [1], non-destructive testing [2], security screening [3], explosives and drug detection [4], monitoring of water content in human tissues [5], and high data-rate communications [6,7]. Moreover, THz communications are gaining an increasing interest, connected to the possibility to have access to Tbps wireless links [8]. In this context, there is a request to develop antennas, which (i) could be fabricated with low-cost materials, (ii) are light and simple to integrate, and (iii) may exhibit reconfigurable properties.

Among the class of planar antennas, leaky-wave antennas (LWAs) are promising devices at THz frequencies. LWAs are travelling-wave antennas and they are well-known radiating systems at microwaves from decades. In LWAs, radiation phenomena are interpreted as an energy leakage of a wave that propagates in a partially open structure. The wave propagating in a LWA progressively loses its energy due to both the (undesirable) losses in the medium and the radiation losses [9]. Because of such energy outflow, leaky waves are described by a generally complex propagation wavenumber.

The main advantages of LWAs are the simple fabrication and design as well as the ubiquity of leaky-wave phenomena from microwaves (i.e., the historical starting point of the leaky-wave theory), to nano-optics (thanks to the current developments in metasurfaces and nanostructured materials). For this reason, they are perfectly suitable for being scaled at THz frequencies.

This review aims to analyze and describe the design of novel reconfigurable Fabry-Perot cavity leaky-wave antennas (FPC-LWAs) layouts for THz applications, focusing on the fabrication constraints that could have an impact on the FPC-LWA layout in the THz range. After introducing in Section 2 the theoretical background of the leaky waves and the leaky-wave antennas, in Sections 3 and 4, we focus on two complementary approaches for the design of tunable THz FPC-LWAs by means of an electro-optical control. In Section 5, we present and discuss the strategies for a wide reconfigurability of the THz radiation pattern by means of two-dimensional (2D) materials with tunable quasi-optical parameters. Conclusions are drawn in Section 6.

2. Leaky-Wave Antennas

LWAs are travelling wave antennas in which the propagating modes responsible for the radiation (i.e., the leaky waves) are characterized by a generally complex propagation wavenumber $k_z = \beta_z - j\alpha_z$, where β_z is the phase constant and α_z is the attenuation constant. Moreover, in order to effectively radiate [10,11], the leaky wave propagating in the structure has to be a fast wave, i.e., characterized by $|\beta_z| < k_0$ where k_0 is the free-space wavenumber. Therefore, the operation of a LWA is quite different from a slow-wave or a surface-wave type of antenna, for which radiation mainly takes place at discontinuities (e.g., at the edge of the structure) [12]. In the absence of ohmic losses, α_z takes into account only power losses due to radiation and it is called "leakage constant": it represents the rate of decreasing of the aperture field due to the leakage mechanism of radiation.

LWAs can be divided into several categories, according to their geometrical structure and their principle of operation [13]. A rigorous description goes beyond the aim of this work and can be found in [14]. A first distinction can be made between one-dimensional (1D) and two-dimensional (2D) LWAs, that differ from each other for the nature of the leaky waves propagating in the structure. In the former case, leaky waves propagate along a longitudinal direction with a planar wavefront; in the latter case, leaky waves radially propagate with a cylindrical wavefront. This important distinction considerably affects the radiation features: a 1D LWA usually produces fan beams, while a 2D LWA mainly radiates pencil beams or fully conical beams.

Then, LWAs can also be divided, according to their structure, in uniform, quasi-uniform and periodic antennas. This distinction can affect the antenna scanning characteristics. In fact, a uniform LWA usually scans inside the forward spatial quadrant by varying the operating frequency, while a periodic LWA can scan the radiation beam in both forward and backward quadrants. However, some exceptions could exist.

2.1. Fabry-Perot Cavity Leaky-Wave Antennas

FPC-LWAs as those discussed here are 2D quasi-uniform LWAs and can be considered as partially-open and partially-guiding structures that support the radial propagation of cylindrical leaky waves [15]. The working principle of these structures is based on the occurrence of a gain-enhancement and was originally formulated, according to a ray-optics modeling, as a phenomenon of multiple beam interference [16]. However, the radiation properties of the FPC-LWAs can also be addressed under the frame of leaky-wave theory [13].

An FPC is able to radiate when the fundamental leaky modes supported by the cavity are excited by the antenna feed. Interestingly, if a horizontal electric or magnetic dipole is selected as excitation for the FPC-LWA, the antenna radiates a directive pencil beam at broadside at its frequency of operation.

For frequencies close to the operating frequency, an FPC-LWA emits a conical beam that has the cone axis along the vertical x-axis (see Figure 1a) [13,17,18].

An FPC-LWA layout consists of a grounded dielectric slab (GDS) covered with a partially reflecting sheet (PRS) [19]. The PRS can assume several shapes, e.g., a distributed Bragg reflector, a simple dielectric layer with a refractive index higher than the substrate, or a metasurface. In any case, the antenna design process is the same, thanks to the leaky-wave approach.

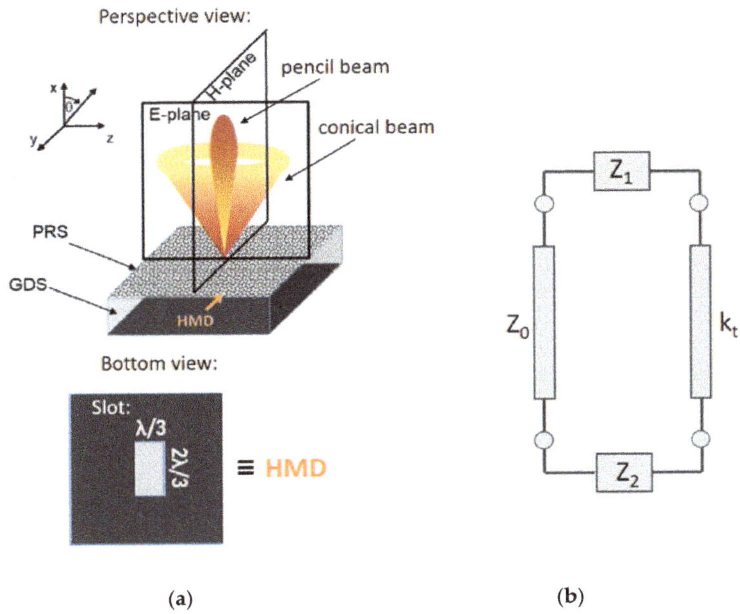

Figure 1. (**a**) an example of FPC-LWA fed by an HMD (horizontal magnetic dipole) oriented along the y-axis [15]; (**b**) terminated equivalent network representing a waveguide cross-section [13]. An FPC-LWA is constituted by a dielectric substrate with a metallization (i.e., a mirror) on the bottom (grounded dielectric slab, GDS) and a PRS (partially reflective surface) on top. A slot for antenna excitation is etched on the metalized side of the substrate. Note that a back-illuminated quasi-resonant slot can be modeled as an equivalent horizontal magnetic dipole (HMD).

2.2. Antenna Parameters and Main Rules for FPC-LWA Design

The radiation properties of an FPC-LWA depend on the source type. FPC-LWAs for THz radiation are usually fed by a horizontal magnetic dipole (HMD) because, differently from a vertical dipole, it allows for emitting THz radiation in the broadside direction. An HMD is able to excite the pair of fundamental TE and TM leaky modes, both characterized by a complex propagation wavenumber, as already discussed in Section 1.

The setting of a proper design of the FPC-LWAs, i.e., a design in which some assumptions are satisfied [17], is important for describing radiation as caused only by the fundamental leaky modes. When these two conditions are met (i.e., an HDM is selected as a feed and the FPC-LWA is designed according to the design rules in [20]), the TM leaky mode affects radiation in the antenna E-plane, while the TE leaky mode is responsible for the radiation pattern in the antenna H-plane. In the condition of broadside radiation, nearly the same attenuation and phase constant values characterize the TE and TM leaky modes: the FPC-LWA radiates an omnidirectional pencil beam.

The pointing angles θ_0^{TE} and θ_0^{TM} of the beam are measured from the vertical axis (x-axis of Figure 1) and their expression is [20].

$$\theta_0^{\mathrm{TE,TM}} \cong \arcsin\sqrt{\left(\frac{\beta_z{}^{\mathrm{TE,TM}}}{k_0}\right)^2 - \left(\frac{\alpha_z{}^{\mathrm{TE,TM}}}{k_0}\right)^2} \tag{1}$$

Equation (1) also defines two regimes of radiation: if $\beta_z \leq \alpha_z$, the FPC-LWA emits radiation at broadside; if $\beta_z \gg \alpha_z$, the FPC-LWA emits a conical beam. When $\beta_z \cong \alpha_z$, the radiation emitted at broadside reaches its maximum value. Such condition is usually set as the operative one and it is the starting point in the FPC-LWA design process. The antenna beamwidth in the E- and H-plane is [15].

$$\begin{cases} \Delta\theta^{\mathrm{TE}} = 2\frac{\alpha_z^{\mathrm{TE}}}{k_0}\sec\theta_0^{\mathrm{TE}}, \ \Delta\theta^{\mathrm{TM}} = 2\frac{\alpha_z^{\mathrm{TM}}}{k_0}\sec\theta_0^{\mathrm{TM}} \ \text{for}\ \beta_z \gg \alpha_z \\ \Delta\theta^{\mathrm{TE}} = 2\sqrt{2}\frac{\alpha_z^{\mathrm{TE}}}{k_0}, \ \Delta\theta^{\mathrm{TM}} = 2\sqrt{2}\frac{\alpha_z^{\mathrm{TM}}}{k_0} \ \text{for}\ \beta_z \cong \alpha_z \end{cases} \tag{2}$$

It is important to point out that the beamwidths and the pointing angles are different on the E- and H-planes when the wavenumber of the TM and TE mode differs, i.e., when the scanning angle increases. For this reason, in several applications in which the pointing angle changes, it could be important to select an antenna PRS that shows the same behavior for both TM and TE polarizations. In this way, the antenna is able to emit an almost circular conical beam when $\beta_z \gg \alpha_z$.

In the leaky-wave theory [13], a leaky-wave antenna is interpreted as a radiating version of a parallel plate waveguide (PPW), due to the presence of a PRS that allows a radiation leakage. In fact, a grounded dielectric slab covered by a PRS (instead of a reflecting surface) supports a pair of fundamental TE-TM leaky modes (instead of the fundamental TE-TM PPW guided modes), that can eventually radiate if properly excited. Two key geometrical parameters in the design of a FPC-LWA [17] are the lateral size of the antenna (i.e., with respect to the axis reported in Figure 1a, the antenna dimensions along the y- and z-axis) and the substrate thickness. The antenna efficiency, i.e., the ratio between the radiated and the incoming power, is related to the lateral size of an FPC-LWA. In order to avoid pattern deterioration due to diffraction effects from edges, the lateral size of the antenna (i.e., lateral truncation) should be set so as the travelling cylindrical modes have radiated at least 90% of the input power before they reach the lateral truncation [21].

Under the condition of broadside radiation and of antenna of infinite lateral extent (i.e., radiation efficiency ≥90% by design), for the TE and TM fundamental leaky modes, the thickness of the antenna substrate is given by a simple formula [15].

$$h = \lambda_0/(2n_s) \tag{3}$$

where n_s is the antenna substrate refractive index at the design frequency, i.e., the frequency at which $\beta_z \cong \alpha_z$.

2.3. Techniques for the Analysis of FPC-LWA Features

Two standard methods are usually employed for the analysis of complex structures: the mode matching technique [22] and the transverse resonance technique [23]. The mode matching technique is a time-consuming technique, especially if the domain of interest is composed by more than one medium. In the situation in which only the determination of the eigenvalues is of interest, the transverse resonance technique represents an appropriate tool [23]. In fact, the eigenvalues of a waveguide problem correspond to pole singularities of a suitable characteristic Green's function in the complex plane of the propagation wavenumber. This Green's function can be identified as the voltage (or current) on a transmission line along one of the transverse direction of the waveguide axial direction [24,25]. The pole singularities correspond to the transverse equivalent network (TEN) model resonances and are usually computed using analytical methods.

A general form of terminated TEN is given in Figure 1b. The transverse wavenumber k_t that corresponds to source-free fields in the waveguide can be derived from the solution of the resonance equation:

$$Z(k_t) = \overrightarrow{Z}(k_t) + \overleftarrow{Z}(k_t) \ \text{or}\ Y(k_t) = \overrightarrow{Y}(k_t) + \overleftarrow{Y}(k_t) \tag{4}$$

where $\overrightarrow{Z}$, $\overrightarrow{Y}$, $\overleftarrow{Z}$, and $\overleftarrow{Y}$ are the impedances or admittances from the left and the right of an arbitrary reference plane. A closed-form analytical solution to this resonance equation does not generally exist, except for very simple canonical examples, such as the parallel-plate waveguide [23,26].

In the simple situation in which the leaky wave can be seen as a perturbation of a mode in a closed waveguide [25], the transverse wavenumber can be considered as:

$$k_{\mathrm{t}} = k_{\mathrm{t}0} + \Delta k_{\mathrm{t}} \tag{5}$$

where $k_{\mathrm{t}0}$ is the wavenumber of the unperturbed mode and Δk_{t} is the perturbation. If the cross-section is homogeneous, the resonance equation can be written in terms of normalized impedances (admittances), expanded in the point $k_{\mathrm{t}} = k_{\mathrm{t}0}$:

$$Z'(k_{\mathrm{t}0}) + \left(\frac{\mathrm{d}Z'(k_{\mathrm{t}})}{\mathrm{d}k_{\mathrm{t}}}\right)_{k_{\mathrm{t}}=k_{\mathrm{t}0}} \Delta k_{\mathrm{t}} \approx 0 \tag{6}$$

$$\Delta k_{\mathrm{t}} \approx -\frac{Z'(k_{\mathrm{t}0})}{\left(\frac{\mathrm{d}Z'(k_{\mathrm{t}})}{\mathrm{d}k_{\mathrm{t}}}\right)_{k_{\mathrm{t}}=k_{\mathrm{t}0}}} \text{ or } \Delta k_{\mathrm{t}} \approx -\frac{Y'(k_{\mathrm{t}0})}{\left(\frac{\mathrm{d}Y'(k_{\mathrm{t}})}{\mathrm{d}k_{\mathrm{t}}}\right)_{k_{\mathrm{t}}=k_{\mathrm{t}0}}} \tag{7}$$

where $Z' = Z/Z_0$ is the normalized impedance, and $Y' = Y/Y_0$ is the normalized admittance.

2.4. Fabry-Perot Cavity Leaky Wave Antennas at THz Frequencies

Leaky-wave phenomena occur at any frequency range. For this reason, LWAs working at THz frequencies are feasible and some examples of devices are already present in the scientific literature [15,27–44]. In this section, only LWAs working in the frequency band between 300 GHz and 3 THz are discussed. The choice of this specific frequency band is linked to a technological challenge that characterizes the so-called "terahertz gap" [45]. In fact, even if the border between millimeters waves and far-IR has not been delimited yet, below 0.3 THz, facilities developed for the millimeter waves are available, while, above 3 THz, the mechanisms connected to the description of far-IR radiation are predominant.

The most studied THz LWA is a leaky waveguide coupled with a hemispherical silicon lens. It is able to operate at a fixed value of frequency [27,28] or in a frequency band of 0.2 THz [29] until 1.4 THz [30–32] wide. When the device works at a single frequency, it can show a very high directivity (i.e., the power density emitted by the antenna versus the power density emitted by an ideal isotropic radiator, when they radiate the same total power); for example [27], of 42.1 dB. Lens-like LWAs able to operate in a frequency band wider than 1 THz can be employed for practical applications, such as THz detectors. However, independently from the operating frequencies of the antenna, the radiation efficiency is always below 75%, due to losses introduced by the coupling with a silicon lens.

With respect to FPC-LWAs, the maximum radiated power at broadside is connected to the resonance condition of the multiple reflections. For this reason, FPC-LWAs usually work around a narrow band centered at the resonance frequency of the FPC. The PRS can assume several forms. In [33–35], the performances of FPC-LWAs with different metasurface geometries are compared. A metasurface is, in general, a metallic patterned surface, in which the geometrical elements composing the pattern have dimensions deeply below the operative wavelength [46]. It means that this surface can be considered as a homogeneous surface with properties depending on the specific geometrical pattern chosen for the metasurface synthesis. Even if several geometries have been investigated in the THz range as PRS for FPC-LWAs, none of them has been fully characterized yet. However, they show advantages common to many LWAs, such as the suitability to miniaturization and integration on complex devices or the scalability at every THz frequency value of operation. The radiation efficiencies, computed for the structures presented in [33–35], do not go beyond the 75%, but the high number of degrees of freedom in the metasurfaces design may allow for higher values. In this context, a fishnet metasurface has been recently introduced as a spatially nondispersive PRS with a design able to synthetize surfaces that allow for directivities ranging from 15 to 30 dB [36–38].

An interesting opportunity offered by FPC-LWAs is the possibility to develop THz antennas with a reconfigurable pointing angle at a fixed frequency. The only two ways, currently present in literature in the frequency band of interest, to steer the beam is the use of a graphene sheet as a PRS [39–41] or the integration in the LWA layout of one or more liquid crystal (LC) layers [15,42]. An alternative way to steer the beam is to change the operating frequency. An example is presented in [43,44], where a microstrip LWA with periodic elements is investigated as a THz radar and shows a beam steering over an angular range of 38°, with a very narrow radiated beam. However, this kind of device has a limited application as radar to a short range object detection [43].

Other forms of FPC-LWAs working in the THz range will be widely analyzed in the following sections, where FPC-LWA structures implementing strategies for reconfiguring the emitted beam pointing angle will be described in terms of design and numerical investigation.

3. Fabry-Perot Cavity Leaky Wave Antennas Based on Tunable Metamaterials

In this section, we start our analysis about the design possibilities offered by different tunable materials by introducing an FPC-LWA structure made with a GDS covered by a metamaterial. The metamaterials are artificial materials obtained from the combination of different materials, present in nature, with spacing and geometrical features below the wavelength dimension. The metamaterial of Figure 2a consists of a multilayered structure in which layers of material with high refractive index are interleaved with LC layers. Because of both the ordinary and the extraordinary refractive index of the LC has a value lower than the other material constituting the multilayer, the PRS can be considered as a distributed Bragg reflector.

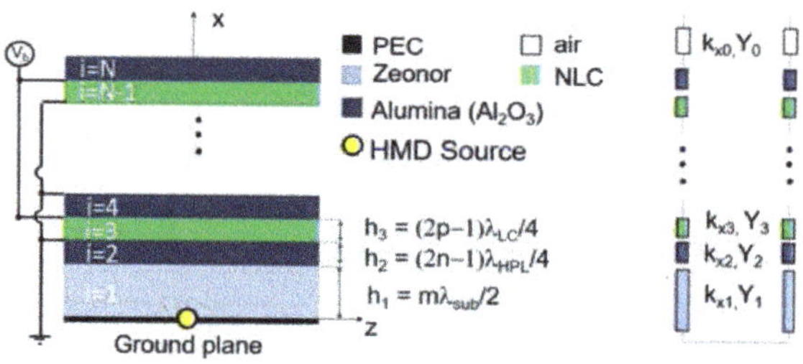

Figure 2. (**a**) cross-section and (**b**) equivalent network of a multilayer FPC-LWA with nematic LC cells (NLC) [42].

3.1. Liquid Crystals for THz Antenna Design

Liquid crystals are materials that exhibit a solid crystalline phase, a liquid phase and intermediate phases in which they flow like a liquid, but show a preferred orientation. These intermediate phases are called mesophases. It is possible to distinguish several mesophases by changing temperature, concentration, constituents, substituents, or other LCs or environmental properties [47]. Materials constituted by molecules with an anisotropic shape could reveal mesophases. For this reason, it is common to refer to LCs as rod-like molecules [48]. Thermotropic are the most extensively studied LCs. They exhibit different mesophases as temperature increases. When the thermotropic LC is in a nematic phase, its rod molecules have a random position. The nematic phase could remind a liquid; however, LC molecules are aligned in the same direction defined by a unit vector called director.

When an external perturbation field is applied on LCs in nematic phase, they are subjected to deformation, as a solid. However, in a solid, the stress produces a translational displacement of molecules while, in LCs, molecules rotate in direction of the force without any displacement in their centre of gravity. In general, in nematic LCs subjected to an external electric field, the director vector experiments a torque force. LCs respond to this external force with a reorientation of their long axis, which is influenced by several factors.

Thermotropic nematic LC layers are usually employed in a confined geometry, called "cell", i.e., a cavity in which the top and bottom walls are covered with thin layers of polymeric material, typically a polyimide or polyamide. In free space optics, the edges are usually neglected as these lie outside the useable area of the device. This thin layer undergoes a mechanical [49] or optical [50] process able to align the polymer chains in a same direction (planar homogeneous alignment). When the nematic LC is in contact with this layer, it is energetically favourable that the LC director has the same direction of polymer chains. In this way, the first layer of rods is considered as anchored to the cell walls. The second layer follows the alignment pattern of the first layer, but experiments a lower anchoring, and so on. When LC molecules are strongly anchored to a boundary, surface interactions are not considered in the evaluation of rod molecules motion in nematic LCs due to external field interactions [47].

Moreover, the electronic response of LCs to an external electric field is characterized by its dielectric constants (or refractive indices), as well as electrical conductivities. These physical parameters are dependent on the direction of the external field and to its frequency: the dielectric permittivity is a tensor and its elements are, in general, complex numbers. Rod-like molecules are usually uniaxial LCs and the permittivity tensor is:

$$\overline{\overline{\varepsilon}} = \varepsilon_0 \begin{bmatrix} \varepsilon_\perp & 0 & 0 \\ 0 & \varepsilon_\perp & 0 \\ 0 & 0 & \varepsilon_\parallel \end{bmatrix} \tag{8}$$

and similarly for the conductivity tensor. In Equation (8), the dielectric permittivity is considered as the product of the vacuum permittivity ε_0 and the extraordinary relative permittivity $\varepsilon_\parallel$ of the LC, when the director is parallel to the applied field, and it is the product of the vacuum permittivity ε_0 and the ordinary relative permittivity $\varepsilon_\perp$ of the LC, when the director is perpendicular to the applied field. Pure organic LCs are highly purified nonconductive materials (i.e., $\sigma = 0$), though they may become conductive by adding ions and impurities. Electrical conductivity influences director orientation, chemical degradation and LC lifetime.

Regarding dielectric permittivity, in most cases, $\varepsilon_\parallel > \varepsilon_\perp$ due to molecular structure and constituents. This condition is called positive anisotropy, i.e., $\Delta\varepsilon = \left(\varepsilon_\parallel - \varepsilon_\perp\right)/S_{\mathrm{eq}} > 0$, where S_{eq} is the order parameter under equilibrium condition. In general, the elements in dielectric permittivity tensor are frequency dispersive and some LCs could change their anisotropy from negative to positive by increasing external field frequency [47].

When an external electric field E is applied, if its strength is more than a critical value E_F, called Freedericksz transition [51], the LC director rotates by an angle θ with respect to its rest position, trying to align itself with the field direction:

$$E_F = \frac{\pi}{d}\left(\frac{K}{\Delta\varepsilon}\right)^{\frac{1}{2}} \text{ or } V_F = \pi\left(\frac{K}{\Delta\varepsilon}\right)^{\frac{1}{2}} \tag{9}$$

where d is the distance between the electrodes which apply the field, and K is a local field factor. K takes into account the Frank elastic constants describing LC director deformation: the splay modulus k_{11}, the twist modulus k_{22}, and the bend modulus k_{33} [48]. Under the condition of strong anchoring, if the external field induces a reorientation of the director from parallel to perpendicular with respect to electrodes plane:

$$K = k_{11} + \frac{1}{4}(k_{33} - 2k_{22}) \tag{10}$$

The rod molecules reorientation determines a change in a LC dielectric permittivity as well as in its refractive index. In fact, as discussed above, in the absence of external fields, LC permittivity is expressed by Equation (8). For values of applied voltage higher than the Freedericksz transition, it becomes [47]:

$$\overline{\overline{\varepsilon}} = \varepsilon_0 \begin{bmatrix} \varepsilon_{\parallel} & 0 & 0 \\ 0 & \varepsilon_{\perp} & 0 \\ 0 & 0 & \varepsilon_{\perp} \end{bmatrix} \tag{11}$$

In the application of Figure 2, we consider the nematic LC mixture 1825 [52,53] due to its high birefringence and moderate losses in the THz frequency range. The selected LC has an ordinary and extraordinary permittivity equal to $\varepsilon_{\perp} = 2.3$ and $\varepsilon_{\parallel} = 2.42$ at 1 THz, Frank elastic constants k_{11} = 12.5 pN, k_{22} = 7.4 pN, k_{33} = 32.1 pN, and viscosity γ = 311.55 mPa·s [52,53]. As every uniaxial LC, the 1825 mixture has a permittivity that, for V = 0, can be described by Equation (8) and, for $V > V_F$, is represented by Equation (11). For intermediate state, i.e., for $0 < V < V_F$, the permittivity tensor is no longer represented by a diagonal matrix. However, if the LC director is aligned along the z-axis, and a low driving voltage is applied, the LC molecules tilt in the xz-plane exhibiting a negligible rotation over both the yz- and the xy-plane [15,42]. With this approximation, the LC can still be locally modelled as a uniaxial crystal. The permittivity tensor can be generically expressed as:

$$\overline{\overline{\varepsilon_r}}(V) = \begin{bmatrix} \varepsilon_x(V) & 0 & 0 \\ 0 & \varepsilon_y(V) & 0 \\ 0 & 0 & \varepsilon_z(V) \end{bmatrix} \tag{12}$$

where $\overline{\overline{\varepsilon}} = \varepsilon_0 \overline{\overline{\varepsilon_r}}(V)$. The assumption of uniaxial LC is important in the design of an FPC-LWA because it allows for modeling the antenna by means of a simple TEN. In the following, we will consider only the two limiting cases V = 0 and $V > V_F$, while the value of the permittivity in the LC layer is considered as a medium value in the selected direction (i.e., it takes into account the local variations due to the anchoring effect).

3.2. Design of the Fabry-Perot Cavity Leaky-Wave Antenna with Liquid Crystals

The innovative elements of the FPC-LWA in Figure 2 are its capability of changing its equivalent impedance by applying an external bias voltage to the LC cells. The change in impedance corresponds to a variation in the FPC-LWA pointing angle at the operative frequency. The FPC-LWA's multilayer is a stack of alumina layers ($\varepsilon_{r,Al} = 9$, $\tan\delta = 0.01$, at 1 THz) and LC cells. The multilayer is placed over a GDS made with Zeonex E48R ($\varepsilon_{r,Ze} = 2.3$, $\tan\delta = 0.006$, at 1 THz, Zeon Corporation, Tokyo, Japan), a cyclic olefin copolymer with a permittivity matched with the ordinary permittivity of the selected LC mixture. In this way, it is possible to enhance the resonance condition of the FPC [54]. The thickness of the layers in the stack is fixed at odd multiples of a quarter wavelength in the corresponding media. This choice guarantees that the FPC-LWA is able to radiate a narrow beam at broadside, i.e., the antenna is highly directive at broadside.

The equivalent circuit model of the multilayer FPC-LWA with LCs is reported in Figure 2b. The TEN takes into account the LC anisotropy and is the starting point of the dispersion analysis. When no voltage is applied to the LC layers, the LC molecules are oriented along the z-axis and $\varepsilon_z(V=0) = \varepsilon_{\parallel}$ (compare Equation (8)). When a voltage $V > V_F$ is applied to the LCs, the rod-like molecules reorient their principal axis parallel to the external E-field and $\varepsilon_x(V > V_F) = \varepsilon_{\parallel}$ (compare Equation (11)). With regard to the transverse transmission line model (Figure 2b), the characteristic admittances, Y_3, and the normal wavenumbers, k_{x3}, of the LC layers for both transverse-electric (TE) and transverse-magnetic (TM) polarizations (with respect to the xz-plane) are functions of the bias voltage. Their expressions are given by:

$$Y_3^{\mathrm{TE}} = \frac{k_x}{\omega\mu_0} \text{ and } Y_3^{\mathrm{TM}} = \frac{\omega\varepsilon_0\varepsilon_z(V)}{k_x} \tag{13}$$

$$k_{x3}^{\mathrm{TE}} = \sqrt{k_0^2 \varepsilon_y - k_z^2} \text{ and } k_{x3}^{\mathrm{TM}} = \sqrt{\frac{\varepsilon_z(V)}{\varepsilon_x(V)} k_0^2 \varepsilon_x(V) \; - k_z^2} \tag{14}$$

where k_0 is the vacuum wavenumber, k_x and k_z is are the wavenumbers in x- and z-directions, respectively, ε_0 is the vacuum dielectric permittivity, and μ_0 is the vacuum permeability. However, the selected LC mixture is uniaxial and ε_y does not depend on the applied voltage: only the fundamental TM leaky mode is affected by the external bias and, thus, the TE leaky mode is not of interest in this investigation. It also allows for showing the LWA radiation features only in the *E*-plane (i.e., the *xz*-plane), that is the plane affected by the TM leaky mode [17].

The dispersion equation relates the frequency and the complex longitudinal wavenumber k_z. The dispersion equation can be obtained by applying the transverse resonance technique to the transverse equivalent network of Figure 2b. Equation (4) evaluated at the cross-section correspondent to $x = h_1$ is

$$Y_{\mathrm{in},i} - jY_0^{\mathrm{TM}} \cot(k_{x,0} h_0) = 0 \tag{15}$$

where the input admittance of the i-th layer can be written as [42]

$$Y_{\mathrm{in},i} = Y_i^{\mathrm{TM}} \frac{Y_{\mathrm{in},i+1} \cos(k_{x,i} h_i) + jY_i^{\mathrm{TM}} \sin(k_{x,i} h_i)}{Y_i^{\mathrm{TM}} \cos(k_{x,i} h_i) + jY_{\mathrm{in},i+1} \sin(k_{x,i} h_i)} \tag{16}$$

where Y_i^{TM} is the characteristic admittance of the i-th layer, $k_{x,i}$ is the transverse wavenumber of the i-th layer, and h_i can be considered as the i-th layer thickness.

We can consider a multilayer FPC-LWA layout working at f_0 = 1 THz, as a case of study able to underline the excellent reconfigurable properties of the proposed structure [42]. The number of layers that constitutes the stack is fixed at 5, i.e., three alumina layers of thickness equal to 75 μm and two LC cells 100 μm thick. Dispersion curves computed between 0.8 and 1.2 THz by means of leaky-wave theory [55,56] are presented in Figure 3a. They are a family of curves obtained gradually changing the voltage between zero (red curve) and the threshold voltage $V_\infty > V_F$ (blue curve). From a numerical viewpoint, the voltage changing corresponds to a dielectric permittivity commutation as reported in Equation (12), where $\varepsilon_x(V=0) = \varepsilon_{\parallel}$, $\varepsilon_y(V=0) = \varepsilon_{\perp}$, $\varepsilon_z(V=0) = \varepsilon_{\perp}$, $\varepsilon_x(V=V_F) = \varepsilon_{\perp}$, $\varepsilon_y(V=V_F) = \varepsilon_{\perp}$, and $\varepsilon_z(V=V_F) = \varepsilon_{\parallel}$. Intermediate results can be obtained for intermediate values of voltage, i.e., intermediate values of dielectric permittivity in the x- and z-directions. For the proposed LC cell thickness, values below 20 V are sufficient to cover almost the complete switching range, as can be derived from LC dynamic complex numerical models that take into account molecules' orientation states at intermediate voltage values [57]. For simplifying the analysis, it is assumed that LC relative permittivity linearly varies with the applied voltage. For this reason, the unbiased and biased states are always correctly predicted. Conversely, the dynamic variation of $\hat{\beta}_z = \beta_z/k_0$ and $\hat{\alpha}_z = \alpha_z/k_0$, corresponding to voltage values between 0 and 20 V, could change if the permittivity tensor is computed for the intermediate biasing states.

The frequency at which the splitting condition $\hat{\beta}_z \cong \hat{\alpha}_z$ is achieved can be considered as the operating frequency. For this case of study, the operative frequency is fixed by the value obtained when the voltage is equal to V_∞ (blue curve), which corresponds to the design frequency. In fact, once the operating frequency is fixed, it is possible to change the value of the normalized phase constant $\hat{\beta}_z$ by simply decreasing the external voltage. On the contrary, the value of the normalized attenuation constant $\hat{\alpha}_z$ remains almost the same, such that $\hat{\beta}_z > \hat{\alpha}_z$. This would allow the steering of the THz beam with a quasi-constant beamwidth (because it is proportional to $\hat{\alpha}_z$ that does not change with the voltage) at a fixed frequency.

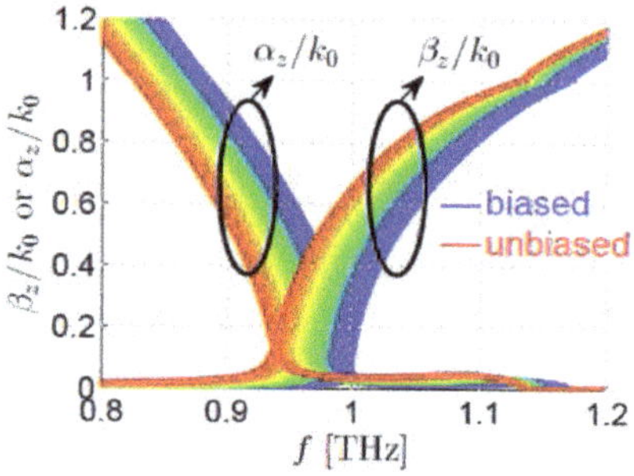

(a)

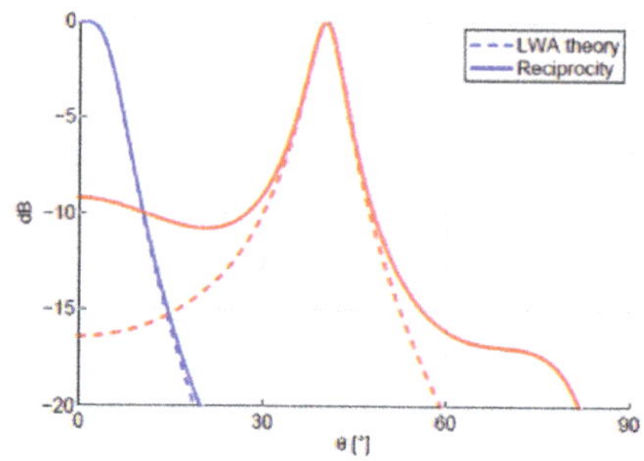

(b)

Figure 3. (**a**) dispersion curves of the fundamental TM mode computed by means of leaky-wave theory [55,56]; (**b**) radiation patterns predicted by leaky-wave theory (dashed lines) or by means of full-wave simulations with CST (solid lines) [55,56,58].

The radiation patterns, computed by means of both leaky-wave theory and full-wave simulations [58], are reported in Figure 3b and confirm the results achieved from the dispersion analysis. The antenna excitation is modelled with an HMD placed on the ground plane. HMD source can be used to model a slot etched in the ground plane and back-illuminated by a coherent THz source, such as QCLs or photomixers. For deriving radiating patterns, two methods are used: (i) taking into account only the contribution of the relevant leaky mode (dash line) and (ii) computing them by means of full-wave simulations (solid line). Radiation pattern for the unbiasing state (blue lines) shows a very good agreement between LWA theory and simulations. However, in the biased state (red lines), the LWA theory does not provide an accurate evaluation. This is probably because a higher-order leaky-wave mode could be present, increasing the sidelobe level. Full-wave simulations, however, correctly predicts this effect. In fact, the thickness of the LC layers extends the range of tunability, but degrades the radiation patterns. In [42], some solutions that take into account different number of layers and different thickness of the layers are introduced and discussed.

4. Liquid Crystal Cell as a Grounded Dielectric Slab

An alternative strategy that can be employed for steering the beam emitted by an FPC-LWA consists in substituting the GDS with a slab of anisotropic material, i.e., with a layer of LC. In fact, an external voltage on the LC cell is able to tune the values of the nematic LC's permittivity tensor, dynamically affecting the propagation properties of the modes supported by the grounded anisotropic dielectric slab (GADS). The starting point in the design of FPC-LWAs with an anisotropic substrate is to carry out a comprehensive analysis of the modal spectrum supported by a GADS. It constitutes a solid background for optimizing the performances of the GADS when it is covered by a PRS, i.e., when the radiation is enabled [59].

4.1. The Grounded Anisotropic Dielectric Slab Model

The analysis of both the leaky and the surface waves excited by an HMD and supported by the GADS is performed in the frequency range between 0.25 and 2 THz. As in the previous section, two limiting cases are numerically studied: the LCs are unbiased ($V = 0$) and a bias voltage that fully reorients the LC molecules is applied ($V > V_F$). Again, the LC mixture 1825 is selected ($n_o = 1.554 - j0.018$ and $n_e = 1.941 - j0.022$ at 1 THz, where $n_o = \sqrt{\varepsilon_\perp}$ and $n_e = \sqrt{\varepsilon_\parallel}$). According to Equation (3) and considering a design frequency of 1 THz, the LC slab thickness is 96.5 μm. In this way, the LC layer is an FPC at 1 THz and the TE mode always experiments the FPC at 1 THz, even if the LC is biased. Figure 4 depicts the model employed in the study of the GADS complex mode spectra. The model consists of an air region, the LC slab and a ground considered as a perfect electric conductor (PEC).

The electrodes for the LC biasing are neglected as a first approximation because they do not remarkably affect the modal analysis.

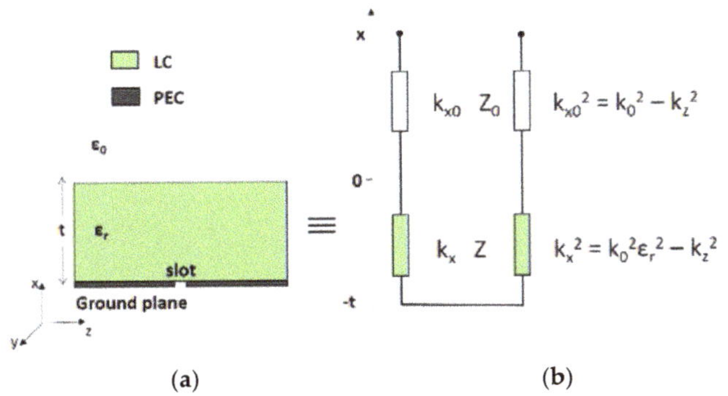

Figure 4. (**a**) cross-section of the model of the GADS and (**b**) its equivalent circuit network [59].

The modal analysis of the GADS can be carried out by means of the TRT applied to the equivalent network of the structure, as described in Section 2.3. However, the simple TEN model of Figure 4a is a good approximation of more rigorous models developed in literature [60,61] when the LCs are uniaxial and their $\varepsilon_{\|}$ and $\varepsilon_{\perp}$ differ less than 50%. The characteristic admittances and the normal wavenumbers of the LC layer for both the TE and the TM polarizations (with respect to the xz-plane) are the same as in Equations (13) and (14). The selected LC mixture is uniaxial and ε_y does not depend on the applied voltage. Equation (4) evaluated at $x = 0$ gives the following dispersion equation for the TM mode of the GADS:

$$jY^{\mathrm{TM}}\cot(k_{x,0}h_0) = Y_0^{\mathrm{TM}} \tag{17}$$

where $Y_0{}^{\mathrm{TM}}$ is the TM characteristic admittance in the air.

4.2. Modal Spectra of Grounded Isotropic and Anisotropic Dielectric Slab

The modal analysis involves two structures: the GDS and the GADS. In the GDS, the permittivity has an isotropic value and it is equal to either $\varepsilon_r = \varepsilon_{\perp}$, or $\varepsilon_r = \varepsilon_{\|}$; in the GADS, the permittivity is the tensor of Equation (12). The complex mode spectra resulting from the dispersion analysis are presented in Figure 5 and are slightly different between the GDS and GADS, confirming the importance to modeling the permittivity of the substrate as a tensor. The normalized phase constant of the surface-wave modes (SW) does not show significant variations between the GDS and the GADS models. The leaky-wave modes (LW) are characterized by values of the normalized attenuation constant that never decrease below 0.75. For this reason, the LW cannot represent a contribution to the radiation and the grounded slabs need to be covered by a PRS for working as antennas. This aspect will be investigated in the future. However, the normalized phase constant (related to the FPC-LWA pointing angle, as expressed in Equation (1)) of the LW shows a significant variation due to the birefringence of the selected LC mixture. This suggests that the GADS once equipped of a PRS may represent an interesting solution for performing the beam steering of the THz radiation.

The analysis presented in Figure 5 is a preliminary numerical study that may constitute a solid background for the experimental implementation of LC-substrate FPC-LWAs with THz beam-steering capabilities.

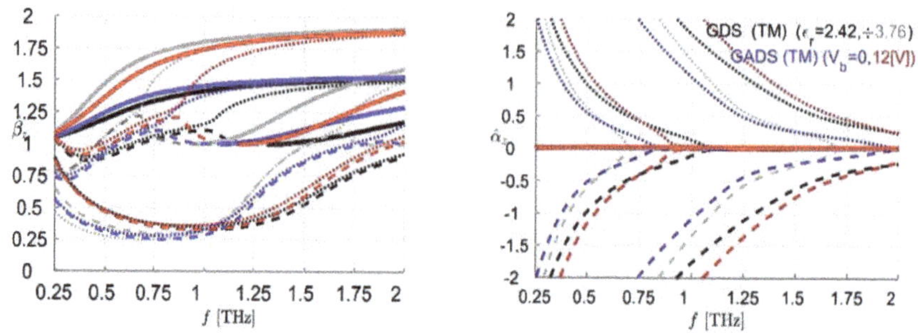

(**a**) (**b**)

Figure 5. (**a**) normalized phase constant and (**b**) normalized attenuation constant spectra of the TM modes of a GDS (black and gray lines) and a GADS (red and blue lines) obtained by changing the values in the LC permittivity tensor. Proper modes are depicted in solid lines, physical and nonphysical improper complex modes are represented in dotted and dashed lines, respectively [59].

5. Graphene Fabry–Perot Cavity Leaky Wave Antennas for THz Beam Steering

In the previous examples, we have seen how it is possible to enhance beam-steering capabilities at fixed frequency in FPC-LWAs by exploiting the tunable features of liquid crystals. However, in such LC-based devices, the beam can only be steered along a principal plane, namely the E-plane, since only TM leaky modes are affected by the axis re-orientation of the LC permittivity sensor. In order to fully switch from a pencil beam to a conical beam, both TM and TE leaky modes need to be affected by the application of a bias voltage. Remarkably, graphene-based FPC-LWAs have recently been shown to effectively provide for such radiating features [62–64]. Indeed, one of the most intriguing features exhibited by graphene (the two-dimensional version of graphite) is the electrical field-effect, i.e., the capability to change its surface conductivity (which completely describes its electrical properties, being graphene a 2Dmaterial) through the application of a bias voltage [65]. In particular, in the THz range, graphene exhibits moderate ohmic losses and a wide variation of reflectivity as the bias is raised from zero (very low reflectivity) to tens of Volts (moderate/high reflectivity) [66]. Interestingly, in the low THz range and for fast waves (as radiating leaky waves), graphene exhibits the same surface conductivity for both TM and TE polarization [67]. As a consequence, when a bias voltage is applied to graphene, both TM and TE leaky modes are affected, and to the radiating features along the principal E- and H-plane, respectively [63,68–70]. Two relatively simple realizations of graphene-based FPC-LWAs are the graphene planar waveguide (GPW) [68] and the graphene substrate superstrate (GSS) [69].

5.1. *Graphene Planar Waveguide*

A GPW is the simplest configuration of graphene-based FPC-LWAs. In practice, the PRS is represented by a uniform graphene sheet that is placed above a $\lambda/2$ thick grounded dielectric slab (usually silicon oxide), fed through a sub-resonant slot etched in the ground-plane and back-illuminated by a THz source (as depicted in Figure 1). The graphene sheet can be biased by placing a thin layer of a low-loss THz conductive polymer beneath the graphene sheet and contacting the layers with two electrodes (see Figure 6a). The greatest the bias voltage, the highest would be the reflectivity of the graphene sheet [68]. However, the value of the maximum bias voltage depends on the voltage breakdown of the dielectric filling the gap between graphene sheet and the conductive polymer (see Section IV in [63] for an in-depth discussion). For antenna applications, a value of bias voltage that raises the chemical potential of graphene up to 1 eV is considered a good trade-off (an approximate relation between the chemical potential and the bias voltage can be found e.g., in [40,66,68]). In particular, in the case of the GPW studied in [68], when the graphene chemical potential is raised to 1 eV, the GPW radiates a pencil beam at broadside. By progressively lowering the bias, the beam evolves in a conical beam reaching a maximum angle that is dictated by both the graphene quality and the substrate material (see Figure 6b–c). In this regard, we should mention that graphene quality considerably affects the antenna radiating features, and very high quality graphene flakes are needed to achieve a satisfactory performance with a GPW. However, even using high-quality graphene, a GPW is not capable of radiating highly-directive beams. This limitation is due to the both the non-negligible ohmic losses of graphene and the moderate values of its reflectivity. Therefore, in order to improve directivity in graphene-based FPC-LWAs, a PRS with a higher reflectivity is needed. This idea is realized in the GSS [69].

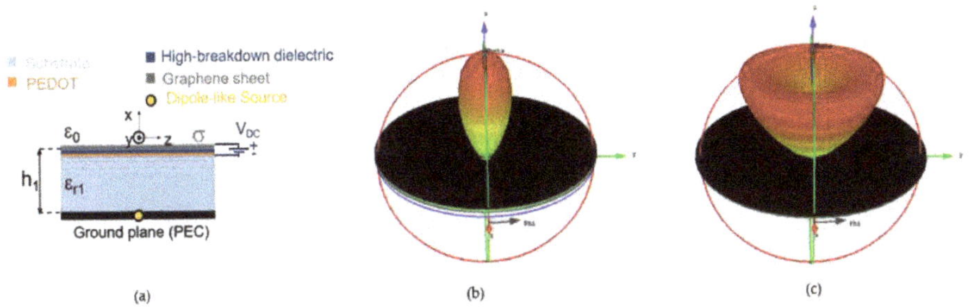

Figure 6. (**a**) a sketch of the GPW described in 5.1 [55]. In (**b**,**c**), the evolution from a pencil beam to a conical beam as the chemical potential is lowered from 1 eV to 0.5 eV, respectively [70]. The radiation patterns are obtained through CST full-wave simulations [58].

5.2. Graphene Substrate Superstrate

In a GSS, a $\lambda/4$ thick superstrate layer of a high refractive index material (e.g., Hafnium Oxide) is placed on top of the low-refractive index dielectric substrate, whereas the graphene sheet is lowered down to an optimal position between the ground plane and the substrate-superstrate interface (see Figure 7a). Such an optimal position can be retrieved from semi-analytical approaches

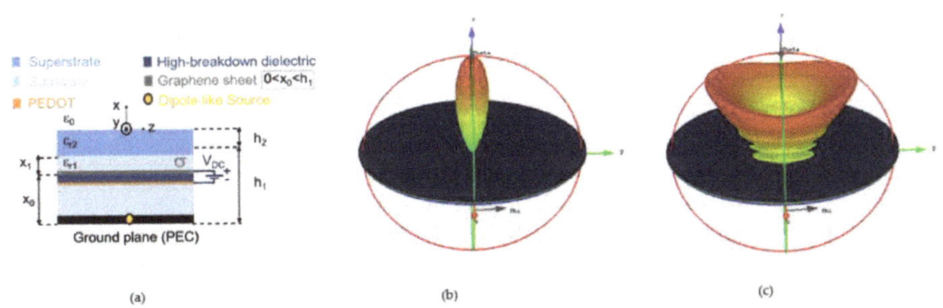

Figure 7. (**a**) a sketch of the GSS described in 5.2 [55]. In (**b**) and (**c**), the evolution from a pencil beam to a conical beam as the chemical potential is lowered from 1 eV to 0.3 eV, respectively [70]. The radiation patterns are obtained through CST full-wave simulations [58].

where the objective function is represented by the search of a position that maximizes directivity at broadside [69]. The result is a position in a middle-of-the-way between the substrate–superstrate interface (where the interaction between graphene and the tangential electric modal field is expected to be minimum) and the middle of the substrate (where the interaction between graphene and the tangential electric modal field is expected to be maximum). The reason for this result lies in the graphene electrical properties: the optimal position has to counterbalance the benefits of graphene tenability and reflectivity with the drawbacks of graphene ohmic losses, whose effects are more prominent in the points of maximum interaction with the electric field (see the relevant discussion in [63]). As a result, the GSS exhibits a higher directivity and a slightly reduced rate of reconfigurability with respect to the GPW (see Figure 7b–c). Nevertheless, the GSS offers more degrees of freedom to design FPC-LWAs with certain radiating features. In both cases, the quality of graphene plays a crucial role, and this aspect is still hindering a first experimental validation of these concepts.

6. Conclusions

Leaky-wave antennas are fast traveling-wave antennas, in which radiation occurs thanks to a power leakage during the wave propagation. They are antennas characterized by a high directivity and by the possibility to be employed as radiating systems at microwaves as well as at optical frequencies. For these reasons, LWAs seem to be particularly suitable for being integrated in THz systems. Among the several existing LWAs typologies, FPC-LWAs are attractive structures because they allow for tuning the radiating features by applying an external voltage. In this review, some strategies for implementing a reconfiguration of the emitting radiation have been discussed and some examples have been analyzed.

Author Contributions: Writing—Original draft preparation, S.T.; writing—review and editing, S.T. and W.F.; supervision, W.F. All authors have read and agreed to the published version of the manuscript.

Funding: This research received no external funding.

Acknowledgments: The authors would like to thank A. Galli, P. Burghignoli, P. Baccarelli, R. Beccherelli, and D.C. Zografopoulos for their valuable contributions on the topic of the current short Review.

Conflicts of Interest: The authors declare no conflict of interest.

References

1. Dhillon, S.S.; Vitiello, M.S.; Linfield, E.H.; Davies, A.G.; Hoffmann, M.C.; Booske, J.; Paoloni, C.; Gensch, M.; Weightman, P.; Williams, G.P.; et al. The 2017 terahertz science and technology roadmap. *J. Phys. D Appl. Phys.* **2017**, *50*, 043001. [CrossRef]
2. Tu, W.; Zhong, S.; Shen, Y.; Incecik, A. Nondestructive testing of marine protective coatings using terahertz waves with stationary wavelet transform. *Ocean Eng.* **2016**, *111*, 582–592. [CrossRef]
3. Heinz, E.; May, T.; Born, D.; Zieger, G.; Anders, S.; Zakosarenko, V.; Meyer, H.-G.; Schäffel, C. Passive 350 GHz video imaging systems for security applications. *J. Infrared Millim. Terahertz Waves* **2015**, *36*, 879–895. [CrossRef]
4. Leahy-Hoppa, M.R.; Fitch, M.J.; Zheng, X.; Hayden, L.M.; Osiander, R. Wideband terahertz spectroscopy of explosives. *Chem. Phys. Lett.* **2007**, *434*, 227–230. [CrossRef]
5. Haglund, R.F.; Neev, J.; Wood, R.F. Commercial and biomedical applications of ultrashort pulse lasers; laser plasma generation and diagnostics. In Proceedings of the SPIE, San Jose, CA, USA, 29 May 2001. [CrossRef]
6. Rappaport, T.S.; Murdock, J.N.; Gutierrez, F. State of the Art in 60-GHz Integrated Circuits and Systems for Wireless Communications. *Proc. IEEE* **2011**, *99*, 1390–1436. [CrossRef]
7. Miles, R.; Zhang, X.-C.; Eisele, H.; Krotkus, A. *Terahertz Frequency Detection and Identification of Materials and Objects*; Springer Science & Business Media: Berlin/Heidelberg, Germany, 2007; ISBN 978-1-4020-6503-3.
8. Akyildiz, I.F.; Jornet, J.M.; Han, C. Terahertz band: Next frontier for wireless communications. *Phys. Com.* **2014**, *12*, 16–32. [CrossRef]
9. Monticone, F.; Alù, A. Leaky-wave theory, techniques, and applications: from microwaves to visible frequencies. *Proc. IEEE* **2015**, *103*, 793–821. [CrossRef]
10. Tamir, T.; Oliner, A.A. Guided complex waves. Part 1: Fields at an interface. *Proc. Inst. Electr. Eng.* **1963**, *110*, 310–324. [CrossRef]
11. Tamir, T.; Oliner, A.A. Guided complex waves. Part 2: Relation to radiation patterns. *Proc. Inst. Electr. Eng.* **1963**, *110*, 325–334. [CrossRef]
12. Croswell, W.; Zucker, F.J. Surface-wave antennas. In *Antenna Engineering Handbook*; Volakis, J.L., Ed.; McGraw-Hill: New York, NY, USA, 2007; Chapter 10.
13. Jackson, D.R.; Oliner, A.A. Leaky-wave antennas. In *Modern Antenna Handbook*; Balanis, C.A., Ed.; John Wiley & Sons, Inc.: Hoboken, NJ, USA, 2008; pp. 325–367, ISBN 978-0-470-29415-4.

14. Galli, A.; Baccarelli, P.; Burghignoli, P. Leaky-Wave Antennas. In *Wiley Encyclopedia of Electrical and Electronics Engineering*; Webster, J.G., Ed.; John Wiley & Sons, Inc.: Hoboken, NJ, USA, 2016; ISBN 978-0-471-34608-1.
15. Fuscaldo, W.; Tofani, S.; Burghignoli, P.; Baccarelli, P.; Galli, A. Terahertz leaky-wave antennas based on metasurfaces and tunable materials. In *Metamaterials and Metasurfaces*; Canet-Ferrer, J., Ed.; IntechOpen: London, UK, 2018; pp. 93–116.
16. von Trentini, G. Partially reflecting sheet arrays. *IEEE Trans. Antennas Propag.* **1956**, *4*, 666–671. [CrossRef]
17. Ip, A.; Jackson, D.R. Radiation from cylindrical leaky waves. *IEEE Trans. Antennas Propag.* **1990**, *38*, 482–488. [CrossRef]
18. Zhao, T.; Jackson, D.R.; Williams, J.T.; Oliner, A.A. General formulas for 2Dleaky-wave antennas. *IEEE Trans. Antennas Propag.* **2005**, *53*, 3525–3533. [CrossRef]
19. Feresidis, A.P.; Vardaxoglou, J.C. High gain planar antenna using optimised partially reflective surfaces. *IEE Proc. Microwaves Antennas Propag.* **2001**, *148*, 345–350. [CrossRef]
20. Lovat, G.; Burghignoli, P.; Jackson, D.R. Fundamental properties and optimization of broadside radiation from uniform leaky-wave antennas. *IEEE Trans. Antennas Propag.* **2006**, *54*, 1442–1452. [CrossRef]
21. Luukkonen, O.; Simovski, C.; Granet, G.; Goussetis, G.; Lioubtchenko, D.; Raisanen, A.V.; Tretyakov, S.A. Simple and accurate analytical model of planar grids and high-impedance surfaces comprising metal strips or patches. *IEEE Trans. Antennas Propag.* **2008**, *56*, 1624–1632. [CrossRef]
22. Itoh, T. *Numerical Techniques for Microwave and Millimeter-Wave Passive Structures*; Wiley-Interscience: Hoboken, NJ, USA, 1989.
23. Sorrentino, R. *Numerical Methods for Passive Microwave and Millimeter Wave Structures*; IEEE Press: New York, NY, USA, 1989.
24. Felsen, L.B.; Marcuvitz, N. *Radiation and Scattering of Waves*; John Wiley & Sons: Hoboken, NJ, USA, 1994; ISBN 978-0-7803-1088-9.
25. Goldstone, L.; Oliner, A. Leaky-wave antennas I: Rectangular waveguides. *IRE Trans. Antennas Propag.* **1959**, *7*, 307–319. [CrossRef]
26. Pozar, D.M. *Microwave Engineering*; Wiley India Pvt. Limited: Delhi, India, 2009; ISBN 978-81-265-1049-8.
27. Alonso-delPino, M.; Reck, T.; Jung-Kubiak, C.; Lee, C.; Chattopadhyay, G. Development of silicon micromachined microlens antennas at 1.9 THz. *IEEE Trans. Terahertz Sci. Technol.* **2017**, *7*, 191–198. [CrossRef]
28. Llombart, N.; Chattopadhyay, G.; Skalare, A.; Mehdi, I. Novel terahertz antenna based on a silicon lens fed by a leaky wave enhanced waveguide. *IEEE Trans. Antennas Propag.* **2011**, *59*, 2160–2168. [CrossRef]
29. Hussain, N.; Park, I. Optimization of a small lens for a leaky-wave slit dipole antenna at the terahertz band. In Proceedings of the 2016 International Symposium on Antennas and Propagation (ISAP), Okinawa, Japan, 24–28 October 2016; pp. 782–783.
30. Neto, A.; Llombart, N.; Baselmans, J.J.A.; Baryshev, A.; Yates, S.J.C. Demonstration of the leaky lens antenna at submillimeter wavelengths. *IEEE Trans. Terahertz Sci. Technol.* **2014**, *4*, 26–32. [CrossRef]
31. Bueno, J.; Yurduseven, O.; Yates, S.J.C.; Llombart, N.; Murugesan, V.; Thoen, D.J.; Baryshev, A.M.; Neto, A.; Baselmans, J.J.A. Photon noise limited performance over an octave of bandwidth of kinetic inductance detectors for sub-millimeter astronomy. In Proceedings of the 2016 41st International Conference on Infrared, Millimeter, and Terahertz waves (IRMMW-THz), Copenhagen, Denmark, 25–30 September 2016; pp. 1–2.
32. Llombart, N.; Bueno, J.; Yurduseven, O.; Baselmans, J.; Neto, A. Dual polarized leaky wave antenna coupled KIDs for THz space applications. In Proceedings of the 2014 39th International Conference on Infrared, Millimeter, and Terahertz waves (IRMMW-THz), Millimeter, Tucson, AZ, USA, 14–19 September 2014; pp. 1–2.
33. Hussain, N.; Park, I. Design of a wide-gain-bandwidth metasurface antenna at terahertz frequency. *AIP Adv.* **2017**, *7*, 055313. [CrossRef]
34. Hussain, N.; Nguyen, T.K.; Park, I. Performance comparison of a planar substrate-integrated Fabry-Perot cavity antenna with different unit cells at terahertz frequency. In Proceedings of the 2016 10th European Conference on Antennas and Propagation (EuCAP), Davos, Switzerland, 10–15 April 2016; pp. 1–4.
35. Nguyen, T.K.; Ta, B.Q.; Park, I. Design of a planar, high-gain, substrate-integrated Fabry–Perot cavity antenna at terahertz frequency. *Curr. Appl. Phy.* **2015**, *15*, 1047–1053. [CrossRef]

36. Fuscaldo, W.; Tofani, S.; Zografopoulos, D.C.; Baccarelli, P.; Burghignoli, P.; Beccherelli, R.; Galli, A. Systematic design of THz leaky-wave antennas based on homogenized metasurfaces. *IEEE Trans. Antennas Propag.* **2018**, *66*, 1169–1178. [CrossRef]
37. Tofani, S.; Fuscaldo, W.; Zografopoulos, D.C.; Burghignoli, P.; Baccarelli, P.; Beccherelli, R.; Galli, A. Design-flow of Fabry-Perot cavity leaky-wave antennas based on homogenized metasurfaces. In Proceedings of the 2019 13th European Conference on Antennas and Propagation (EuCAP), Krakow, Poland, 31 March–5 April 2019; pp. 1–4.
38. Tofani, S.; Fuscaldo, W.; Zografopoulos, D.C.; Baccarelli, P.; Burghignoli, P.; Beccherelli, R.; Galli, A. Spatial Dispersion Analysis of Homogenized Metafurfaces for Terahertz Leaky-wave Antennas. In Proceedings of the 12th European Conference on Antennas and Propagation (EuCAP 2018), London, UK, 9–13 April 2018.
39. Li, J.; He, M.; Wu, C.; Zhang, C. Radiation-pattern-reconfigurable graphene leaky-wave antenna at terahertz band based on dielectric grating structure. *IEEE Antennas Wirel. Propag. Lett.* **2017**, *16*, 1771–1775. [CrossRef]
40. Esquius-Morote, M.; Gómez-Diaz, J.S.; Perruisseau-Carrier, J. Sinusoidally modulated graphene leaky-wave antenna for electronic beamscanning at THz. *IEEE Trans. Terahertz Sci. Technol.* **2014**, *4*, 116–122. [CrossRef]
41. Cheng, J.; Jafar-Zanjani, S.; Mosallaei, H. Real-time two-dimensional beam steering with gate-tunable materials: a theoretical investigation. *Appl. Opt.* **2016**, *55*, 6137–6144. [CrossRef]
42. Fuscaldo, W.; Tofani, S.; Zografopoulos, D.C.; Baccarelli, P.; Burghignoli, P.; Beccherelli, R.; Galli, A. Tunable Fabry-Perot cavity THz antenna based on leaky-wave propagation in nematic liquid crystals. *IEEE Antennas Wirel. Propag. Lett.* **2017**, *16*, 2046–2049. [CrossRef]
43. Murano, K.; Watanabe, I.; Kasamatsu, A.; Suzuki, S.; Asada, M.; Withayachumnankul, W.; Tanaka, T.; Monnai, Y. Demonstration of short-range terahertz radar using high-gain leaky-wave antenna. In Proceedings of the 2016 41st International Conference on Infrared, Millimeter, and Terahertz waves (IRMMW-THz), Copenhagen, Denmark, 25–30 September 2016; pp. 1–2.
44. Murano, K.; Watanabe, I.; Kasamatsu, A.; Suzuki, S.; Asada, M.; Withayachumnankul, W.; Tanaka, T.; Monnai, Y. Low-profile terahertz radar based on broadband leaky-wave beam steering. *IEEE Trans. Terahertz Sci. Technol.* **2017**, *7*, 60–69. [CrossRef]
45. Chamberlain, J.M. Where optics meets electronics: recent progress in decreasing the terahertz gap. *Philos. Trans. R. Soc. A* **2004**, *362*, 199–213. [CrossRef]
46. Holloway, C.L.; Kuester, E.F.; Gordon, J.A.; O'Hara, J.; Booth, J.; Smith, D.R. An overview of the theory and applications of metasurfaces: The two-dimensional equivalents of metamaterials. *IEEE Antennas Propag. Mag.* **2012**, *54*, 10–35. [CrossRef]
47. Khoo, I.-C. *Liquid Crystals: Physical Properties and Nonlinear Optical Phenomena*; John Wiley & Sons: Hoboken, NJ, USA, 2007; ISBN 978-0-470-08402-1.
48. Chandrasekhar, S. *Liquid Crystals*; Cambridge University Press: Cambridge, UK, 1992; ISBN 978-0-521-42741-8.
49. Berreman, D.W. Solid surface shape and the alignment of an adjacent nematic liquid crystal. *Phys. Rev. Lett.* **1972**, *28*, 1683–1686. [CrossRef]
50. Yaroshchuk, O.; Reznikov, Y. Photoalignment of liquid crystals: basics and current trends. *J. Mater. Chem.* **2011**, *22*, 286–300. [CrossRef]
51. Fréedericksz, V.; Zolina, V. Forces causing the orientation of an anisotropic liquid. *Trans. Faraday Soc.* **1933**, *29*, 919–930. [CrossRef]
52. Dąbrowski, R.; Kula, P.; Herman, J. High birefringence liquid crystals. *Crystals* **2013**, *3*, 443–482. [CrossRef]
53. Reuter, M.; Vieweg, N.; Fischer, B.M.; Mikulicz, M.; Koch, M.; Garbat, K.; Dąbrowski, R. Highly birefringent, low-loss liquid crystals for terahertz applications. *APL Mater.* **2013**, *1*, 012107. [CrossRef]
54. Jackson, D.R.; Oliner, A.A.; Ip, A. Leaky-wave propagation and radiation for a narrow-beam multiple-layer dielectric structure. *IEEE Trans. Antennas Propag.* **1993**, *41*, 344–348. [CrossRef]
55. Fuscaldo, W. Advanced Radiating Systems Based on Leaky Waves and Nondiffracting Waves. Ph.D. thesis, 2017.
56. Tofani, S. Static and reconfigurable devices for near-field and far-field terahertz applications. Ph.D. thesis, Sapienza University of Rome, Rome, Italy, 2018.
57. Zografopoulos, D.C.; Beccherelli, R.; Kriezis, E.E. Beam-splitter switches based on zenithal bistable liquid-crystal gratings. *Phys. Rev. E* **2014**, *90*, 042503. [CrossRef] [PubMed]
58. CST. Available online: https://www.cst.com/ (accessed on 9 May 2017).

59. Tofani, S.; Fuscaldo, W.; Zografopoulos, D.C.; Burghignoli, P.; Baccarelli, P.; Beccherelli, R.; Galli, A. Terahertz modal analysis of a grounded liquid-crystal cell and its application as a tunable cavity antenna. In Proceedings of the 2019 41st Photonics and Electromagnetics Research Symposium (PIERS), Rome, Italy, 17–20 June 2019.
60. Yakovlev, A.B.; Hanson, G.W. Fundamental modal phenomena on isotropic and anisotropic planar slab dielectric waveguides. *IEEE Trans. Antennas Propag.* **2003**, *51*, 888–897. [CrossRef]
61. Pavone, S.C.; Martini, E.; Caminita, F.; Albani, M.; Maci, S. Surface Wave Dispersion for a Tunable Grounded Liquid Crystal Substrate Without and With Metasurface on Top. *IEEE Trans. Antennas Propag.* **2017**, *65*, 3540–3548. [CrossRef]
62. Wang, X.-C.; Zhao, W.-S.; Hu, J.; Yin, W.-Y. Reconfigurable Terahertz Leaky-Wave Antenna Using Graphene-Based High-Impedance Surface. *IEEE Trans. Nanotechnol.* **2015**, *14*, 62–69. [CrossRef]
63. Fuscaldo, W.; Burghignoli, P.; Baccarelli, P.; Galli, A. Graphene Fabry-Perot cavity leaky-wave antennas: Plasmonic versus nonplasmonic solutions. *IEEE Trans. Antennas Propag.* **2017**, *65*, 1651–1660. [CrossRef]
64. Fuscaldo, W.; Tofani, S.; Burghignoli, P.; Baccarelli, P.; Notargiacomo, A.; Cibella, S.; Pea, M.; Carelli, P.; Mishra, N.; Coletti, C.; et al. Graphene-Based Fabry-Perot Cavity Leaky-Wave Antennas: Towards an Experimental Validation. In Proceedings of the 2018 48th European Microwave Conference (EuMC), Madrid, Spain, 25–27 September 2018; pp. 276–279.
65. Geim, A.K.; Novoselov, K.S. The rise of graphene. *Nat. Mater.* **2007**, *6*, 183–191. [CrossRef] [PubMed]
66. Hanson, G.W. Dyadic Green's functions for an anisotropic, non-local model of biased graphene. *IEEE Trans. Antennas Propag.* **2008**, *56*, 747–757. [CrossRef]
67. Lovat, G.; Hanson, G.W.; Araneo, R.; Burghignoli, P. Semiclassical spatially dispersive intraband conductivity tensor and quantum capacitance of graphene. *Phys. Rev. B* **2013**, *87*, 115429. [CrossRef]
68. Fuscaldo, W.; Burghignoli, P.; Baccarelli, P.; Galli, A. Complex mode spectra of graphene-based planar structures for THz applications. *J. Infrared Milli. Terahz Waves* **2015**, *36*, 720–733. [CrossRef]
69. Fuscaldo, W.; Burghignoli, P.; Baccarelli, P.; Galli, A. A Reconfigurable Substrate-Superstrate Graphene-Based Leaky-Wave THz Antenna. *IEEE Antennas Wirel. Propag. Lett.* **2016**, *15*, 1545–1548. [CrossRef]
70. Fuscaldo, W.; Tofani, S.; Burghignoli, P.; Baccarelli, P.; Galli, A. Terahertz Fabry-Perot cavity leaky-wave antennas. In Proceedings of the 2019 41st Photonics and Electromagnetics Research Symposium (PIERS), Rome, Italy, 17–20 June 2019.

MDPI
St. Alban-Anlage 66
4052 Basel
Switzerland
Tel. +41 61 683 77 34
Fax +41 61 302 89 18
www.mdpi.com

Condensed Matter Editorial Office
E-mail: condensedmatter@mdpi.com
www.mdpi.com/journal/condensedmatter

www.ingramcontent.com/pod-product-compliance
Lightning Source LLC
Chambersburg PA
CBHW061559210926
51850CB00009B/2155